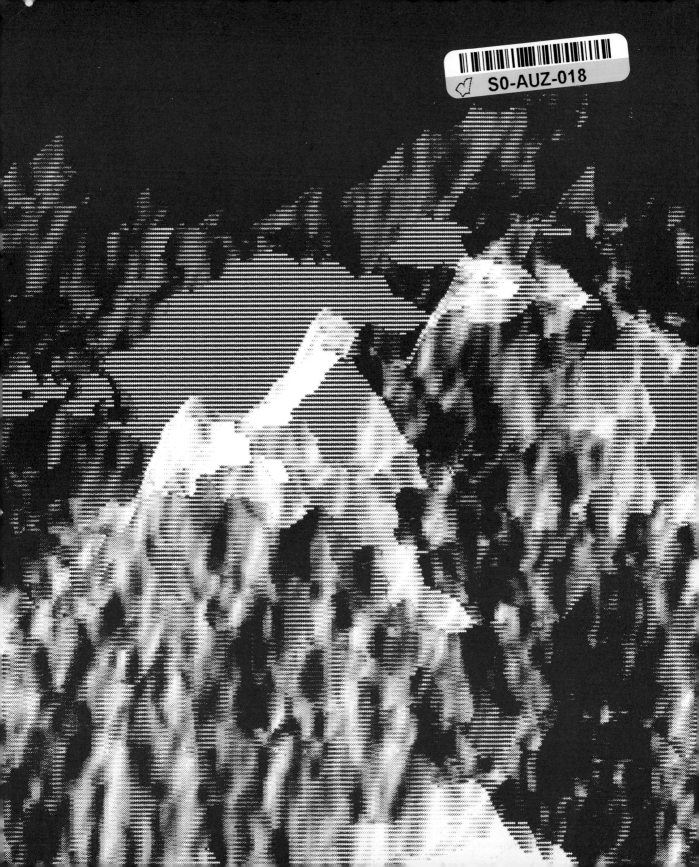

FRACTALS

FORM, CHANCE, AND DIMENSION

FRACTALS
FORM, CHANCE, AND DIMENSION

Benoit B. Mandelbrot

INTERNATIONAL BUSINESS MACHINES
THOMAS J. WATSON RESEARCH CENTER

W. H. FREEMAN AND COMPANY
San Francisco

ABOUT THE AUTHOR: A graduate of Ecole Polytechnique; Caltech M.S. and Ae.E. in Aeronautics; Docteur ès Sciences Mathématiques U. Paris. Before joining the IBM Thomas J. Watson Research Center, where he is now an IBM Fellow, Dr. Mandelbrot was with the French Research Council (CNRS), the School of Mathematics at the Institute for Advanced Study, the University of Geneva, and a Junior Professor of Applied Mathematics at the University of Lille and of Mathematical Analysis at Ecole Polytechnique. On leave from IBM, he has been a Visiting Professor of Economics and later of Applied Mathematics at Harvard, of Engineering at Yale, of Physiology at the Albert Einstein College of Medicine, and of Mathematics at the University of Paris-Sud, and has been with M.I.T., first in the Electrical Engineering Department, and most recently as an Institute Lecturer. A Fellow of the Guggenheim Foundation, Trumbull Lecturer at Yale, Samuel Wilks Lecturer at Princeton, and Abraham Wald Lecturer at Columbia, it was as Lecturer at Collège de France that he gave in 1973 and 1974 the lessons which eventually were developed into the present Essay.

AMS Classifications: Primary, 5001, 2875, 60D05; Secondary, 42A36, 54F45, 60J65, 76F05, 85A35, 86A05.

Current Physics Index Classification (ICSU): o5.40.+j, 45.30.−b, 98.20.Qw, 91.65.Br.

Library of Congress Cataloging in Publication Data

Mandelbrot, Benoit B
    Fractals

    Translation of Les objets fractals.
    Bibliography: p.
    Includes index.
    1. Geometry. 2. Mathematical models. 3. Stochastic processes. I. Title.
QA447.M3613   516′.15   76-57947
ISBN 0-7167-0473-0
ISBN 0-7167-0474-9 pbk.

Printed in the United States of America

1 2 3 4 5 6 7 8 9

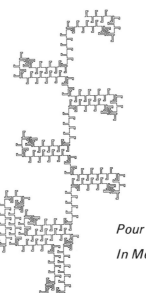

Pour Aliette

In Memoriam, B. et C.

FOREWORD

This book is a much modified and augmented second version of *Les objets fractals: forme, hasard et dimension* (Paris & Montreal: Flammarion, 1975).

In writing and designing it, my associates were Sigmund W. Handelman for computation and graphics, and H. Catharine Dietrich for editing.

Dr. Richard F. Voss was of great help in many ways and is coauthor of this version's illustrations.

The end papers, frontispiece and other caption-less patterns scattered here and there are explained on page 25.

The work was performed at the IBM Thomas J. Watson Research Center in Yorktown Heights, New York. Detailed acknowledgments are collected toward the end of the book.

CONTENTS

coastlines ◆ global effects in Brownian space-to-line functions ◆ Brownian relief on a spherical Earth ◆ the horizon ◆ fractional Brownian relief on a flat Earth ◆ fractional Brownian model of river discharge, motivated ◆ fractional Brownian model of the relief, motivated ◆ projective island surfaces and the Korčak law ◆ deadvalleys and lakes ◆ antipersistent fractional Brownian motions ◆ isosurfaces of turbulent scalars: the problem stated ◆ the Kolmogorov and Burgers delta variances ◆ in homogeneous turbulence, isosurfaces are fractals

LIST OF PLATES AND PATTERNS

FRACTALS
FORM, CHANCE, AND DIMENSION

CHAPTER I

Introduction

Many important spatial patterns of Nature are either irregular or fragmented to such an extreme degree that *Euclid* – a term used in this Essay to denote all of classical geometry – is hardly of any help in describing their form. The coastline of a typical oceanic island, to take an example, is neither straight, nor circular, nor elliptic, and no other classical curve can serve without undue artificiality in the presentation and the organization of empirical measurements and in the search for explanations. Similarly, no surface in Euclid represents adequately the boundaries of clouds or of rough turbulent wakes. More generally, many patterns of Nature that had already attracted the attention of primitive Man involve, in comparison to Euclid, not only a higher degree but an altogether different level of complexity.

In the present Essay I hope to show that it is possible in many cases to remedy this absence of geometric representation by using a family of shapes I propose to call *fractals* – or *fractal sets*. The most useful among them involve *chance*, and their irregularities are statistical in nature. A central role is played in this study by the concept of *fractal* (or Hausdorff-Besicovitch) *dimension*.

The best-known fractals arise from the geometric model of Brownian motion. Among the phenomena of higher complexity to be discussed in this book, Brownian motion is exemplary, and it is also exceptional in having been very fully described and explained. My claim is therefore that fractal kin of Brownian sets can be brought usefully to bear not only on microscopic phenomena but also on phenomena on or above Man's scale, including those listed in the first paragraph.

Some fractal sets are curves, others are surfaces, still others are clouds of disconnected points, and yet others are so oddly shaped that there are no good terms for them in either the sciences or the arts. The variety of these forms should be sampled by browsing through the illustrations. After the fact, their application seems so natural and so easy that one may well argue that the study of fractals should henceforth be added as a full-fledged chapter to the collection of shapes and techniques one calls elementary geometry.

This work is referred to throughout as a scientific Essay, and it conforms indeed strictly (other than by its length) to an old dictionary's definition as a "composition dealing with a subject from a personal point of view and without attempting completeness." One may find it also (to quote the eleventh edition of the *Encyclopaedia Britannica*) to be "irregular, with constant digressions and interruptions."

It should help unify what has been until now a collection of mathematical odds and ends, many of them classical but somewhat obscure, and make them known to nonmathematicians and mathematicians alike. Also, the mathematical questions it raises are compelling on their own merits. However, this Essay is neither a treatise nor a textbook in mathematics.

It is a collection of theories in diverse branches of natural science, that should be judged by the prospective users on the basis of their powers of organization and explanation (and their esthetic quality) rather than by their attractiveness as examples of a mathematical structure.

Unfortunately, each of these theories has to be cut short before it becomes too technical; the reader is referred to other publications for detailed development. One might say (to echo d'Arcy Thompson) that this Essay is all preface from beginning to end. Anyone who expects more from it will be disappointed.

GOALS

The term *fractal* (the origin of which will soon be described) is new, but individual fractals have been known to mathematicians for quite a long time. The basic ones arose during the crisis that mathematics underwent between (roughly) 1875 and 1922 and that remains associated with names like Weierstrass, Cantor, Peano, Lebesgue, and Hausdorff. Other notable contributors were Koch, Sierpiński, and Besicovitch. The preceding names, to put it mildly, are not ordinarily encountered with any frequency in the empirical study of Nature. Yet it will be seen again and again that the impact of the work of these giants far transcends its intended scope. Once again we are surprised by what we should have come to expect, that "the language of mathematics reveals itself unreasonably effective in the natural sciences ... a wonderful gift which we neither understand nor deserve. We should be grateful for it and hope that it will remain valid in future research and that it will extend, for better or for worse, to our pleasure even though perhaps also to our bafflement, to wide branches of learning" (Wigner 1960).

In sum, the present Essay proposes new solutions to very old concrete problems with the help of mathematics that is very old too, but that (to my knowledge) had not been used in this fashion – with the exception of Brownian motion.

Mathematicians will (I hope) be surprised and delighted to find that concepts thus far reputed pathological should evolve naturally from very concrete problems. And that the study of Nature should involve so many new mathematical problems. However, the bulk of the material to be presented involves few technical difficulties.

As a consequence, it is better to address the bulk of the text of this Essay not to the providers of the tools but rather to their would-be users – a mixed group of scientists. A few asides or digressions will be addressed to the mathematician (the first of which will be a definition of fractals later in the Introduction), but the presentation will proceed mostly from the concrete and specific to mild formalism and generalities. Almost every chapter will focus on a single subject or several closely related ones. But each will also serve to introduce either a limited species, so to speak, or a broader genus of fractals, and to explore some aspects of fractals in general. Formal mathematical details will be postponed as much as feasible – all but the simplest ones not appearing until Chapter XII. The nature of fractals is meant to be gradually discovered by the reader, not revealed in a flash by the author.

ETYMOLOGY OF "FRACTAL"

The remainder of the Introduction will expand on the topics touched in the above opening sections: motivation, method of approach, and manner of presentation. First a long quotation from Perrin 1906 gives a more complete description of some of the problems to be attacked. A later Section is addressed to mathematicians or other readers who might welcome or even demand at an early stage a general definition of fractal sets and dimension; this is provided. The chapter ends on some points of style, purpose, and organization.

In any event, the remaining sections of the Introduction are written to be sampled according to one's taste. One can also proceed from here to Chapter II without knowing anything beyond the etymology of the term *fractal*. This etymology can be asserted with full authority because I am responsible for coining the term to denote a collection of concepts and techniques that seems finally to acquire a clear-cut identity.

Fractal comes from the Latin adjective *fractus*, which has the same root as *fraction* and *fragment* and means "irregular or fragmented;" it is related to *frangere*, which means "to break." The stress should logically be on the first syllable, as in fraction. Our term will primarily be applied to mathematical fractal sets but will also be used as part of the expression *natural fractal,* which will designate a natural pattern representable by a fractal set. For example, since the curves relative to Brownian motion are prime examples of fractal sets, physical Brownian motion is the prime example of a natural fractal.

Postscript. Just before finishing this English version, I came to realize that the idea that Cantor sets may be of use in physics had already occurred briefly around 1883 to Henri Poincaré! The episode led Hadamard 1912, p. 212, to comment that "examples of curves without tangent are indeed classical since Riemann and Weierstrass. Anyone can grasp, however, the deep differences that exist between, on the one hand, a fact established under circumstances arranged for the enjoyment of the mind, with no other aim and no interest other than to show its possibility, an exhibit in a gallery of monsters, and on the other hand, the same fact as encountered in a theory that is rooted in the most usual and the most essential problems of analysis." It may be added that differences deepen when such facts enter into the theory of Brownian motion, and deepen further when they enter into the studies to be reported in this Essay. A more detailed story of the Poincaré episode is told under CANTOR SETS in Chapter XI and is followed by a mention of recent tantalizing suggestions on the same lines due to Ulam and to Ruelle & Takens.

J. PERRIN ON IRREGULARITY AND FRAGMENTATION IN NATURE

The irregular or fragmented character of coastlines, Brownian trajectories, and many other patterns of Nature, mentioned in the introductory paragraph of this chapter, will now be elaborated upon by reproducing in free translation some excerpts from Perrin 1906. Jean Perrin's subsequent work on Brownian motion eventually won him the Nobel Prize in physics and (as we shall see later in the

Introduction) had a direct influence upon Norbert Wiener and hence on probability theory. However, Perrin's authority will be of service in indirect fashion. The work to be "quoted" was not a scientific treatise, rather a philosophical manifesto published in a general intellectual monthly. It reveals that Perrin's thinking was remarkable not only for the deep but narrow aspects that were to be immediately influential but also for broader aspects that failed to gain any attention. That is, not only for insights about Brownian motion itself but also for the recognition that Brownian motion and coastlines and their kin have basic features in common.

It may be added that I am fascinated by this text because of the interest of the ideas and the eloquence of the style, and not because of any actual influence it had upon the present work, since it came very late to my attention, when the French version was nearly complete.

The translation is based on a slight revision of Perrin 1906, used in the preface of *Les atomes* (Perrin 1913).

"It is well known that, before giving a rigorous definition of continuity, a good teacher would show beginners that they already possess the idea which underlies this concept. He will draw a well-defined curve and say, holding a ruler, 'You see that there is a tangent at every point.' Or again, in order to impart the more abstract notion of the true velocity of a moving object at a point in its trajectory, we say, 'You see, of course, that the mean velocity between two neighboring points on this trajectory does not vary appreciably as these points approach infinitely near to each other.' And many minds, aware that for certain familiar motions this view appears true enough, do not see that it involves considerable difficulties.

"Mathematicians, however, are well aware of the childishness of trying to show by drawing curves that every continuous function has a derivative. Though functions for which such is the case are the simplest and the easiest to deal with, they are nevertheless exceptional. Using geometrical language, curves that have no tangents are the rule, and regular curves, such as the circle, are interesting but quite special.

"At first sight the consideration of the general case seems merely an intellectual exercise, ingenious but artificial, the desire for absolute accuracy carried to a ridiculous pitch. Those who hear of curves without tangents, or of functions without derivatives, often think at first that Nature presents no such complications, nor even suggests them.

"The contrary, however, is true, and

the logic of the mathematicians has kept them nearer to reality than the practical representations employed by physicists. This assertion may be illustrated by considering certain experimental data without preconception.

"Consider, for instance, one of the white flakes that are obtained by salting a solution of soap. At a distance its contour may appear sharply defined, but as we draw nearer its sharpness disappears. The eye no longer succeeds in drawing a tangent at any point. A line that at first sight would seem to be satisfactory appears on close scrutiny to be perpendicular or oblique. The use of a magnifying glass or microscope leaves us just as uncertain, for fresh irregularities appear every time we increase the magnification, and we never succeed in getting a sharp, smooth impression, as given, for example, by a steel ball. So, if we accept the latter as giving a useful illustration of the classical form of continuity, our flake could just as logically suggest the more general notion of a continuous function without a derivative.

An interruption is necessary to draw the reader's attention to Plate 9. (*Note:* The numbers on the plates are the same as the pages on which they occur.) The "quote" resumes.

"We must bear in mind that the uncertainty as to the position of the tangent at a point on the contour is by no means the same as the uncertainty observed on a map of Brittany. Although it would be different according to the map's scale, a tangent could always be found, for a map is a conventional diagram. On the contrary, an essential characteristic of our flake (and, indeed, of the coast itself) is, that we *suspect*, without seeing them clearly, that any scale involves details that absolutely prohibit the fixing of a tangent.

"We are still in the realm of experimental reality when we observe under the microscope the Brownian motion agitating each small particle suspended in a fluid [this Essay's Plate 11]. The direction of the straight line joining the positions occupied at two instants very close in time is found to vary absolutely irregularly as the time between the two instants is decreased. An unprejudiced observer would therefore conclude that he is dealing with a function without derivative, instead of a curve to which a tangent could be drawn."

"It must be borne in mind that, although closer observation of any object generally leads to the discovery of a highly irregular structure, we often can with advantage approximate its properties by continuous functions. Although wood may be indefinitely porous, it is useful to speak of a beam that has been sawed and planed as having a finite area. In other words, at certain scales and for certain methods of investigation, many phenomena may be represented by regular continuous functions, somewhat in the same way that a sheet of tinfoil may be wrapped round a sponge without following accurately the latter's complicated contour.

"If, to go further, we refuse to limit our considerations to the part of the universe we actually see, and if we attribute to matter the *infinitely* granular structure that is in the spirit of atomic theory, our power to apply to reality the *rigorous* mathematical concept of continuity will suffer a very remarkable diminution.

"Let us consider, for instance, the way in which we define the density of air at a given point and at a given moment. We picture a sphere of volume v centered at that point and including the mass m. The quotient m/v is the mean density within the sphere, and by *true* density we denote some limiting value of this quotient. This notion, however, implies that at the given moment the mean density is practically constant for spheres below a certain volume. This mean density may be notably different for spheres

containing 1,000 cubic meters and 1 cubic centimeter respectively, but it is expected to only vary by 1 in 1,000,000 when comparing 1 cubic centimeter to one-thousandth of a cubic millimeter.

"Suppose the volume becomes continually smaller. Instead of becoming less and less important, these fluctuations come to increase. For scales at which the Brownian motion shows great activity, fluctuations may attain 1 part in 1,000, and they become of the order of 1 part in 5 when the radius of the hypothetical spherule becomes of the order of a hundredth of a micron.

"One step further and our spherule becomes of the order of a molecule radius. Then, in a gas, it will generally lie in intermolecular space, where its mean density will henceforth *vanish*. At our point the *true* density will also *vanish*. But about once in a thousand times that point will lie within a molecule, and the mean density will be a thousand times higher than the value we usually take to be the true density of the gas.

"Let our spherule grow steadily smaller. Soon, except under exceptional circumstances, it will become empty and remain so henceforth owing to the emptiness of intra-atomic space; the true density *vanishes* almost everywhere, except at an infinite number of isolated

Plates 8 and 9

FRACTAL FLAKE AND ITS SHADOW

In an inspiring text quoted in the Introduction of this Essay, Jean Perrin commented on the form of the "white flakes that are obtained by salting a solution of soap." The shape illustrated on these Figures is meant to accompany Perrin's remarks.

First of all, these Figures are neither photographs nor computer reconstitutions of a real flake.

Neither does the shape illustrated here claim to result from a theory embodying the diverse aspects of a real flake's formation – chemical, physico-chemical, and hydrodynamical.

A fortiori, this flake does not claim to be directly related to any irreducible scientific principle.

The flake is merely a computer-generated shape meant to illustrate as simply as the author could manage the geometric characteristics he believes to be embodied in Perrin's description and that he proposes to model using the notion of fractal.

Portrait lights are used, one on the viewer's side and one to the left, but of course this flake never existed as a hard object, only in a computer's memory.

Furthermore, precisely the same drawing could have been inspired by the description of a small asteroid. An earlier version of a portion of this Figure was oddly reminiscent of one of the better known recent "photographs" of the Loch Ness monster. Could this conver-

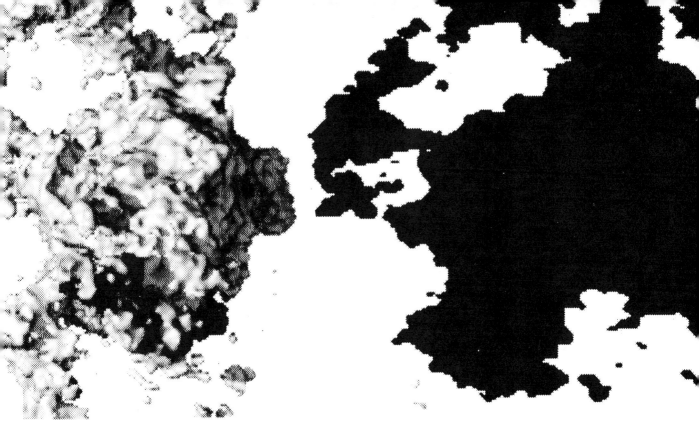

gence of form be merely coincidental?

Readers who would argue that real flakes are even farther removed from a polished ball of steel ought to inspect the alternative flake shown in Plates 222 and 223. Their caption also performs a task that could not be tackled at this stage of this Essay; it explains how the present flake has been obtained.

The bottom Figure of Plate 9 represents a view from a distance, so it shows very little detail. The top Figures represent a close view containing much additional detail. At the top of Plate 9, we have included (after having enhanced it strongly) the shadow cast by the side source of light. At the top of Plate 8, we have a side view of the same flake.

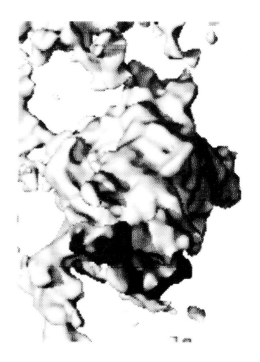

Plate 11

CLASSICAL EXAMPLES OF
PHYSICAL BROWNIAN MOTION

The phenomenon of Brownian motion is described in Perrin 1909 as follows: "When we consider a fluid mass in equilibrium, for example, water in a glass, all the parts appear completely motionless. If we put into it an object of greater density, it falls, and if it is spherical, it falls vertically. The fall, it is true, is the slower the smaller the object; but a visible object always ends by reaching the bottom of the vessel. When at the bottom, it does not tend again to rise. However, it would be difficult to examine for long a preparation of very fine particles in a liquid without observing that instead of assuming a regular movement of fall or ascent, they are animated with a perfectly irregular movement. They go and come, stop, start again, *mount*, descend, *mount again*, without in the least tending toward immobility."

Deliberately, the present Plate is the only one in this book to picture a natural phenomenon. It is reproduced from Jean Perrin's *Atoms*. The bottom Figure combines three detailed tracings, and the top Figure is one bigger piece of the motion of a colloidal particle of radius 0.53μ, as seen under the microscope. The successive positions were marked every 30 seconds (the grid size being 3.2μ), then joined by straight segments having no physical reality whatsoever.

To resume our free translation from Perrin, one may be tempted to "define an 'average velocity of agitation' by following a particle as accurately as possible. But such evaluations are *grossly wrong*. The apparent average velocity varies crazily in magnitude and direction. [This Plate] gives only a much weakened idea of the prodigious entanglement of the real trajectory. If indeed this particle's positions were marked down 100 times more frequently, each segment would be replaced by a polygon relatively just as complicated as the whole drawing, and so on. It is easy to see that in practice the notion of tangent is meaningless for such curves."

This Essay shares Perrin's concern with irregularity but attacks it from a different angle. The impossibility of defining a tangent (and the associated presence of nondifferentiable continuous function) will play a subsidiary role. We shall instead stress the fact that when a Brownian trajectory is examined increasingly closely (Chapter III), its length increases without bound.

In fact it ends up by filling the whole plane. Is it not tempting to conclude that in some sense still to be defined, this peculiar curve's dimension is the same as the plane's?

A principal aim of this Essay will be to show that the loose notion of dimension indeed splits into several distinct components. Brownian motion is *topologically* of dimension 1. However, being practically plane filling, it is *fractally* of dimension 2. The discrepancy between these two values will, in the terminology introduced in this Essay, qualify Brownian motion as being a fractal.

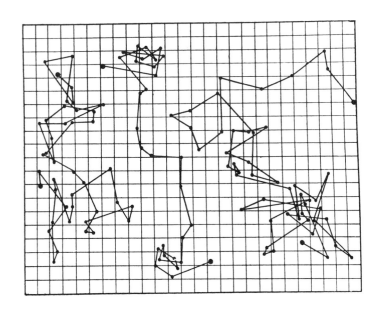

points, where it reaches an infinite value.

"Analogous considerations are applicable to properties such as velocity, pressure or temperature. We find them growing more and more irregular as we increase the magnification of our necessarily imperfect image of the universe. The function that represents any physical property will form in intermaterial space a *continuum* that presents an infinite number of singular points.

"An infinitely discontinuous matter, a continuous ether studded with minute stars, is also the picture represented by the cosmic universe. Indeed, the conclusion we have reached above can also be arrived at by imagining a sphere that successively embraces planets, solar system, stars, and nebulae."

FROM PERRIN
THROUGH WIENER

As a first comment on the preceding excerpts, it should be stressed that physicist-philosopher Perrin wrote in an informal style but gave every evidence of having had a precise knowledge of the most advanced mathematics of his day.

Thus the version of our quote published in *Atoms* is prefaced by the words "I wish to offer a few remarks designed to give objective justification to certain logical demands of the mathematicians." Furthermore, a short fragment from the 1906 paper not included in the preface of *Atoms* reads as follows: "Allow us now a hypothesis that is arbitrary but not self contradictory. One might encounter instances where using a function without a derivative would be simpler than using one which can be differentiated. When this happens, the mathematical study of irregular continua will have proven its practical value." Then, starting a whole new section for emphasis, "However, this is as yet nothing but a daydream."

Perrin had the good fortune of see *a part* of this daydream become reality in his lifetime. He initiated this implementation with the actual experiments on Brownian motion which were to bring him fame. Second, the restatement of his thoughts in Perrin 1909 chanced to catch the attention of Norbert Wiener around 1921 (Wiener 1956, pp. 38-39, or 1964, pp. 2-3) and, to his "surprise and delight" moved him to define and study rigorously a mathematical model of Brownian motion. Through Wiener's intermediacy, some of Perrin's ideas have become extraordinarily influential

However (insofar as I can tell), the overwhelming bulk of the problems sketched in Perrin 1906 were never studied from the viewpoint he had foreseen.

It seems that before his daydream could start being implemented, it had, so to speak, to be replayed independently in someone else's mind (sixty years later) under more favorable circumstances.

Much delay in a full implementation was probably inevitable because many of the necessary tools were not available. Today, nothing exceeds in usefulness the process which Perrin inspired Wiener to define, the Bachelier-Wiener-Lévy Brownian motion and the countless variants of it due to the ingenuity of mathematicians. Another tool of a different kind now available, which was not invented until 1919, is *fractal dimension*, but it will be better not to discuss this notion until later on.

THE FRACTAL PALACE: MUSEUM AND TOOLSHED

Nevertheless, other tools of fractal geometry reach back before Brownian motion. For the two basic ones, it seems clear that they were already known to Perrin. The oldest recorded nondifferentiable continuous function is due to Weierstrass, and its publication in duBois Reymond 1875 is chosen in this Essay as the starting point of the 1875-1922 crisis of mathematics. As to the "infinite number of singular points in the continuum representing a physical property," this was doubtless meant to refer to the Cantor set. And the following of Weierstrass and Cantor soon created a huge collection of other sets of the same sort.

They might, since they are not ordinarily called classical, be viewed as modern, but it would be silly to continue applying the word modern to mathematics that, in fact or in spirit, belong to the period from 1875 to 1922. However, they do remain nonclassical in the sense that they had never been taught properly. From being scarecrows, they went on all too quickly to be viewed as special examples hardly worthy of individual consideration by mathematicians. Many of them are best known to practitioners of mathematical games. This is how, to take on example, Koch's simple variant of the Weierstrass function came to be called *snowflake curve*. This collection of sets went so far afield as to be called a "Mathematical Art Museum" (in N. Ya. Vilenkin 1965, a charming little book). Others (beginning with Henri Poincaré) have called it a "Gallery of Monsters." One of the aims of the present Essay will be to show that these descriptions were either inadequate or wrong, and that this collection of sets may also be visited as a "Museum of Science." My teacher Paul Lévy (great even when anachronistic;

see Chapter XI) contributed a good deal to building it by stressing the role of chance. I have also enriched it a little. Mathematics is to be praised for having put these sets at our disposal long ago and scolded for having discouraged us from using them.

This splendid collection of mathematical sets should prove invaluable to the concrete minded scientist, and I hope, by using it in diverse contexts, to help make it known and used by others. Under the formalistic shell that has insulated it, its basic ideas are extremely simple and intuitive. One can begin to work with fractals very quickly, with very light technical apparatus. There is almost no need to enter into those formal preliminaries that all too often tend to be viewed as either an impenetrable jungle or an Eden one does not wish to leave behind.

The ideal is to avoid either distortion. Deliberate over-generality and abstraction, the star billing given to "proper form," and the proliferation of concepts and terms have often done more harm than good in the sciences. It should be a matter of regret that the least exact among them, sciences whose very principles are the least certain, tend to be the most concerned with rigor, generality, and axiomatics. Thus I am delighted to feel I may have identified fresh examples in which, in classical fashion, form seems to be intimately related to substance.

MULTIPLE FACETS OF DIMENSION; DEFINITION OF FRACTALS

It seems useful at this point to insert a section that breaks the sequence from substance to form and uses mathematical terms without dwelling on their definitions. This section is unavoidably addressed to mathematicians specializing in these matters, but other readers may also find it helpful or at least reassuring, and anybody can skip it.

It is the first of many digressions in this Essay, designated by the symbols ◁ and ▶. The latter is very bold, so as to be readily found by anyone who becomes lost in a digression and wants to skim ahead. The beginning sign is less bold, to prevent digressions from receiving excessive attention. And there will even be a few digressions within digressions. It should be mentioned that material to be discussed in the body of the Essay at a later stage will often receive advance mention in digressions. Such is the case in the present instance. And of course all the material in this section is discussed in greater detail (including references) in Chapter XII.

◁ During the crisis of mathematics to which we alluded, 1875 to 1922, it became increasingly clear that a proper understanding of irregularity or fragmentation (as of their opposites, regularity and connectedness) must among others involve an analysis of the intuitive notion of dimension. Such an analysis eventually revealed that said notion involves a whole "bundle" of facets that not only are conceptually distinct but may lead to different numerical values. Analysis also revealed that Euclid is limited to sets for which all the useful dimensions coincide, so that one may choose to call them *dimensionally concordant* sets.

◁ On the other hand, many of the paradoxical aspects of the Cantor set could be traced to the fact that it is *dimensionally discordant*. So is the Koch curve. Furthermore, the discordant character of the basic fractals is not at all a minor nuisance. Rather, it is such a basic feature that we shall now use it to give a tentative definition of the concept of fractal.

◁ To be specific, we shall consider two definitions of dimension, each of which, to paraphrase Hurewicz & Wallman 1941, assigns to every set of points in ordinary Euclidean space, no matter how "pathological," a real number which on intuitive and formal grounds strongly deserves to be called its dimension. The more intuitive of the two is the topological dimension, which was put in final shape in 1922 by Menger and Urysohn – hence the date selected as closing the 1875-1922 crisis. We shall denote it by D_T. The second dimension had been formulated a bit earlier, in Hausdorff 1919 and was put in final form by Besicovitch. We shall denote it by D.

◁ The two dimensions equal at least 0 and at most the dimension E of the Euclidean space in which we work. But the similarity ends here: the two dimensions need not coincide, and they only satisfy the inequality

$$D \geq D_T.$$

The cases where $D = D_T$ include all of Euclid, and the cases where $D > D_T$ include every set I was ever tempted to call fractal. ($D - D_T$ is for them a kind of dimensional excess.) Hence, there is no harm in proposing the following definition:

A fractal will be defined as a set for which the Hausdorff-Besicovitch dimension strictly exceeds the topological dimension.

◁ For example, a set to be described in Chapter III as the trail of Brownian motion is a fractal because $D = 2$, as we shall see in said chapter, while on the other hand $D_T = 1$.

◁ As to the original Cantor set, it is a fractal because

$$D_T = 0 ,$$

and we shall see in Chapter IV that

$$D = \log 2 / \log 3 > 0 .$$

And a Cantor set in E-dimensional space can be tailored so that

$$D_T = 0 ,$$

and D takes on any desired value between D_T (excluded) and E (included).

◁ Furthermore, the Koch curve is a fractal because

$$D_T = 1 ,$$

and we shall see in Chapter II that

$$D = \log 4 / \log 3 > 1 .$$

◁ As seen in the preceding examples, D_T is *always an integer*, but D *need not be an integer*. This most striking characteristic deserves a terminological aside. If one is to use the term *fraction* broadly as synonymous of a noninteger real number, several of the above-listed values of D are fractional, and indeed Hausdorff-Besicovitch dimension is often called *fractional dimension*.

◁ However, this term is awkward because (a) it disturbs those who view *fraction* as a synonym of *rational number* and (b) nothing prevents D from being an integer (not greater than E but strictly greater than D_T). The widely used contracted form, *Hausdorff dimension*, is historically inaccurate and happens to encounter an unexpected difficulty among certain nonmathematicians. Since Felix Hausdorff is best known for Hausdorff topological spaces, the term *Hausdorff dimension* seems to have undertones of the "dimension of a Hausdorff space," thus suggesting it is a topological concept – which emphatically is not the case.

◁ When this Essay was underway, I found to my surprise that there was no term to designate the sets for which $D > D_T$ and to distinguish the mathematical structures relative to these sets from the topological structures with which they tend to be confused because both concern different facets of form. This need led to coining the term *fractal*. Now that it exists, one might as well refer to the Hausdorff-Besicovitch dimension as the fractal dimension. ►

SPATIAL HOMOGENEITY AND SELF SIMILARITY

While adding fractals to the model maker's tool kit to increase its versatility, one must not increase its weight too much. The most useful shapes in Euclid were the simplest, such as lines, planes, or spaces, and the simplest physics arises when some quantity such as density, temperature, pressure, or velocity is distributed upon them uniformly, in a homogeneous manner.

When we go on to fractals, we must also seek to keep as long as feasible to those which can somehow be viewed as homogeneous, in the sense that they are *invariant by certain translations*. Such is, for example, the case for various curves associated with Brownian motion. If it is divided into parts, said parts cannot be precisely superposed on each other – as can be done with equal parts of a straight line – nevertheless the division can be such that the parts are alike in a statistical sense. Nearly all the fractals introduced by the great precursors and nearly all the fractals for which we shall find an actual use will also be either strictly or statistically homogeneous.

◁ When homogeneity is allowed to be no more than statistical, it becomes possible to make it more demanding, so to speak, from other viewpoints. This is one reason why chance is so valuable a tool in the study of fractals. ▶

Furthermore, each piece of a straight segment is like a reduced scale version of the whole, and the same will be true for nearly all the fractals we shall encountered in this Essay. Either in a strict or a statistical sense, nearly all are *invariant by certain transformations of scale*.

Many are *self similar fractals*. In this compound term, the adjective serves to mitigate the noun. While the primary term *fractal* points to disorder and is compatible with intractable irregularity, the modifier *self similar* points to a kind of strict order. The term *self similar fractals* can alternatively be explained by taking *self similar* as the primary term, thus pointing first to strict order, and then adding *fractal* as a modifier meant to exclude lines, planes, and so on.

One must stress that the generalization of self similarity beyond lines and planes is not new in mathematics. Also, it is not new in science, since Lewis F. Richardson 1926 postulated that over a wide range of scales turbulence is made of self similar eddies. Furthermore, striking *analytical* consequences of this idea were drawn in Kolmogorov 1941. A novelty of this Essay is that it is addressed to *geometric* aspects of self similarity in Nature.

◁ In addition, the notion of self similarity seems tantalizingly close to the physicists' much more recent and still unsystematic notions of *scaling* and of renormalization groups. ▶

LIMITATIONS OF TOPOLOGY IN THE STUDY OF FORM

If a mathematician is asked which is the well-defined branch of mathematics that studies form, he is very likely to mention topology. Clearly, the present Essay is predicated upon the belief that this answer is incomplete: the loose notion of form possesses mathematical aspects other than topological ones.

It is necessary to elaborate, in order to avoid any misunderstanding on this matter. Topology is the branch of mathematics that teaches us, for example, that all pots with two handles are of the same form because, if they were made of an infinitely flexible clay, each could be molded into any other continuously, without tearing any new opening or closing up any old one.

Obviously, this particular aspect of the notion of form is not useful in the study of individual coastlines, since it simply indicates that they are all topologically identical to each other – and to a circle! And the identity is underlined by the fact that the topological dimension is in each case equal to 1. Furthermore, if one includes in an island all its offshore "satellites," the cumulative coastline is topologically identical to "many" circles. Exactly how many is hard to tell, as the number depends on the precise definition of the terms *island* and *satellites*. In practice, the number of islands is always so large that it can be viewed as infinite.

Thus (whether or not one counts the satellite islands), when it comes to discriminating between different coastlines,

topological form is too general an instrument and hence too blunt. In other analogous instances topology does contribute something, but rarely as much as one may want.

By way of contrast, it will be seen that coastlines of different "degrees of irregularity" tend to have different fractal dimensions. Differences in fractal dimension express differences in a nontopological aspect of *form,* which can be called the *fractal form.*

PHYSICAL DIMENSION

Before discussing dimensions that may be fractions, we must say a few more words about *physical dimension.* It is an intuitive notion that appears to go back to an archaic state before Greek geometry, yet deserves to be taken up again. One aspect of this Essay is that each of the chapters that follow will study one category of concrete objects for which a physical dimension is a fraction or is otherwise nonstandard.

Physical dimension concerns the relation of sets to objects, where the two terms will apply respectively to mathematical concepts and to patterns from reality. Strictly speaking, objects such as a small ball, a veil, or a thread – thin though they may be – should all have to be represented by three-dimensional sets. In practice, however, every physicist knows that one cannot handle these objects the way one handles large balls. It is much more useful to think of a veil, a thread, or a ball – if they are fine enough – as closer in dimension to 2, 1, and 0, respectively.

Let us restate and clarify this last assertion. It expresses that in order to describe a thread, the theories relating to sets of dimension 1 or 3 require corrective terms. Therefore, after the fact, the better geometrical model is the one in which those corrections are smaller. If our luck holds, this model will continue to give a good idea of what is being studied even when corrections are omitted. In other words, physical dimension inevitably has a subjective basis. It is a matter of approximation and therefore of degree of resolution.

MANY DIFFERENT DIMENSIONS IMPLICIT IN A BALL OF THREAD

To confirm this last hunch, we will take up an object more complex than a single thread, namely, a ball of 10 cm diameter made of a thick thread of 1 mm diameter. Depending on one's viewpoint, it possesses (in a latent fashion) several distinct physical dimensions.

Indeed, at the resolution possible to an observer placed 10 m away, it appears as a point, that is, as a zero-dimensional figure. At 10 cm it is a ball, that is, a three-dimensional figure. At 10 mm it is a mess of threads, that is, a one-dimensional figure. At 0.1 mm each thread becomes a sort of column and the whole becomes a three-dimensional figure again. At 0.01 mm resolution, each column is dissolved into filiform fibers, and the ball again becomes one-dimensional, and so on, with the dimension jumping repeatedly from one value to another. And below a certain level of analysis, the ball of thread is represented by a finite number of atomlike pinpoints, and it becomes zero-dimensional again.

That a numerical result should thus depend on the relation of object to observer is of course in the very spirit of physics in this century and is even a particularly exemplary illustration of it. The inevitable result is that depending on the criteria used, different observers may disagree as to the number of distinct dimensions latent in the same object. Where one observer sees a zone having its characteristic D, others are likely to see only a gradual transition which may not deserve separate study.

In the same way, the objects we are going to consider will offer a succession of different dimensions. The novelty will be that certain zones of transition will be reinterpreted as being fractal zones within which the fractal dimension is greater than the topological dimension.

The reality of none of these zones will be fully established until it is associated with a true deductive theory. In general this goal will lie beyond this Essay's ambition. It is furthermore indisputable that, just as the entities according to William of Occam, the dimensions must not be multiplied beyond necessity. In particular, certain fractal zones may be so narrow that they need not be truly worth distinguishing. The best, however, is to delay the examination of these doubts until such time when their object will have been duly described.

OF "NOTIONS THAT ARE NEW, BUT"

Before agreeing to become acquainted with strange tools of thought, the reader may want to know what is in my view their present contribution to Natural Science. The ideal would be to evade the question and to let the reader judge, but a few words are inescapable.

Most emphatically, I do not view fractals as a panacea, therefore I shall state my claims in a low key.

It must be acknowledged that in a few applications all I have done thus far is to focus on a concept that had already been isolated and to interpret it geometrically

as being a fractal dimension. Such is, for example, the case of some exponents encountered in the study of polymers. The usefulness of such relabeling must be evaluated on its own merits. In the case of polymers, the new label is apparently stimulating, but one must beware of terms that are merely of esthetic, or simply even cosmetic, interest. Being a language (J. W. Gibbs), mathematics may be used not only to inform but also, among other things, to seduce. We must watch out – in the wise words of Henri Lebesgue – for "notions that are new, to be sure, but of which no use can be made after they have been defined."

However, the self criticism expressed in the preceding paragraph applies to an insignificant portion of this book. I believe that in every case to be discussed in detail, the concepts of fractal and fractal dimension contribute much more than new terms to denote old ideas.

For example, it is characteristic of many of the phenomena tackled in this Essay that observations or experiments concerning the same object but employing different methods appear to contradict each other. If each of them is indisputable, the unconscious conceptual framework within which the data were originally interpreted must be radically inappropriate. In conclusion it will emerge, painlessly and almost without notice, that the simplest solution to each of these paradoxes happens to be the introduction of a fractal, and that the intrinsic parameter one has been seeking is a fractal dimension.

To relabel an old parameter was neither a good nor a bad thing, but to identify a new one is viewed by almost everyone in a favorable light. Once a proper parameter is identified, all mystery seems to dissipate: some measurements begin to make sense (typically they turn out to hasten a numerical evaluation of the fractal dimension), while other measurements remain meaningless because they turn out to be attempts to answer an "ill-posed problem."

This last notion is of course to be contrasted with that of "well-posed problem." The distinction between the two is a step any science must take sooner or later. This step is vital and often nontrivial. ◁ Mechanics did not take it until Hadamard's work at the beginning of this century. ▶

The models referred to in the preceding paragraphs are descriptive and no more than "phenomenological." However, phenomenology must not be despised. One might well argue that many theoretical endeavors lag because they attempt to skip a very useful stage of scientific analysis. Many of the most successful theories of the past had not consisted in deriving a new formula to account for raw data but in accounting for descriptive laws that had already been derived or guessed previously using entirely different means.

In the same way, the phenomenological description of coastlines in Chapter II as fractal curves will be the basis of theorizing in Chapter IX. The theory will only partly succeed, since it will confirm that coastlines ought indeed to be fractal curves but will yield the "wrong" dimension: $D = 1.5$ instead of the observed $D \sim 1.2$ to $D \sim 1.3$. In order to obtain the correct D, it is still necessary to resort to an artificial "fix," nevertheless I believe that phenomenology has already fully justified itself in this case. Similarly, the buildup of a theory of stellar distribution has already been helped by the availability of a phenomenology backed by rich mathematics; see Chapter V. However, the present theory – also to be tackled in Chapter V – predicts $D = 1$ instead of the observed $D \sim 1.2$ to $D \sim 1.3$.

More generally, it will be possible in many cases to show that a fractal that I had picked out merely as a solution to a puzzle of Nature can be explained through the interaction of more natural elementary mechanisms. And even those phenomena for which there is as yet no explanation are likely to be modified qualitatively after many riddles that seem to be of limited impact are pooled in one that is very widely encountered.

◁ Incidentally, D may be related to several notions that physicists call *dimensions*. Some have been described as useful devices without theoretical significance, but nothing proves that such is indeed the case. ▶

THIS ESSAY MIXES STYLES; IS SEMIPOPULAR *AND* SCHOLARLY

Now that the objective of the present Essay has been outlined, we must examine its manner. The above heading is a confession that carries no regret, for in my opinion the walls that separate literary styles are not brought down as frequently as they should be. In science, the same remark applies also to disciplines. Anyone who chooses to disregard these walls resigns himself in advance to losing many readers at an early stage. On the other hand, it is evident that such an Essay would not be undertaken by someone who did not think that other readers would react differently and in particular will show tolerance regarding the numerous compromises that are unavoidable whenever one mixes styles.

A first characteristic of the text is that pains are taken to develop all problems and techniques from the outset, so as to make the work accessible to a public not necessarily specializing in the various subjects tackled. In other words, *this work is in part expository. But this is not its main purpose.*

Further, an attempt is made not to frighten away needlessly those who are not interested in mathematical precision per se. Despite the fact that rigorous mathematical backup is available throughout (and is sounder than in much of theoretical physics), the style is kept informal, the detail being set aside to Chapter XII, the Mathematical Lexicon. Serious original work is not expected or even supposed to show such concerns. From these points of view, therefore, this Essay has taken on some appearances of a *work of popularization. But this also is not its main purpose.*

In addition, this Essay takes on some appearances of a *work of erudition* because it follows many paths back into history, even some that are solely of antiquarian interest and became known too late to influence in any manner whatsoever the development of my own views. These paths are followed because I delight in the history of ideas, and because the incredulity some theses seemed to encounter made me seek to implant them by finding antecedents.

And yet any search for origins is controversial. For each author who a hundred years ago expressed an idea for which we have a good use, we run the risk of finding a contemporary — sometimes the same person — who developed the opposite idea, the arguments being weak in both directions. If these authors were neglected, need we consign one or both to oblivion? Or shall we grant a bit of posthumous glory to the one whom we approve now? Need we bring back to life personalities whose trace had disappeared because we lend only to the rich (the phenomenon Robert Merton has called Matthew Effect), and because often the work of one individual is only accepted on the basis of the superior authority of another who adopts it and makes it live on under his name?

We must beware of excessive erudition in relation to the history of ideas. We know how difficult it is to speak of unrecognized geniuses without falling into the romantic trap which implies that there are no others. On the other hand, Stent 1972 might lead us to the conclusion that to be ahead of one's time is to be premature, a failing that deserves nothing but compassionate oblivion. In any event, I did wish to assert these links with the past, stressing them further in the biographical sketches in Chapter XI.

The main purpose of this Essay, however, lies elsewhere. As has been stressed in the introductory paragraphs, this work is above all a description (both monographic and synthetic) of theories and theses which I believe to be new.

So far, they are either incompletely described in my specialized articles, or even unpublished, and I do not propose to develop any of them to the full detail desired by the specialists. Is it right to gather miscellaneous "prefaces" in this fashion? Does science in the making lend itself to such an Essay? Is it wise to attempt to give wide readership to the basic aspects of theories that have scarcely seen the light of day — which implies that scientific opinion has not had time to express itself fully about them? One must hope that the reader will judge on the evidence.

THE ROLE OF GRAPHICS

Before proceeding further, the reader is again advised, if he has not yet done so, to browse through the illustrations. This Essay was designed as a picture book to help make its contents accessible in various degrees to a wide variety of actual readers and of mere lookers.

Nearly all these pictures are computer generated and, because of their importance to my goals, one must digress a moment to comment on their precise interpretation. There is no question that any attempt to illustrate geometry involves a basic fallacy. For example, a straight line is, strictly speaking, un-bounded and infinitely thin and smooth, while any illustration is unavoidably of finite length, of positive thickness, and rough edged. Nevertheless, a rough evocative drawing of lines is felt by many to be useful and by some to be necessary in order to build up intuition and help in the search for proof. And of course, when it comes to providing a geometric model of a thread, a rough drawing is in fact more adequate than the mathematical line itself. In other words, it suffices for all practical purposes that a geometric concept and its image should fit within a certain range of characteristic sizes that lies between a sufficiently large but finite outer scale and a sufficiently small but positive inner scale.

Thanks to advances in computer-controlled graphics, the same kind of evocative illustration has become practical in the case of fractals. It is, for example, characteristic of a self similar fractal curve that it be (again) unbounded and infinitely thin. Also, each has a very specific kind of *un*smoothness, which makes it more complicated than anything in Euclid. The best representation, therefore, can only hold within a limited range, on the principles we have already encountered. From the viewpoint of applications, however, cutting off the very large and the very small de-tail is not only quite acceptable but even eminently appropriate. Almost everywhere in Nature, both cutoffs are either present or suspected. Thus the typical fractal curve can be evoked satisfactorily by combining elementary strokes in finite number.

The larger the number of strokes and the accuracy of the process, the more attractive the representation, and also the more useful. It is safe, indeed, to use very rough outlines in the study of topological concepts that are by definition deformation invariant. To the contrary, fractal concepts refer to the mutual placement of strokes in space, and it is vital in illustrating them to keep as close as feasible to precise scale. Hand drawing would have been prohibitive, but I was fortunate in having access to an experimental computer graphics device that produces camera-ready copy. This Essay provides a sample of its output.

Graphics is a wonderful tool for matching models with reality. When a chance mechanism agrees with the data from some analytic viewpoint but simulations of the model do not at all "look real," the analytic agreement should become suspect. Any formula can relate to only a small aspect of the relationship between model and reality, while the eye has enormous powers of integration and discrimination. In other words, one should not be worried here by the fact that statisticians tend to stress the opposite pitfall. It is true that the eye can sometimes see spurious relationships which analysis later negates, but this problem arises mostly in areas of science where samples are very small. In the areas we shall explore, samples are huge.

In addition, graphics helps find new applications for existing models. I first experienced this possibility with the Brownian motion that is honored by an illustration in Volume I of William Feller's treatise, *An Introduction to Probability Theory and Its Applications,* Feller 1950. The whole curve looked to me like a mountain's profile, or perhaps a cross-section, and its zeroset looked like a record of telephone errors. The ensuing hunches each led eventually to a full theory, to be presented in Chapters IX and IV, respectively. Since then, my own computer-generated illustrations provided similar inspiration, both to me and to others who were kind enough to act as my "scouts" in more sciences than I knew existed.

Naturally, graphics is extended by cinematography, and it so happens that a film concerned with some classical fractals is available (see Max 1971).

Finally, the book includes a different kind of graphics. The end papers, the frontispiece, and several captionless patterns used as filling or decoration are examples of semi-abstract "art." Some are fractals of which no use is made in this book. Others were created by preserving from destruction some attractive outcomes of faulty computer programming. Ordinarily (though the fact may not be readily apparent) the fault consists merely in a misplacing of a feature or two.

POINTS OF LOGISTICS

The order adopted in Chapters II through X is primarily ruled by convenience of presentation, in particularly by a desire to rank topics by increasing complexity. It is easiest to begin with curves, continue with points, and end by introducing surfaces.

Also, it seems best to begin with problems to which the reader has most likely given little thought, a fact that will have saved him from prejudice. The end of the discussion begun in Chapters II and III is, however, put off until Chapter IX, because it should be more approachable after the reader has become accustomed to the fractal way of thinking.

To complement the text, a great deal of information is included in the captions and in Chapter XII, which has already been described as containing definitions, addenda, and references. On the other hand, the amount of built-in repetition is such that the reader is unlikely to lose the thrust of the early stages of the argument if he skips the passages he feels to be either repetitious or too complicated (in particular, those that go beyond the most elementary mathematics).

As has already been mentioned, the writer will every so often feel the need to engage in private conversation, so to speak, with some readers who might be overly troubled if some specific point were left unmentioned or unexplained. To avoid the use of footnotes, the digressions are marked by the brackets ◁ and ►, which should make them easier to skip. Some digressions are of a different character. They are devoted to incidental remarks I have had no time to explore

fully. To avoid disruption, digressions other than very short ones are concentrated at the ends of sections.

An attempt has been made to indicate at first glance whether the discussion is concerned with theoretical or empirical dimensions D. The latter are mostly known to two decimals and will therefore be written as 1.2 or 1.37, ...). The former will be written as integers, ratios of integers, ratios of logarithms of integers, or in decimal form to *four* decimals. The Apollonian gasket's dimension (Chapter VIII) does not fit in this scheme, being empirical but known to more than two decimals; it is written with six decimals.

CHAPTER II

How Long Is the Coast of Britain?

In this chapter we shall begin with one particularly obvious aspect of the surface of Earth and shall introduce a first category of fractals, namely, connected physical curves for which the fractal dimension is greater than 1.

In the manner of Jean Perrin (see Chapter I), we shall consider a stretch of rugged coastline. Whereas Perrin concerned himself with defining its tangents, we shall attempt to measure its length.

It is evident that this length is at least equal to the distance along a straight line between the beginning and the end of our stretch of curve. Moreover, had the coastline been straight, the problem of length measurement would have been solved by this first step. However, a truly rugged coastline is so irregular and winding as to be much longer than the straight-line distance mentioned above. There are various ways of evaluating it more accurately, and we shall momentarily proceed to analyze them. We will

see that they all lead to the conclusion that the final estimated length is not only extremely large but in fact so large that it is best considered infinite. Hence, if one wishes to compare different coasts from the viewpoint of their "extent," length will be inadequate. We shall seek an improved substitute, and in doing so shall find it impossible to avoid the introduction of various forms of the concept of fractal dimension and measure.

After this task is performed, the scope of this chapter will broaden gradually to rivers, then to river networks. Eventually we shall wander far out and examine certain facets of Man's lung and blood vessels.

MULTIPLICITY OF ALTERNATIVE METHODS OF MEASUREMENT

Method A: Set dividers to a prescribed opening η, the yardstick, and walk them along the coastline, in the sense that each new step starts where the previous step leaves off. The number of steps multiplied by η is an approximate length $L(\eta)$. If we make the opening of the dividers smaller and smaller and repeat the operation, we expect this $L(\eta)$ to soon cease to change and to take a well-defined value to be called the *true length*. In most cases, however, the observed $L(\eta)$ tends to increase without limit.

Before discussing this finding, let us note that the principle of the above-mentioned procedure consists, first of all, in acknowledging that a coastline is too irregular to be measured directly by reading it off in a catalog of lengths of simple geometric curves. Therefore, our Method A replaces the coastline by a sequence of polygons, each of which is easier to handle because it is arbitrarily "smoothed out." We are reminded of Perrin's image of a sheet of tinfoil one may put around a sponge without really following its outline. Such "smoothing out" can also be accomplished in other ways.

Method B: We can imagine a man walking along the coastline, keeping away from it by no more than the prescribed distance η, and taking the shortest possible path. Then he resumes his walk, after having reduced his yardstick; then again, with another reduction, and again, and again, until the value of η

reaches, say, 1 m. Any finer detail is inaccessible to this sort of analysis. One may go further and argue (a) that it is of no direct interest to Man and (b) that very fine detail varies with the seasons and the tides so much that it is altogether meaningless. We shall take up argument (a) later on in this chapter. In the meantime, we shall neutralize argument (b) by restricting our attention to a rocky coastline observed when the tide is low and the waves are negligible. To follow it in even smaller detail, our man must be replaced by a mouse, then by an ant, and so forth. Again, as our walker wishes to stay increasingly closer to the coastline, the distance to be covered will continue to increase with no limit in sight.

Method C: To avoid the asymmetry between land and water created by Method B, there is a good method due to G. Cantor. It views the coastline, so to speak, with an out-of-focus camera that transforms every point into a circular blotch of radius η. In other words, it considers all the points of both land and water for which the distance to the coastline is no more than η. They form a kind of sausage or snake that, in effect, provides an intrinsic method of covering the coastline by a tape of width 2η. An example cannot be drawn until later on, but one can peek ahead to Plate 33 (without bothering to read the caption yet). We then measure the surface of the tape and divide it by 2η. If the coastline were rectilinear, then the tape would be a rectangle, and the above ratio would be the actual length. With actual coastlines, this ratio is an estimated length. As η decreases, this estimate is, again, found to increase without unarguable limit.

Method D: Imagine a map drawn by a pointillist painter who uses circular blotches of radius η. It is clear that the coastline can be covered completely with such circles. Had we demanded that their centers be all the points of the coastline, we would be brought back to Method C. By contrast, the present method will require nothing of the circle centers, and will require that the number of blotches be as small as possible. As a result, they may well lie mostly inland near the capes and mostly in the sea near the bays. The total area of the circles, divided by 2η, is another estimate of the length.

We will soon discuss the results yielded by these diverse methods. However, the main finding is always the same. As η is made smaller and smaller, every one of the approximate lengths tends to become larger and larger without bound. Insofar as one can tell, each seems to tend toward infinity.

ARBITRARINESS OF THE RESULTS OF MEASUREMENT

Before we investigate the empirical rule that governs the rates of increase of these various approximate coastline lengths, we should stop for a few words about the possible existence of intrinsic values for the yardstick η. (A discussion of the practical role of the mathematical passage to the limit – a procedure that physicists need but prefer to handle with caution – will be carried out later in the chapter.)

In order to ascertain the meaning of what has just been established, let us imagine we perform the same measurements on a coastline that man has tamed, say, the coast at Chelsea as it is today. It might be foreseen that every method of measuring the length with a yardstick equal to η will again yield a result that increases until η becomes 200 or perhaps only 20 meters. Beyond that, however, there is likely to be a zone in which $L(\eta)$ varies only a little. $L(\eta)$ does not begin to increase again until the η becomes less than, say, 20 centimeters long, that is, until the corresponding approximate length begins to take into account the irregularity of the stones in the harbor. Thus, if we trace the curves representing $L(\eta)$ as a function of η, the length measured today will most likely exhibit, in the zone of η's between $\eta = 20$ meters and $\eta = 20$ centimeters, a flat portion that was not observable in the past. (The values of 20 meters and 20 centimeters should not be taken too seriously.)

Measurements made in this zone are obviously of greater practical usefulness than any other measurements. It follows that to the extent that boundaries between the different scientific disciplines are only a matter of conventional division of labor between scientists, one might restrict geography to the study of phenomena well above Man's reach (for example, above 20-200 meters). This restriction would yield a well-defined value of geographical length. The Coast Guard may well choose to use the same η for untamed coasts, and encyclopedias and almanacs could adopt the corresponding $L(\eta)$.

However, the adoption of the same η by all the agencies of a government is hard to imagine, and its adoption by all countries is all but inconceivable. For example, Richardson 1961 reports that the recorded lengths of the common frontier between Spain and Portugal are entirely different in Spanish and Portuguese encyclopedias. The discrepancy must at least in part result from different choices of η.

More important, Nature does exist apart from Man and to give too much weight to any specific η and $L(\eta)$ is to let the study of Nature be dominated by Man, either through his typical step size or his highly variable technical reach. The basic uncertainty concerning the value of a coastline's length cannot be legislated away in this fashion. In one manner or another, the concept of geographic length is not as inoffensive as it seems. It is not now, nor has it ever been, entirely "objective." The observer inevitably intervenes in its definition.

RICHARDSON'S EMPIRICAL DATA

It so happens that the variations of the approximate length $L(\eta)$ obtained by Method A have been studied empirically, in Richardson 1961. Lewis Fry Richardson was a great scientist whose originality, as we shall see in Chapter XI, mixed with eccentricity, and who did not in his lifetime achieve the fame he deserves. As we shall see in Chapter VI, we are indebted to him for some of the most profound and most durable ideas regarding the nature of turbulence, a field into which he introduced a notion of self similarity. He also concerned himself with other difficult problems, such as the nature of armed conflict between states. His work is characterized by the conception and execution of experiments of classic simplicity, and by respect for the facts thus revealed. He never hesitated to use precise and refined concepts when he deemed them necessary.

The diagrams reproduced in Plate 32 were found among his papers after he died. They all lead to the conclusion that there are two constants, which we shall call λ and D, such that a polygon of side η approximating a coastline has a number of sides roughly of the form $\lambda\eta^{-D}$. Therefore the length $L(\eta)$ is roughly $\lambda\eta^{1-D}$. The value of the exponent D seems to depend upon the coastline that is chosen and it happens that different pieces of the same coastline, if considered separately, produce different values of D. To Richardson, the D in question was a simple exponent of no particular significance. However, its value seems to be independent of the method chosen to estimate its length and thus seems to warrant attention. What can be said about it?

FRACTAL DIMENSION

Having, so to speak, "unearthed" Richardson's work, my initial contribution to this field (in Mandelbrot 1967s) was to suggest that D, even though it is not an integer, be interpreted as a dimension, namely, as a fractal dimension. I had indeed recognized that each of the many methods of measuring $L(\eta)$, as enumerated above, corresponds to one of a variety of nonstandard generalized definitions of dimension already used in pure mathematics. For instance, the definition based on the coastline being covered by the smallest number of large points of radius η is precisely that used by Pontrjagin & Schnirelman 1932 to define the covering dimension. (A brief mathematical exposition of this and the following two methods may be found under COVERING BY SPHERES in Chapter XII.) The definition based on a tape of width 2η is a perfectly natural extension of an idea of Cantor and Minkowski (Plate 33), and is due to Bouligand. Other definitions are linked to the epsilon entropy of Kolmogorov & Tihomirov 1959-1961. But these definitions are both of insufficient generality and too formal to be truly expressive. It is better to examine in more detail two other points of view that are very analogous but geometrically "richer." The best and oldest one dates back to Felix Hausdorff and will serve to define fractal dimension.

The most fundamental task, that of representing and explaining the shape of the coastlines and deducing the value of D from other considerations, will be put off until Chapter IX. It is sufficient at this point to announce that this task will lead us, in a first approximation, to D = 1.5. This value is much too large to

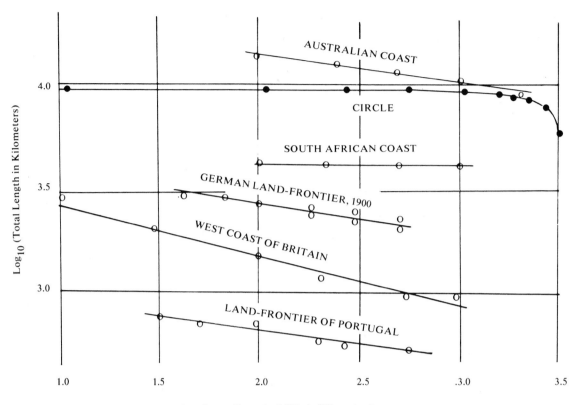

Plate 32

RICHARDSON'S EMPIRICAL DATA
ON THE RATE OF INCREASE
OF COASTLINES' LENGTHS

This Figure reproduces Richardson's experimental measurements of length performed on various curves using polygons of increasingly short side η. As expected, increasingly precise measurements made on a circle stabilize very rapidly near a well-determined value.

In the case of coastlines, on the contrary, the approximate lengths do not stabilize at all. As the yardstick η tends to zero, the approximate lengths, as plotted on doubly logarithmic paper, fall on a straight line of negative slope.

To Richardson, the slope had no theoretical interpretation. The present Essay, on the other hand, interprets coastlines as approximate fractal curves and uses the present diagram's slope as an estimate of 1–D, where D is the fractal dimension.

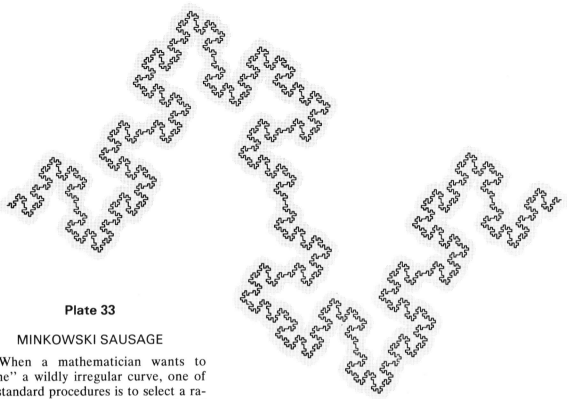

Plate 33

MINKOWSKI SAUSAGE

When a mathematician wants to "tame" a wildly irregular curve, one of the standard procedures is to select a radius η and to draw around each point of the curve a disc of radius η. This procedure, dating back at least to Herman Minkowski, is brutal but very effective. (As to the term *sausage,* rumor has it that it is a leftover of the application of this procedure to curves associated with Norbert Wiener.)

In the present illustration such smoothing has been applied not to an actual coastline but rather to a theoretical curve to be constructed later by continual addition of ever smaller detail (Plate 49). Comparing the piece of sausage to the right with the corresponding piece of sausage above, we see that the construction of the curve passes a critical stage when it begins to involve details of size smaller than η. Later stages of construction leave the sausage essentially unaffected.

describe the facts but more than sufficient to establish that it is, so to speak, natural for a coastline's dimension to exceed the classical value $D = 1$.

The preceding arguments do not constitute a proof that $D > 1$ for coastlines. Such a proof is inconceivable in Natural Science. But at least they should have created doubt. Now that fractal dimension has been injected into the study of coastlines, and even if specific reasons for claiming $D > 1$ come to be challenged, we are unlikely ever to return to the stage when $D = 1$ was accepted thoughtlessly and naively. He who thinks that after all it is useful to consider that $D = 1$ will have to argue his case.

HAUSDORFF FRACTAL DIMENSION

If we accept that various natural coasts are really of infinite length and that anthropocentric length can give but an extremely partial idea of reality, how can different coastlines be compared to each other? Since infinity equals four times infinity, we are still permitted to say that every coastline is four times longer than each of its quarters, but such statements are clearly of no interest. How can we better express the solidly anchored idea that the entire curve has a "measure" that is four times greater than each of its fourths?

A most ingenious method of reaching

this goal has been provided by Felix Hausdorff. Its intuitive motivation starts out from the fact that the linear measure of a polygon is calculated by adding its sides' lengths without transforming them in any way. One may say (the reason for doing so will soon become apparent) that these sides are raised to the power one (which happens to be the Euclidean dimension of a straight line). The surface measure of a closed polygon's interior is similarly calculated by adding the sides of squares that pave it raised to the power two (which is the dimension of a plane).

Let us do likewise for a polygonal approximation of a coastline made up of small segments of length η. If their lengths are raised to the power D, we obtain a quantity we may call tentatively an "approximate measure in the dimension D." Since according to Richardson the number of sides is $N = \lambda \eta^{-D}$, said approximate measure takes the value $\lambda \eta^D \eta^{-D} = \lambda$.

This result is important: the approximate measure in the dimension D is independent of η. With actual data, we simply find that this approximate measure varies little with η.

In addition, there is a simple counterpart and a generalization to the classical fact that by attempting to calculate the length of a square, one finds it to be infinite. Indeed, the limit for $\eta \to 0$ of the coastline's approximate measure evaluated in any dimension d smaller than D is infinite. Similarly, there is a counterpart and a generalization to the fact that the area of a straight line is zero. Indeed, whenever d is larger than D, the corresponding approximate measure tends to 0 as $\eta \to 0$. The approximate measure behaves reasonably if and only if d = D.

See HAUSDORFF ... in Chapter XII for definitions of Hausdorff measure and Hausdorff and Hausdorff Besicovitch dimensions.

A CURVE'S FRACTAL DIMENSION MAY BE > 1; FRACTAL CURVES

By design, the Hausdorff dimension is defined to preserve some aspect of the ordinary dimension, namely, its role as exponent in defining a *measure*. But from another viewpoint, D behaves very oddly indeed. It is a fraction! In particular it exceeds 1, which is the intuitive dimension of all curves and which may be shown rigorously to be their topological dimension. I propose that continuous curves for which the fractal dimension exceeds the topological dimension 1 be called *fractal curves*.

INTUITIVE NOTIONS OF SELF SIMILARITY AND CASCADE

Until now we have stressed the complication that is characteristic of coast-lines considered as geometric figures, but there is also a great degree of order in their structure.

It is indeed striking that when a bay or peninsula first noticed on a map scaled to 1/100,000 is reexamined on a map at 1/10,000, innumerable sub-bays and subpeninsulas become visible. On a 1/1,000 scale map, sub-sub-bays and sub-subpeninsulas appear, and so forth. We cannot go on to infinity, but we can go very far indeed. We find that although the various maps are very different in their specific details, they are of the same overall character and have the same generic features. In other words, it appears that the specific mechanisms that brought about both small and large details of coastlines are geometrically identical except for scale.

The resulting overall generic mechanism might be thought of as a sort of cascade, or perhaps of Italian fireworks, with each stage creating details smaller than those of the preceding stages. When in addition it so happens that each piece of coastline, statistically speaking, is similar to the whole, with the exception of specific details statistics can choose to disregard, the coast will be said to be self similar.

Self similarity, as it has been introduced, is complicated by the need to define the notion of "details one chooses to disregard." Therefore it is useful, first of

Plate 36

TRIADIC KOCH ISLAND
(COASTLINE DIMENSION
D=log4/log3=1.2618):
THE ORIGINAL METHOD
OF CONSTRUCTION

The present construction begins with an "inner island," namely, a black equilateral triangle with sides of unit length. Then the points in the middle third of each side are displaced perpendicularly to the side in question, until they become aligned along a V-shaped peninsula, both sides of which are straight and of length 1/3. This second stage ends with a Star of David. The same process of formation of peninsulas is repeated with the Star's sides, and then again and again, ad infinitum. Some points – such as the triangle's vertices – never move. Other points – such as the nine vertices of the Star of David, other than those it has in common with the triangle – achieve their final positions after a finite number of stages. Still other points continue to be displaced without end. However, they move by decreasing amounts and eventually converge to limit points, which define the island's contour.

The island itself is the limit of a sequence of areas bounded by polygons, each of which contains all the preceding polygons in the sequence. This limit is represented by Plate 39.

NOTE. It seems desirable to make it obvious at first glance whether the value of some D is derived from a theory or from measurements. With this goal in mind, all empirical values are written as decimal fractions carried to one or two decimals. Theoretical values are written either as integers, or as ratios of integers, or as ratios of logarithm integers, or as decimal fractions carried to four decimals.

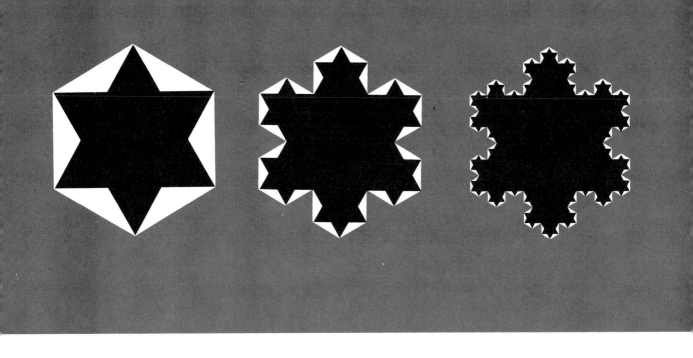

Plate 37

TRIADIC KOCH ISLAND
(COASTLINE DIMENSION
D=log4/log 3):
AN ALTERNATIVE METHOD
OF CONSTRUCTION

An alternative method of constructing the Koch curve devised by Cesàro 1905 begins with an "outer island," namely, a regular hexagon with sides of length $\sqrt{3}/2$. The surrounding ocean is colored in gray. Then one squeezes in bays ad infinitum, the Koch island being the limit of *decreasing* approximations.

This method of construction and the one described in Plate 36 are carried out in parallel in Plate 37. In this way, the contour of the Koch island is squeezed between an inner and an outer approximation that are made increasingly close to each other. One can think of the resulting two-sided cascade process as starting with a shape made of three successive rings: solid land (in black), swamp (in white), and water (in gray). Each stage of the construction transfers chunks of territory from the swamp to either solid land or water.

This Figure leaves no doubt that at the limit the swamp becomes completely exhausted; that is, it narrows down from a "surface" to a line.

all, to mention a notion that can be viewed as the opposite of self similarity, and, second, to probe self similarity in the context of very regular figures.

The most extreme contrasts to a self similar coastline with its characteristic structures at all scales are provided by curves that (a) have no scale, like the straight line, or (b) have a single scale, like the circle, or (c) have two clearly-separated scales, like a scalloped circle in which the zigzags are much smaller than the radius. In other words, the contrary of self similar is finitely structured, where the term *finite* is thought of as implying "small."

A ROUGH MODEL OF A COASTLINE: THE TRIADIC KOCH CURVE

As to the geometric cascade underlying the shape of a coastline, it may be simplified as indicated in Plate 36, particularly in the top thirds of the successive Figures. (See also the bottom Figure of Plate 41.) Let us assume that a bit of the coastline drawn roughly to a scale of 1/1,000,000 is a straight segment of length 1. Then assume that the detail that becomes visible on a map at 3/1,000,000 results from the replacement of the segment's middle third by a promontory in the shape of an equilateral triangle. The resulting second approx-

imation is an open polygon formed of four segments of equal lengths, and it will be called *standard polygon*. Assume further that the new detail that appears at 9/1,000,000 results from the replacement of each of these four segments by four subsegments of the same shape but smaller by a ratio of one-third, forming subpromontories.

Proceeding in this fashion, we add an increasing number of corners. And, if we continue to infinity, we reach a limit first considered by von Koch 1904, an idea of which is provided by Plate 39. It is ordinarily called a Koch curve, but we have to be more specific and shall call it a triadic Koch curve. One of the earliest discussions of this curve is to be found in Cesàro 1905, a work of such charm as to make one almost forget the hard search needed to locate it. We shall return to it later (especially in Chapter XII under PEANO CURVES), but we must stop for a free translation of a few ecstatic lines now. "This endless imbedding of the shape into itself gives us indeed an idea of what Tennyson describes somewhere as the *inner* infinity, which is after all the only one we could conceive in Nature. Such similarity between the whole and its parts, even its infinitesimal parts, leads us to consider the triadic Koch curve as truly marvelous among all lines. Had it been given life, it would not be

Plate 39

TRIADIC KOCH ISLAND
(COASTLINE DIMENSION D=1.2618)

The construction used to generate this island is described in the captions of Plates 36 and 37. One should mention why this and many other Plates choose to represent islands rather than coastlines and other "solid areas" rather than their contours. This method's superiority is that it takes much fuller advantage of the fine resolution of our graphics system. Even when the diameter of the smallest discernible dot takes the same value η as the width of the thinnest line, islands can be represented down to the diameter η but coastlines can only be represented down to a diameter above 3η. The actual difference in precision is even greater, since the finest perceivable irregularity on the contour of a big black area is smaller in scale than assumed in the preceding sentence.

possible to do away with it without destroying it altogether for it would rise again and again from the depths of its triangles, as life does in the Universe."

It is indeed a curve. Its area vanishes, as is obvious with the alternative method of construction illustrated on Plate 37, but on the other hand each stage of its construction increases its total length by a ratio of $4/3$. Thus the limit curve is of infinite length. It is also continuous, but it has no definite tangent in almost all of its points because it has, so to speak, a corner almost everywhere. Its nature borders on that of a continuous function without a derivative. The physicist (other than Perrin) who meets such constructs in a mathematical treatise cannot help but conclude that they could only be monsters devoid of any concrete interest. In the present case, however, such a conclusion is impossible, since the triadic Koch curve was introduced as a simplified model of a coastline. True, it is at best a first approximation, but not because it is too irregular, rather because, in comparison with a coastline, its irregularity is by far too systematic.

In this connection it is good to quote two great mathematicians who did not personally contribute to science but had an acute sense of the concrete. Lévy 1970 observes that "the absence of tangents and the infinite length of the Koch curve are intuitively linked to the presence of infinitely small deviations which one could not dream of tracing. I have always been surprised to hear it said that geometric intuition inevitably leads one to think that all continued functions are differentiable. From my first encounter with the notion of derivative, my personal experience proved that the contrary is true." Similarly Steinhaus 1954, a deeply perceptive study of coastlines and river banks (that, however, stopped short of the notion of dimension), observes that "a statement nearly approaching reality would be to call most arcs encountered in nature nonrectifiable. This statement is contrary to the belief that nonrectifiable arcs are an invention of mathematicians and that natural arcs are rectifiable: it is the opposite that is true." Note that the last part of the sentence is almost identical to a phrase in the Perrin quote in Chapter I.

One must also comment on the contrast between the above words and the famous invective of Charles Hermite, who on May 20, 1893, wrote to Stieltjes of "turning away in fear and horror from this lamentable plague of functions with no derivatives." (See Hermite & Stieltjes 1905, **II**, p. 318.) One likes to believe that great men are perfect, but a recollection by H. Lebesgue (in the 1922 vita, *Notice,* reproduced in Lebesgue 1972–, **I**) suggests that this sentence was by no means ironic. Having written a paper on

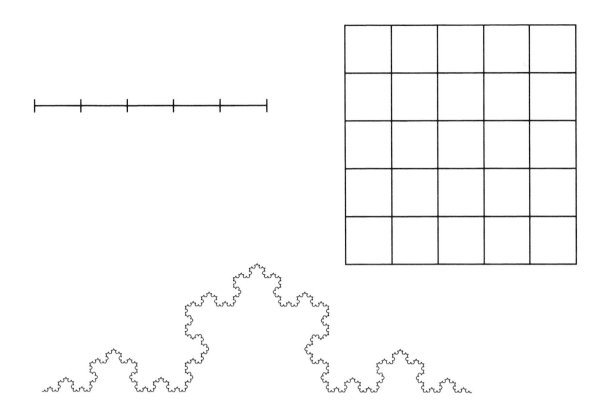

Plate 41

SELF SIMILARITY,
CLASSICAL AND FRACTAL

The top Figures recall how, given an integer γ, a segment of straight line of unit length may be divided into $N = \gamma$ sub-segments of length $r = 1/\gamma$. Similarly, a unit square can be divided into $N = \gamma^2$ squares of side $r = 1/\gamma^2$. In either case, $\log N / \log(1/r)$ could be called the shape's similarity dimension, but school geometry has not felt the need of pin-pointing this concept, since it merely reduces to the Euclidean dimension.

The bottom Figure is a triadic Koch curve. It is obvious that it, too, can be decomposed into reduced-size pieces,

and that $N = 4$ while $r = 1/3$. The resulting formal dimension $D = \log N / \log(1/r)$ is not an integer and corresponds to nothing classical. Hausdorff showed that it is of use in mathematics, and I hope to show it is also of use in natural science.

As to the whole coastline of the Koch island of Plate 39, it could not be self similar because a loop-free closed line cannot be decomposed into the union of other loop-free closed lines. ◁ We shall see, however, that the notion of self similarity can apply within infinite collections of islands. ▶

very irregular surfaces applicable on a plane similar in structure to "thoroughly crumpled handerchiefs," Lebesgue wanted it published by the Académie dès Sciences, but "Hermite for a moment opposed its inclusion in the *Comptes Rendus* ; this was about the time when he wrote ..." – and this is followed by the sentence quoted above.

THE SIMILARITY DIMENSION D FOR THE TRIADIC KOCH CURVE, D=log 4/log 3 = 1.2618

The length of each successive approximation of the triadic Koch curve can be measured exactly, and the result is most curious. It has the same form as Richardson's empiric law relative to the coast of Britain, namely,

$$L(\eta) = \eta^{1-D},$$

with $D = \log 4/\log 3 \sim 1.2618$

◁ Proof: Clearly, $L(1) = 1$ and

$$L(\eta/3) = (4/3)L(\eta).$$

To achieve $L(\eta) = \eta^{A}$, we must have

$$3^{-A} = 4/3,$$

hence $A = 1 - \log 4/\log 3.$ ▶

Naturally, this D is no longer to be empirically estimated but a mathematical constant. Hence, Hausdorff's reason for calling D a dimension is even more persuasive in the case of the Koch curve than in the case of coastlines.

Unfortunately, (a) the Hausdorff definition is disappointingly difficult to handle rigorously, and (b) had it been easy to handle, the generalization of dimension beyond integers is so far-reaching that one would still welcome further motivation for it.

It happens that in the case of self similar shapes a very easy further motivation is available in the notion of similarity dimension, to be explored soon. Although no *writing* on its account could be located, one often *hears* mathematicians use the similarity dimension to guess the Hausdorff dimension, and the present Essay will only encounter cases where this guess is correct. In their context, there can be no harm in thinking of fractal dimension as being synonymous to similarity dimension. We have here a counterpart to the use of topological dimension as synonymous to "intuitive" dimension.

As a motivating prelude on similarity, let us examine the very simplest shapes: line segments, rectangles in the plane, and the like; see Plate 41.

Because a straight line's Euclidean dimension is 1, it follows for every integer γ that the "whole" made up of the segment of straight line $0 \leq x < X$ may be "paved" (each point being covered once and only once) by $N = \gamma$ "parts." These "parts" are segments of the form

$(k-1)X/\gamma \leq x < kX/\gamma$, where k goes from 1 to γ. Each part can be deduced from the whole by a similarity of ratio $r(N) = 1/N$.

Likewise, because a plane's Euclidean dimension is 2, it follows that whatever the value of γ, the "whole" made up of a rectangle $0 \leq x < X$; $0 \leq y < Y$ can be "paved" exactly by $N = \gamma^2$ parts. These parts are rectangles defined by

$$(k-1)X/\gamma \leq x < kX/\gamma$$

and

$$(h-1)Y/\gamma \leq y < hY/\gamma,$$

wherein k and h go from 1 to γ. Each part can now be deduced from the whole by a similarity of ratio

$$r(N) = 1/\gamma = 1/N^{1/2}.$$

For a right-angled parallelepiped, the same argument gives us $r(N) = 1/N^{1/3}$.

Finally, we know that there is no serious problem in defining spaces whose Euclidean dimension is $E > 3$. (The Euclidean dimension of the space in which we work will always be denoted by E.) In that case, D-dimensional parallelepipeds defined for $D \leq E$ satisfy

$$r(N) = 1/N^{1/D}$$

that is,

$$Nr^D = 1 .$$

Alternatively written,

$$\log r(N) = \log(1/N^{1/D}) = -(\log N)/D,$$

and

$$D = -\log N/\log r(N) = \log N/\log(1/r).$$

It is the latter equality that will now be generalized. In order to do so, note that the exponent of self similarity continues to have formal meaning for some nonstandard shapes. The main requirement is that the whole may be split up into N parts deducible from it by self similarity having the ratio r (followed by displacement or by symmetry). The D obtained in this fashion always satisfies $0 \leq D \leq E$. In the example of the triadic Koch curve, $N = 4$ and $r = 1/3$, hence $D = \log 4/\log 3$. In other words, we do fall back upon the dimension obtained previously on the basis of the Hausdorff argument.

INTUITIVE MEANING OF FRACTAL DIMENSION WITHOUT PASSAGE TO THE LIMIT

The already quoted memoir, Cesàro 1905, begins with the motto,

The will is infinite
*   and the execution confined,*
the desire is boundless
*   and the act a slave to limit.*

Indeed, limits apply to scientists no less than to Shakespeare's Troilus and Cressida. In order to obtain any Koch curve, the mechanism of adding smaller and smaller new promontories has to be pushed to infinity, but in reality this process soon becomes meaningless. While in the case of actual coastlines the assumption of endless promontories may conceivably be defended, the notion that they are self similar can only apply between certain limits. Below the lower limit, the concept of coastline ceases to belong to geography.

It is useful therefore to stop the cascade a little earlier and reasonable to suppose that the real coastline is in fact characterized by two scales. Its outer scale might be measured in tens or hundreds of kilometers and could be defined as the diameter of the smallest circle encompassing an island, the inner scale being harder to pinpoint.

Yet, even after cutting off both the very big and the very small, D would still continue to stand for a *physical dimension* as described in Chapter I. Intuitively as well as pragmatically (from the point of view of the simplicity and naturalness of the corrective terms required), it is reasonable to consider a very close approximation to a Koch curve as closer to a curve of dimension $\log 4/\log 3$ than to a curve of dimension 1. As for a coastline, it is likely to have several separate dimensions (remember the balls of thread in Chapter I). From the point of view of geography (i.e., in the zone of scales going from, say, one meter to 100 kilometers), it is reasonable to say that its dimension is the D estimated by Richardson. This does not exclude that in the range of sizes of interest in physics, the coastline may have a different dimension – one associated with the concept of interface between water, air, and sand.

GENERALIZED KOCH CURVES AND AVOIDANCE OF DOUBLE POINTS

The Koch construction is readily changed by modifying the shape of the standard polygon, in particular by combining promontories with bays, as indicated in the upcoming portfolio of Plates (Plate 46 ff.). In this way we obtain, so to speak, cousins of the triadic curve, with dimensions equal to $\log 9/\log 7$ ~1.1291, then $\log 8/\log 4 = 1.5$, then higher still, but always at least 1 and less than 2. The details concerning these curves are given in the captions.

It is clear by inspection that the generalized Koch curves up to Plate 55 have no double point. This is why their wholes can be divided into disjoint parts with no ambiguity. However, double points are desirable in certain cases (Plate 57). Moreover, a Koch construction using carelessly chosen standard polygons runs a high risk of involving double points such as self contacts or self intersections, or worse. For example, a Koch curve may include portions that are covered repeatedly. When the desired D is small, it is easy to avoid double points by careful choice of the standard polygon. This becomes increasingly difficult as one pushes the principle of the Koch construction in the hope of increasing the value of D.

Finally, when the desired similarity dimension reaches $D=2$, points of self contact become unavoidable, at least asymptotically. This is not a matter of corrigible lack of imagination, but rather a fundamental question of principle, the discovery of which contributed mightily to the solution of the 1875–1922 crisis of mathematics. For example, Plates 59 to 63 show what happens in case one selects $r=1/3$, $N=9$, then $r=1/2$ and $N=4$, and finally $r=1/\sqrt{7}$ and $N=7$.

The corresponding limits are studied further in Chapter XII under PEANO CURVES. These limits fill "solid chunks" of the plane that include squares or circles, and as a result the formal definition of similarity dimension as given by $\log N/\log(1/r)=2$ is no longer justified. However, this value turns out to be perfectly correct simply because a Peano curve is really a way of looking at a piece of plane for which all the classical definitions yield 2. ◁ And the last section of

the chapter will describe a procedure that yields a bona fide curve that has no double point but is a fractal with $D=2$. ▶

When it attempts to read a dimension beyond $D=2$, the Koch construction inevitably leads to curves that cover the plane infinitely many times.

THE CONTRIBUTION OF OFFSHORE ISLANDS TO THE DIMENSION

An important but simple further generalization of the original Koch construction is performed now. The main new feature (which was allowed for without fanfare in an earlier section) consists in letting the *standard polygon* split into several disconnected portions. However, to insure that the limit Koch curve remains interpretable in terms of coastlines, a first portion of the standard polygon continues to be a connected open polygon joining the end points of the segment [0,1]. It is to contain $N_0 < N$ links and is called *continental polygon*. The remaining $N-N_0$ links are to be shared between one or several closed polygonal "islands" or "lakes." (The distinction between "island" and "lake" polygons is obvious but cumbersome to state.) A shape of this sort is shown to the right of Plate 50.

The novelty in this case is that the

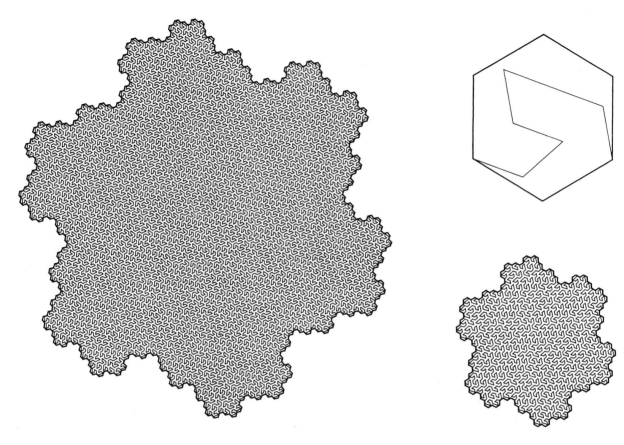

Plates 46 and 47

A SMOOTHER KOCH ISLAND
(COASTLINE DIMENSION D=1.1291)

The classic construction of the triadic Koch curve or island involves two ingredients. The first is a *standard polygon* which is oriented from one end to the other and has equal sides. The second ingredient is an *initial polygon*, which also is oriented and has equal sides. For the triadic Koch curve, it is a segment set to [0,1], and for the triadic Koch island, it is a triangle. Each stage of the construction starts with an oriented equal-sided polygon and replaces each side with a standard polygon reduced and displaced so as to have the same end points. The straight distance between the end points of the standard polygon being set to 1, the lengths of its sides are designat-ed by r (Koch selected r=1/3), and the number of sides by N (Koch selected N=4). The fractal dimension is

$$D = \log N / \log (1/r)$$

(Koch obtained $\log 4 / \log 3$).

Plate 46 takes a regular hexagon as its initial polygon, and it uses a standard polygon shown to the side, made of N=3 legs with $1/r = \sqrt{7}$. The Koch construction then results in a coastline of dimension

$$D = \log 3 / \log (\sqrt{7}) = 1.1291$$

Observe that contrary to Koch's original, the present standard polygon is symmetric with respect to its center point. As a result, it combines peninsulas and bays in such a way that the island's area remains constant throughout the construction. The same will be

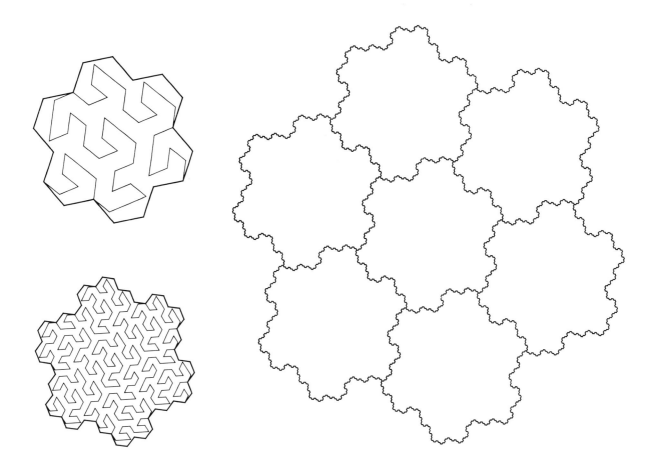

true of all the Koch curves in the plates that follow. The peculiar cross-hatching used here will be explained in the caption of Plate 62.

Tiling. The present values of N and r were singled out by W. Gosper (Gardner 1976), and the present set of Plates was substituted for its counterpart in the French version of this book because it has the following virtue. Not only is each sixth of the coastline self similar but so is the island itself. (The snowflake shares the first property but not the second.) Indeed, as shown by Plate 47, the Gosper island is divisible into seven "provinces" deducible from the whole by similarity of ratio r = 1/√7. To numerous authors, such as the geographer W. Christaller (see Haggett 1972), it had long been a cause for irritation that sev-

en hexagons put together do not quite make up a bigger hexagon. The present shape fudges the hexagon just enough to correct this defect! It is readily checked that no other value of D would do.

France. The dimension D = 1.1291 is too small to represent a "typical" coast-line, but the present overall shape is by no means unreasonable. As a matter of fact, a certain geographical outline of unusual regularity that is often described as "the Hexagon," namely the outline of France, is closer in form to the present shape. (Of course, Brittany is sadly undernourished here. By contrast, the subsequent Plates in the portfolio will illustrate the appearance of curves for which the fractal dimension is excessively high for the application to geography, but it is very appropriate for others.

Plate 49

A QUADRIC KOCH ISLAND
(COASTLINE DIMENSION D=1.5000)

This Plate shows four steps of the Koch construction that is initiated with a square (hence the term *quadric*) and uses a standard polygon characterized by N=8, r=1/4 and D=1.5000. Once again, the total island area remains constant throughout the succession of stages.

In the last stage, the detail of the curve is reduced to very fine whiskers, but it is clear that much would be lost perceptually if the graphics had to omit this detail.

In the present construction, not only the finite approximations but also the limit curve involve no self overlap, no self intersection, and no self contact. The same is true through Plate 55. This series of examples is followed on Plate 57 by a Koch curve of dimension log3/log2 having an infinity of self contacts, then on Plates 59 through 63 by Koch curves of dimension D=2 in which self contacts are everywhere dense.

Tiling. This Figure continues a sequence of Koch islands that can be used for tiling. For example, the present island is decomposable into 16 islands reduced in the ratio of r=1/4. Each of them can be imagined to be the Koch island built on one of the 16 squares forming the first stage of the construc-

tion. More generally, tiles made to replicate the present Koch islands can be used to cover the plane.

◁ We shall find in Chapters II and IX that the dimension D=1.5000 is also encountered for various Brownian functions. Hence their value is especially easy to obtain with random curves and surfaces.

◁ *Maximality.* The quadric Koch curves represented in Plates 49 to 55 possess an interesting property of maximality. Consider all Koch curves devoid of double points and such that the standard polygon is traced on a grid of lines parallel and perpendicular to [0,1] and spaced by $r=1/\gamma$. Among them, we denote as maximal the curves that attain the highest possible value of N and hence of D. When γ is even, it turns out that $N_{max}=\gamma^2/2$, and when γ is odd, that $N_{max}=(\gamma^2+1)/2$. In all cases, the first approximation of a maximal Koch island is made of black subsquares having white neighbors and vice versa. Moreover, each lattice vertex (with the exception, when γ is even, of the lattice's center) lies on the first approximate coastline. A combinatorial problem (seemingly not a trivial one) arises: counting the number of distinct Koch curves with given γ and maximal N. ▶

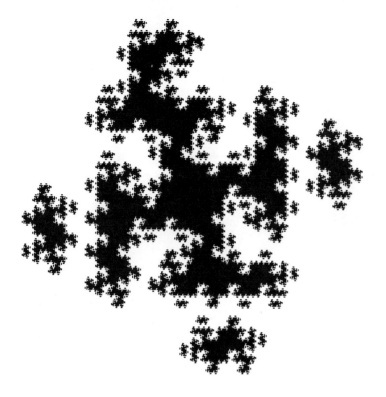

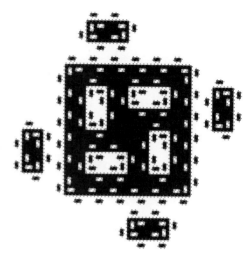

Plates 50 and 51

A KOCH ISLAND AND
A KOCH ARCHIPELAGO
(COASTLINE DIMENSION D=1.6131)

The construction of Plate 51 proceeds as in Plate 49 except that $\gamma=6$ and $N=18$ ◁ a maximal value ▶, so that

$$D = \log 18/\log 6 = 1.6131.$$

The dull Figure to the right of Plate 50 is a Koch archipelago made mostly of rectangular islands or lakes. Topologically and in terms of intuitive "form" it is very different from Plate 51, but its cumulative coastline's fractal dimension is precisely the same. Incidentally, a full fractal description of its shape involves at least three fractal dimensions: that of the individual islands or lakes, which is $D=1$, that of the cumulative coastline assuming the lakes to be filled in, and that of the total cumulative coastline.

The Figure to the left on Plate 50 is a pseudo random mixture explained in the caption of Plate 85.

The fact that the form of the quadric Koch islands in the present portfolio of illustrations depends very markedly upon D is highly significant. However, their having roughly the same overall outline is not significant, being due to the fact that the initial polygon is in all cases a square ◁ and the standard polygons are maximal for this initial shape. ▶ When one starts with an M-sided regular polygon (M>4), the overall shape looks smoother, increasingly so as M itself increases ◁ and the standard polygons become farther removed from being maximal. ▶ A genuine link between overall form and the value of D will not enter in until Chapter IX, which deals with random coastlines that effectively determine the standard polygon and the initial shape at the same time.

Incidentally, if the disconnected shape to the right of Plate 50 could be made into a tile, such tiles could also cover the plane.

Plate 53

A KOCH ISLAND
(COASTLINE DIMENSION D=1.6667)

Now the same construction as in Plate 49 is carried out with $r=1/8$ and $N=32$ ◁ a maximal value ▶, so that

$$D = \log 32/\log 8 = 5/3 = 1.6667.$$

The widths of the causeways and of the channels in the nightmarish marinas in the present series of Plates become increasingly small as one proceeds toward the peninsulas' tips or the bays' deepest points. In addition, these widths tend to decrease as the fractal dimension increases, and "wasp waists" appear around $D \sim 5/3$.

◁ *Digression concerning turbulence.* I see an uncanny resemblance between the sequence of approximate fractals drawn on this Plate and the successive stages of turbulent dispersion of black ink in milk. Actual dispersion is of course very irregular, but this defect can be fixed by invoking chance.

◁ One can almost see a Richardsonian cascade at work. A finite pinch of energy, having been injected as a large scale eddy, spreads a square ink blob around. Then the original eddy splits into smaller scale eddies, the effects of which are more local. The initial energy continues to cascade down to ever smaller typical sizes, eventually contributing nothing but slight fuzziness to the outline of the final ink blob. For comparison, the reader may look up the diagram in Monin & Yaglom 1971, Vol. I, p. 592, reproduced from Corrsin 1959b.

◁ The broad conclusion that a Richardsonian cascade necessarily leads to a shape bounded by a fractal seems inescapable, but the specific conclusion that $D=5/3$ is based on shakier evidence. A student of turbulence is, so to speak, predisposed to pay special attention to this value of D because it corresponds to planar sections of spatial surfaces with $D=8/3$. This last value plays other well-confirmed roles in turbulence. In one case (end of Chapter IX), $D=8/3$ is reducible to the Kolmogorov theory and to the empirical spectra. In another case (Chapter V), $D=8/3$ is only a rough estimate.

◁ Nevertheless, analogies of this sort are not very convincing. In fact, the value of D may well depend on the energy with which the cascade begins and on the size of the vessel in which dispersion is observed. A high initial energy in a small vessel might possibly lead to more thorough dispersion, with planar sections reminiscent of Plate 55 ($D=1.7373$) rather than of Plate 53.

◁ In any event, one cannot give too much weight to the appearance of the shapes on this series of Plates because all are nonrandom and most are connected. Planar sections of blobs dispersed by turbulence are on the contrary very likely to be disconnected, whether or not the blob itself is connected.

◁ The fractal approach to the study of turbulent dispersion is mentioned in Mandelbrot 1976c. ▶

Plate 55

A KOCH ISLAND
(COASTLINE DIMENSION D=1.7373)

Marina lots become positively labyrinthine. We take $\gamma=14$ and $N=98$ ◁ a maximal value ▶, so that

$$D = \log 98/\log 14 = 1.7373.$$

Looking back on the sequence of Koch islands concluded by this Figure, we note that each can be surrounded by an outline of total area roughly equal to 2, the interior of which is roughly half land (black) and half ocean (white). Furthermore, as one proceeds to islands with coastlines of increasing dimension, one encounters a qualitative jump at a point where black and white areas become separated by areas in various shades of gray. On the other hand, a uniform "printer's gray" is only obtained in the limit case $D=2$, to be considered in later Plates. Indeed, as long as $D<2$, a construction avoiding self contact and self intersection necessarily involves large contiguous areas that are interior to land or to the sea.

◁ One should not forget that the fractal in this Plate is the coastline; the land and sea are conventional shapes, in the sense that they have positive and finite areas. Later chapters will consider cases in which the "sea" alone has a well-defined area, being again the union of simple-shaped "cutouts," while the land has no interior point.

◁ *More on maximality.* As the value of γ increases, so does the maximal N, and so also does the number of alternative maximal polygons. Therefore, the limit Koch curve becomes increasingly influenced by the original standard polygon, and hence begins to look increasingly contrived. In other words, the wish to achieve a maximal dimension without contact point imposes a degree of discipline that must increase with D. It will reach its paroxysm when we reach the Peano limit $D=2$. ▶

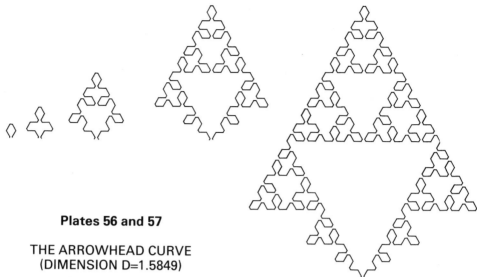

THE ARROWHEAD CURVE
(DIMENSION D=1.5849)

The first reason for including this curve from Mandelbrot 1975m is that it is attractive (turned sideways, it becomes a tropical fish). A second reason resides in its being an example of a generalization of the Koch construction; it allows the standard polygon to be transformed by various symmetries before it is used. In other words, some algorithm is used to attach to every side of each approximation a two-digit binary *orientation index.* The index 00 says the side must simply be replaced by the standard polygon. The orientation index 01 says the standard polygon must first be replaced by its mirror image with respect to [0,1], and so on.

In the present case, the standard polygon is made of the segment [0,0.5] of the x-axis and of two segments of length 0.5 which would combine with the segment [0.5,1] of the x-axis to form an equilateral triangle. Hence, $N=3$, $r=1/2$ and

$$D = \log 3 / \log 2 = 1.5849.$$

The initial polygon is a rhombus, the lower half of which is the biggest black triangle in this "island." The early stages of construction are shown in Plate 56.

Self contacts. Contrary to the preceding Plates, the present one does not attempt to avoid self contacts. (A different contrast is offered by the pattern on page 80, which is a Koch curve in which contacts and overlaps are due to faulty programming.)

Tiling. Arrowheads tile the plane. Neighboring tiles are linked together by a nightmarish extrapolation of Velcro (to mix metaphors, one fish's fins fit exactly those of two other fish). Furthermore, by fusing together four appropriately chosen neighboring tiles, one gets a tile increased in the ratio of 2.

The Sierpiński gasket. The boundary between white and black in the upper portion of this drawing (above the biggest black triangle) goes back to Sierpiński 1915 and will be encountered again in Chapter VII. Sierpiński also described an alternative construction that relies upon cutouts, a method to be used extensively in Chapters IV and VII. He subtracted from an equilateral triangle pointing up a sequence of triangles pointing down, seen here as solid white or solid black.

Drainage divides. Yet another construction of the arrowhead curve, which has the advantage of showing the order in which it ought to be followed, was advanced in Mandelbrot 1975m, which obtained it as a drainage divide between different rivers' drainage basins – a concept to which we will return in later Plates.

57

Plate 59

QUADRIC KOCH CONSTRUCTION
(DIMENSION D=2)
AND FINITE PEANO ISLANDS

Peano curve is a synonym of *plane-filling curve*. The one represented here is also a Koch curve in the strict sense: the standard polygon is always placed in the same way on the sides of the polygon obtained at the preceding stage.

After rotation by 45°, this curve turns out to be a slight variant of one due to Moore 1900, p. 77. (See Chapter XII under PEANO CURVES, 3.)

The initial polygon here is the unit square (boundary of the black box) and the standard polygon is made of nine segments of length 1/3: the outer thirds of [0,1] hug between them a shape reminiscent of the figure eight as seen on digital electronic displays. The resulting finite Koch islands are merely sets of black squares on a distorted chessboard. And the nth finite Koch curve is a grid of lines, a distance of $\eta = 3^{-n}$ apart, all of them contained within a square of area equal to 2 that becomes covered increasingly tightly as $n \to \infty$. It suffices to show one example of this dull design (above the black box).

The present standard polygon self contacts and hence cannot be followed unambiguously without additional guidance. To make such guidance unnecessary, the first stage of this construction is exhibited in a modified form on the second Figure clockwise from the black box; all the corners have been cut off using a scheme that leaves the total area invariant. The second and third stages are also drawn with corners cut off.

As to the fourth stage, its representation on the scale of this Plate would have merged into gray, but a larger drawing of one fourth of the coastline can be followed unambiguously (at some risk of becoming seasick). It shows graphically what is meant by saying that with or without cut off corners, the limit of the present Koch curve fills the plane – as should be the case when D = 2.

It would have been nice to be able to define a limit island in analogy to the Koch islands of the preceding Plates in this portfolio, but in the present case it is impossible. (Indeed, a point chosen at random will almost surely flip between being inland and in the ocean, without end.) As to advanced finite approximations, they are strange indeed. The outline is penetrated by "bays" so deeply and uniformly that a square of middling side x – that is, such that $\eta \ll x \ll 1$ – is divided between dry land and water in about equal proportions!

The limit Peano-Moore curve establishes a continuous correspondence between the straight line and the plane, but this correspondence is not one-to-one since the limit curve has an infinite number of self contacts. The fact that they are mathematically unavoidable is classical, and is discussed in Chapter XII under DIMENSION (TOPOLOGICAL). The fact that they are valuable in modeling certain natural phenomena is new, and is discussed in the caption of Plate 61.

Plate 61

A DIFFERENT
FINITE PEANO ISLAND

This Plate introduces a different Peano curve, that is due to Cesàro 1905 and was originally based on a limit case of the Koch construction (Chapter XII, PEANO CURVES, 3). We shall derive it here in an alternative fashion, then use it to argue that Peano curves, far from being mathematical monsters with no concrete interpretation, are one step removed from very good first-order models of such natural phenomena as river networks or human vascular systems. This caption will concentrate on the former phenomenon and thereby limit itself to the portion of the Figure shown in gray.

In building this Peano curve, each stage splits into three substages. First substage: draw a "skeleton," so to speak, along the appropriate lines in the facing page; begin with the thickest lines of length $1/2$, then add thinner lines of length $1/4$, then even thinner lines of length $1/8$, and so on. Second substage: define the mth finite Peano island as the Minkowski sausage (Plate 33) drawn along the mth skeleton, the value of η being selected so that the island's area is 1. Third substage: define the mth Peano curve as the "skin" of the mth sausage. As $m \to \infty$, the sausage spreads around over the same ultimate square as in Plate 59, without overlap and with increasing uniformity, and the sausage skin asymptotically fills the square.

Rivers. To inject river networks, one interprets the area within the sausage as being filled with water. It can no longer be viewed as a marina, however nightmarish, but it does give a rough idea of a network of rivers branching off ad infinitum! One can furthermore, if one wishes, modify the total area of rivers, but the cumulative shore will continue to tend to a Peano curve. (The sausage area can take any value between 0 and a certain upper threshold above which the sausage would self overlap.)

Double points. The necessity of double points in a Peano curve now comes to be associated inevitably with the following facet of Nature. Assume one starts anywhere along a Peano river's shore and moves upstream or downstream, making a detour for the slightest branch in the network (moving ever faster as one gets to finer branches). It is clear one will eventually find oneself facing one's point of departure from across the river. And since at the limit the river is taken to be infinitely narrow, one will effectively come back where one started. This interpretation shows that the double points in a Peano curve are unavoidable, not only from a logical but also from an intuitive viewpoint. (Concerning triple points, see Chapter XII, PEANO CURVES, 1.)

Returning to Plate 59, the reader will have no difficulty finding that it, too, involves river basins. The networks are less obvious and, by the same token, more interesting. Hilbert's version of the Peano curve (Hilbert 1891; T. Hawkins 1970, pp. 189ff; Gardner 1976) involves still another interesting river network.

In the Peano-Moore and the Peano-Cesàro curves, each individual river is of finite length, hence of dimension 1. To achieve the dimension $D=2$, all rivers have to be taken together. The next set of Plates will exhibit a Peano river network of more realistic form.

Conversely, any actual river network that drains a region with reasonable uni-

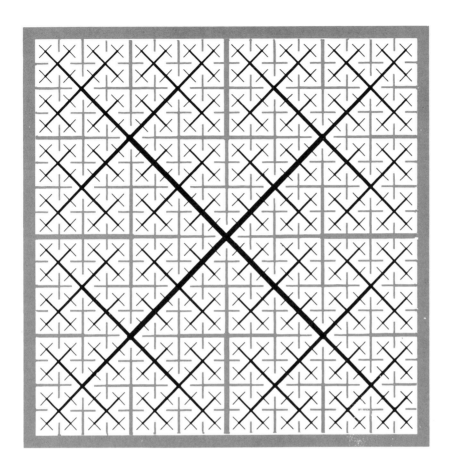

formity automatically defines an approximate Peano curve.

However, if one did not know of the descending cascades that built our various finite Peano curves, one would have had to marvel at the extraordinary discipline and long-range order that allow them to avoid not only self intersection but also self contact. Any lapse in discipline would make the latter very likely.

◁ And total breakdown of discipline would make multiple self intersection almost certain, since it leads to Brownian motion – to which we have already allud-

ed. We shall explore it in Chapter II to show that its dimension takes the Peano value $D = 2$.

◁ The random self avoiding walk will be investigated in Chapter VIII. Obviously, its chance of following a Peano curve is strictly zero. It surrounds many big or small chunks of the plane and makes them inaccessible, so its future course must run over a much broader portion of the plane. The surrounded areas look like Koch islands of dimension $D < 2$, and indeed (Chapter VIII) the self avoiding walk's dimension is below 2. ▶

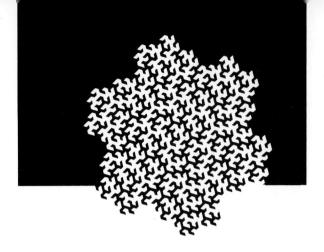
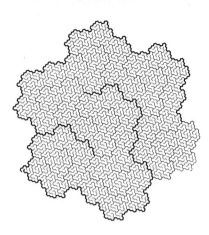

Plates 62 and 63

STILL ANOTHER PEANO MODEL WATERSHED (DRAINAGE DIVIDE AND RIVER DIMENSION D=1.1291)

The Figure on the left of Plate 62 represents the third stage of construction for a Peano curve due to W. Gosper and described in Gardner 1976.

The initial polygon is $[0,1]$, and the first two stages are shown to the right on Plate 63. The limit is a generalized Koch curve in the sense used on page 56, with $N=7$ and $1/r=\sqrt{7}$, so that $D=2$. This Peano-Koch curve has the novel virtue that its finite approximations are free of self contact. (See Chapter XII under PEANO CURVES.) The limit fills the interior of Gosper's curve of Plate 46. The hatching used there was a riper stage of the present construction.

Each Peano curve involves, so to speak latently, the fractal dimension of its own boundary. The same had been true already in Plates 59 and 61, but it had not been worth pointing out because said boundary was merely a square. Here said dimension is noteworthy because $D=1.1291$.

Interpretation in terms of river and drainage divide networks. As suggested in Plate 61, such an interpretation is possible with any Peano curve. On the Figure to the right on Plate 62, the rivers and drainage divides relative to the third stage of the Peano-Gosper curve have been drawn according to the relative importance of each link as measured in a scheme due to Horton and revised by Strahler (Leopold 1962). In addition, their widths are made positive in such a way that links are mutually similar in the sense of having the same ratio of the width to the length measured as the crow flies.

The top Figure of Plate 63 restates the same point differently by drawing the rivers in black and the drainage divides in gray.

If the links' widths are selected judiciously, each river and drainage divide is a fractal, with the dimension $D=1.1291$.

One who as a schoolboy kept gazing on a map showing the rivers Loire and Garonne does not feel far from home.

On the bottom Figure of Plate 63, the widths of the rivers and divide links are no longer proportional to their lengths measured as the crow flies. They thin down faster as one proceeds upstream and soon become invisible. This alternative method leaves out a simple curve of dimension 2. Its application is further explored in the caption to Plates 66 and 67 and in the text's section devoted to the width of the Missouri River. The method becomes especially attractive in the modeling of arteries (black) and veins (gray).

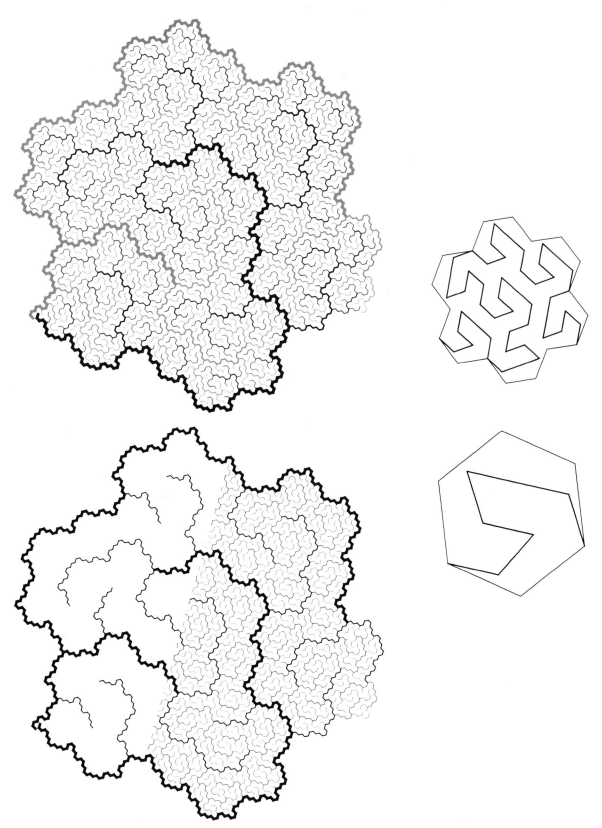

Plates 64 and 65

GENERALIZED KOCH CURVES
AND SELF SIMILARITY
WITH UNEQUAL PARTS
(D=1.4490, D=1.8797, D~1)

These Plates exhibit curves constructed in the manner of Koch except that they involve yet another generalization: the sides of the standard polygon are allowed to take different values r_m. Until now, we had assumed that the N "parts" into which our "whole" is divided all involve the same similarity ratio r.

An incidental benefit when the r_m are made unequal is that the Koch curve becomes less relentlessly regular. Thus Plate 64 adds variety to the triadic Koch curve, and the Figure at the top of Plate 65 improves on Plate 61. Note that the truncation rule is the same as in all earlier Plates: carry out the construction until it reaches details of a predetermined small size. This rule, however, has new implications in the present context. It used to be synonymous with "perform a predetermined number of construction stages," but such is no longer the case.

Contrary to earlier generalizations of the Koch method, the present one requires a corresponding generalization of the notion of similarity dimension. In a search for suggestions, let us begin with pavings of ordinary Euclidean shapes with parts reduced in the respective ratios r_m. When D=1, the r_m must satisfy $\Sigma r_m = 1$, and, more generally, one must have $\Sigma r_m^D = 1$. Furthermore, in the case of fractals splitable into equal parts, the already familiar condition $Nr^D = 1$ can

also be written as $\Sigma r_m{}^D = 1$. Thus we have at least two reasons for using this last equation to define the similarity dimension D in all cases when N and the r_m are given in advance. It remains to investigate whether or not said D coincides with the Hausdorff-Besicovitch dimension. It does.

Examples. Plate 64 has a D above Koch's original $\log 4/\log 3$. The top of Plate 65 has a D slightly below 2. As $D \to 2$, the coastline on this Figure tends toward the so-called Peano-Polya curve. The resemblance between this Figure and a row of trees is not accidental; see PARADIMENSION, APPLICATIONS in Chapter XII. Finally, at the bottom of Plate 65 we have a D slightly above 1.

◁ And when N is random and the r_m are independent random variables with identical distributions, we have the generalized definition $\langle N \rangle \langle r^D \rangle = 1$. ►

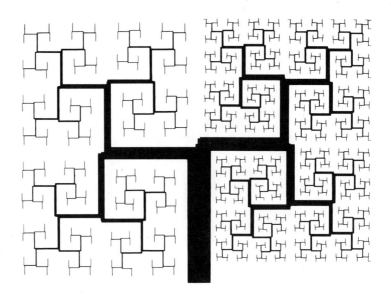

Plates 66 and 67

SPACE-FILLING TREES AS GENERALIZED KOCH CURVES

In the Koch method, every straight-line segment in a finite approximation is eventually broken up into shorter pieces ad infinitum. In many applications it is useful to generalize this aspect of the procedure, simply by allowing certain segments to remain untouched in later stages. To achieve this aim, it is sufficient to take up again the notion of orientation index introduced in the caption on page 56 and allow it to take yet another value, say, −1, with the understanding that a polygon side marked −1 is left alone in later stages.

In Plate 67 this generalized procedure is used to grow a "tree." One starts with a trunk on which the top "bud" is the only side marked otherwise than by −1. The standard polygon involves two "branches" in which again only two terminal "buds" are marked otherwise than by −1. And so on ad infinitum. The algorithm was made asymmetric and its detail was set purposefully to insure that the "tree" fills a roughly rectangular portion of the plane with no gap and no overlap. However, we did not try to

avoid asymptotic self contact due to the fact that every point on the "bark" line can also be obtained as a limit branch tip.

◁ These figures are examples of the plane version of a model of the lung, but space precludes dwelling on this topic in this Essay. ▶

The "subtrees" constructed starting with the main leaders are similar to the whole tree in two different ratios of r_1 and r_2. The whole tree is not itself self similar because in addition to the subtrees it includes a trunk. On the other hand, the set of asymptotic branch tips is indeed self similar. From Plates 64-65, the similarity dimension is the root of the equation $r_1^D + r_2^D = 1$, and one has $D < 2$. In the top Figure of Plate 67, the tips are nearly plane filling and $2-D$ is small; in the bottom Figure of Plate 67, D is much below 2.

The Figure of Plate 66 is very different. It results from a Koch construction in which the standard polygon is changed at each stage, so that the ratio of width to length decreases without end. The result is that the branch tips are no longer self similar. However, if one adds the conditions of no overlap and no gap, the tips achieve the dimension $D = 2$.

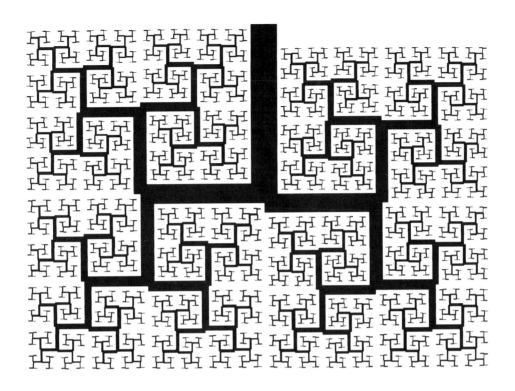

limit curve has several distinct dimensions. The whole curve including the islands and lakes continues to have the dimension $D = \log N / \log(1/r)$, but the "continental outline" taken by itself has the dimension $D_0 = \log N_0 / \log(1/r)$, which satisfies $1 \leq D_0 < D$. (The index 0 evokes an island's shape.)

Because $D_0 < D$, one is tempted to say that the continental outline is negligible in comparison to the whole coastline, in the same way that a straight line $(D = 1)$ is negligible in comparison to a plane $(D = 2)$. Such is indeed the case. Supposing that the term *at random* is given the proper intrinsic meaning (on which we shall not dwell), the classical result is that a point chosen at random on the plane almost never falls on the x-axis. Similarly, it is indeed true that a point chosen at random on a coastline with islands almost never falls on the continental outline.

We shall now examine closer first the islands, then the continental outline.

THE KORČAK LAW

The most interesting single empirical regularity relative to islands concerns the distribution of their areas, more precisely of the areas S measured after projection on an ideal globe. (This roundabout definition is needed because, just like a coastline's length, an island's true area is doubtless infinite, whereas the projective area S is well defined.) The distribution of relative projective areas of the islands is highly visible, and it may even contribute more to what we perceive as the form of a region of the globe than any of the individual coastlines. For example, it is difficult to think of the Aegean's Greek and Turkish shores without the Greek islands. The distribution of S is ordinarily called the islands' number-frequency law. According to Korčak 1938, it is hyperbolic, that is, B and σ being two positive constants, S and the characteristic length $\sqrt{S}$ satisfy

$$\Pr(S > s) = \sigma s^{-B} \text{ and } \Pr(\sqrt{S} > \lambda) = \sigma \lambda^{-2B}.$$

Pr is the abbreviation used for *probability* throughout this Essay.

Korčak's striking and elementary result was not forgotten thanks to Maurice Fréchet, who described it in Fréchet 1941 and often mentioned it in lectures. It has long been left filed in the back of my mind, and I am finally ready to take a first stab at it.

When the standard polygon involves a single island, it specifies the projective area of this island, to be designated by s_1. The first stage of the Koch construction creates a single island of area s_1, the next stage creates N islands of area $r^2 s$, and so on, and in general the mth stage creates N^m islands of area $r^{2m} s_1$. Altogether, as s multiplies by r^{-2}, $Pr(S > s)$ multiplies by $1/N$; hence the distribution must (for all values of s of the form $r^{2m}s$) take the hyperbolic form σs^{-B}, with σ some constant and

$$2B = \log N / \log (1/r) = D.$$

This result is independent of N_0 and D_0. In terms of the characteristic length $\sqrt{S}$, the crucial exponent is merely the whole coastline's fractal dimension!

When the standard polygon involves two islands, there are altogether two parallel series of islands, both of which follow hyperbolic distributions with $2B = D$. It turns out that the merged distribution is also essentially hyperbolic with a different σ but identical $2B = D$.

Without going into further detail, we will simply note that the empirical B regarding the whole Earth is of the order of 0.6 to 0.65. It is really very close to one half of the D measured from the length of coastlines.

◁ While earlier sections of this chapter involve new science but well-known mathematics, the present result appears to be mathematically new as well. ▶

FRACTAL DIMENSION AS A MEASURE OF FRAGMENTATION

Now let us turn to the continental outline. The preceding derivation allows $D_0 = 1$. Thus a Koch continent is allowed to have a rectifiable coastline! And, as a corollary, since each island is a "continent unto itself," its coastline is also rectifiable. In this extreme instance, illustrated by the Figure at the right on Plate 50, the overall dimension is no longer a measure of irregularity but instead a measure of fragmentation. (How perceptive of Perrin to have associated the notions of irregularity and of fragmentation! See Chapter I.) In other words, D ceases to measure the wiggliness of individual curves and measures instead the number-area relationship for an infinite family of rectangular islands.

As to the property that the length tends to infinity as $\eta \rightarrow 0$, it continues to hold, but the reason for it changes completely. Measurements with a yardstick of η can only include islands with a perimeter of at least 2η. However, the number of islands to be measured increases without bound as $\eta \rightarrow 0$, and the measured length is easily seen to behave like η^{1-D}. This means it tends to infinity exactly as it did in the absence of islands.

◁ The cumulative *area* of the very small islands has to be negligible, and it is indeed the tail sum of a convergent series. On the other hand, the cumulative coastline length of small islands is infinite because raising said convergent series' terms to the power $1/2$ yields a divergent series. ▶

We end with two remarks. The fact that D measures irregularity and/or fragmentation in combination is one specific instance of a theme we encounter again and again. The same fractal property is compatible with entirely different sorts of topological connectedness. As to the good stochastic models, such as the final model we shall propose for the relief in Chapter IX, their special virtue will be that they generate a specific mix of fractal and topologic features.

DEPENDENCE OF MEASURE ON THE RADIUS WHEN D IS A FRACTION

Let us now examine the extension from Euclidean to fractal dimensions of another classical result in Euclid. For idealized physical objects of uniform density ρ, it is well known that apart from numerical multiplying constants, the weight of a rod of length 2R, of a disc of radius R, or of a ball of radius R is proportional to ρR^D. For E = 1, 2, and 3, the multiplying constants are respectively equal to 2, 2π, and $4\pi/3$. *Weight* is here a standby for length, area, or volume, whichever is intrinsic to the Euclidean dimension in question.

The same happens to be true — with complications we choose for the moment to evade — for shapes of fractional dimension.

In the particular case of the triadic Koch curve, the proof is especially easy when one takes as origin the midpoint of the curve (that is, the top vertex of the middle tentlike portion). Call the distance from 0 to the tips of the curve $R_0 = 1/\sqrt{3}$. The circle of radius $R = R_0/3$ is seen to contain two-eighths of the curve, that is, a measure equal to

$$M(R) = M(R_0)/4,$$

which can be rewritten as

$$M(R) = M(R_0)(R/R_0)^D = [M(R_0)R_0^{-D}]R^D.$$

Consequently, the ratio $M(R)/R^D$ turns out to be independent of R and, up to a numerical factor dependent on D, can serve to define a "density" ρ.

THE LENGTH-AREA RELATIONSHIP

The preceding result about $M(R)/R^D$ can be given an alternative form that is illuminating and easy to use directly. We begin again with the classical fact that the circumferential length of a circle of radius R is equal to $2\pi R$, and the area of the disc bounded by the circle is πR^2. And so it follows that

$$(\text{length}) = 2\pi^{1/2}(\text{area})^{1/2}.$$

Among squares,

$$(\text{length}) = 4(\text{area})^{1/2}.$$

More generally, within each family of standard planar shapes that are similar to each other and have different linear extents, the ratio (length)/(area)$^{1/2}$ is some constant determined by their common shape. Hence length and (area)$^{1/2}$ provide alternative evaluations of the linear extent of the shape, and in particular their ratio is independent of the units of measurement.

Now consider islands with coastlines of dimension $D > 1$. As we know already, the fact that $D > 1$ does not in any way prevent their areas from being fully defined, that is, positive and finite. The purpose of this section is to perform some comparisons within a family of shapes that are similar to each other and have a well-defined area. A very natural way of generalizing to them the classical ratio (length)/(area)$^{1/2}$ will be identified. The generalized ratio, however, depends upon the unit of measurement. When this unit is $G_0 \sim$ 6 ft., the numerator will be said to involve human-length, defined as the coast length measured with a yardstick of G_0, and the denominator will involve human-area, defined as the area measured in units of $G_0{}^2$.

The generalized relationship asserts that the ratio

$$(\text{human-length})^{1/D}/(\text{human-area})^{1/2}$$

takes the same value for all mutually similar islands. As a result, the linear extent of each island can be evaluated, not only by the near-classical expression (human-area)$^{1/2}$ but also by the non-standard expression (human-length)$^{1/D}$. The novel feature is that if G_0 is replaced by a different yardstick or divider opening G, the ratio of the alternative linear extents is replaced by

$$(G\text{-length})^{1/D}/(G\text{-area})^{1/2}$$

which is larger in the factor $(G/G_0)^{1-D}$.

A careless examination of my claim, omitting the presence of the term human before the term length, would lead back to the ratio (length)$^{1/D}$/(area)$^{1/2}$, which is of course infinite, so its being a constant is trivial. Therefore the term human is essential.

To derive the generalized relationship, the first step is to observe that, measured in units of G_0, each island has the intrinsic overall linear extent equal to (human-area)$^{1/2}$. This quantity equals the diameter of the least circumscribed circle, multiplied by a shape dependent factor. In order to measure each coastline with the variable yardstick

$$G = (\text{human-area})^{1/2}/1000,$$

we can approximate it by polygons of side G. Since these polygons are mutually similar, their length is proportional to (human-area)$^{1/2}$.

Now replace the yardstick by the prescribed G_0. We know that the measured length changes in the ratio $(G_0/G)^{1-D}$.

Hence, the original human-length is proportional to

$$(\text{human-area})^{1/2}(G_0/G)^{1-D}$$
$$= (\text{human-area})^{1/2-(1-D)/2}G_0^{1-D}1000^{D-1},$$

in other words, proportional to

$$(\text{human-area})^{D/2}G_0^{1-D}.$$

Finally, by raising each side to the power $1/D$, we obtain the relationship we have claimed.

Note that as in the classical cases, the ratio of linear extents varies between one family of mutually similar shapes and another (hence it can be said to depend upon the shapes' form). The same is true here, but in addition this ratio now depends on G_0.

APPLICATION

The length-area relationship may be used to obtain an empirical estimate of the dimension of a fractal curve that bounds a classically defined area.

GENERALIZATION TO SPACE

It is obvious that everything we have said about closed planar curves surrounding a well-defined area extends to spatial surfaces surrounding a well-defined volume. For example,

$$(\text{human-area})^{1/D}\sim(\text{volume})^{1/3}.$$

HOW WINDING IS THE MISSOURI RIVER? DATA

This chapter so far is worded in terms of coastlines, which are closed curves, but the argument that geographic curves are nonrectifiable applies equally well to river banks. In fact Steinhaus 1954 – from which we have quoted – is primarily concerned with rivers. It observes that "the left bank of the Vistula, when measured with increased precision, would furnish lengths ten, hundred or even thousand times as great as the length read off the school map." I would again amplify this statement by asserting that the degree of wiggliness of river banks can be tamed further with the help of the concept of fractal dimension.

When rivers give rise to a proper science, it should of course be called potamology – Maurice Pardé was fond of this term – but current usage merges their study into the broader science of hydrology, into which this Essay will make many incursions. One is brought back to closed curves by considering drainage basins of rivers. They are closely akin to island coastlines, but have a richer structure, since each basin is the juxtaposition of partial basins and is crisscrossed by the rivers themselves. A priori, it would seem that Koch curves cannot handle this complication, but in fact we shall show that the generalization one needs is less extensive than was anticipated, because space-filling limit cases of the Koch curves involve all the structure needed to take care of the new phenomena. As usual, however, it is best to begin with concrete considerations.

Let us therefore first investigate the notion of length in the case of the leading rivers of various basins. This investigation disregards the leading river's width and approximates its course by a wiggly self similar line of dimension $D > 1$ going from a point called source to a point called mouth. The argument described earlier in this chapter adapts itself readily to the new problem. Insofar as all rivers as well as their basins can be assumed to be mutually similar, the ratio

$$(\text{human-length})^{1/D}/(\text{basin area})^{1/2}$$

should be the same for all rivers. Moreover, $(\text{basin area})^{1/2}$ should be proportional to

(distance from source to mouth
as the crow flies)

Combining the two results, the ratio of

$$(\text{human-length})^{1/D}$$

divided by

(distance from source to mouth
as the crow flies)

should be expected to be the same for all rivers. And in this case, human-length would no longer be measured with a six-foot-long yardstick but rather with some other unit more appropriate to this field (and which we need not attempt to determine precisely).

Most remarkably, Hack 1957, which studies the relationship between length and basin area empirically, finds that the ratio

$$(\text{human-length})/(\text{area})^{.6}$$

is indeed common to all rivers. It can be viewed as yet another method of estimating D and the value it yields turns out to be $D = 1.2$ – right in the ballpark of the values inferred from coastlines. Forgetting a river has two shores, it appears that if one measures the degrees of irregularity by D, local wiggles of the banks and enormously global bends are equally irregular!

◁ Only for extremely large basins $(\text{area} > 10^4 \text{km}^2)$ and correspondingly long leading rivers does this extrapolation become invalid. J. E. Mueller observes that the value of D estimated in this fashion goes down to 1. The juxtaposition of two different values of D suggests that if one maps all basins on sheets of paper of the same size, maps of short rivers look about the same as maps of long rivers but maps of very long rivers are more nearly straight. This finding seems to suggest that self similarity breaks around an outer scale of the order of 100 km.

Large-scale wiggles in the river's outline are a self similar extrapolation of the small wiggles, but very large-scale wiggles seem less accentuated. ▶

THEORY: THE SHORES OF A RIVER NETWORK AS A PEANO CURVE

Now let me advance a theory of river shores. The key factor is as follows: if a river together with its tributaries is to drain an area thoroughly, it must, insofar as other constraints allow, penetrate everywhere. If one disregards all those constraints and assumes the river's width to vanish, the river's bank might be a plane-filling curve joining the points situated on opposite sides of the river's mouth. Although crude, the space-filling model should suffice to lay to rest any notion that the Peano curve is *necessarily* pathological!

This potential application is already exemplified by the finite Peano-Moore island of Plate 59. The thin fingers of water that penetrate it through and through can no longer be viewed as forming a marina, however exaggerated, but can be viewed as branching rivers. On the other hand, we obtain with no special effort the two features that have seemed a priori to require an ad hoc enrichment of the Koch islands, namely, natural splitting into separate sub-basins and crisscrossing by rivers.

Nevertheless, the model based on Plate 59 is crude. In particular, its regularity is excessive in the sense that (a) the basin's overall boundary and (b) the longest river's course are both Euclidean curves of dimension 1 (respectively, a square and a polygon). The same objections apply to the model of Plate 61. Most fortunately, we geometers of Nature have recently been blessed with a new Peano curve due to R. W. Gosper and made available in Gardner 1976. Plates 62 and 63 show how it can be interpreted to eliminate defects a and b. Because $D > 1$, this Peano curve suffices to provide a first model of the Hack relationship. The predicted $D = 1.1291$ is on the low side, but for a first model we are doing remarkably well, I think.

Incidentally, we are now in a position to acknowledge that the Richardson data had included, mixed among coastlines, several land frontiers between countries. Many of them, however, are either rivers or water divides, so after all there had been no harm in failing to dwell on this aspect of the data.

Nevertheless, the precise identity between the dimensions of the rivers and of the drainage divides is not a logical necessity. It is a feature of certain specific models such as those based on the Peano curves of Moore, Cesàro, and Gosper. By way of contrast, a river network linked with the arrowhead curve and described in Mandelbrot 1975m involves rivers of dimension $D = 1$, which is too small, and water divides of dimension $D \sim 1.5849$, which is too large.

"The most famous of all hexagonal configurations, and one of the most beautiful, is the bee's cell." (Thompson 1917–) and many other authors have marvelled at this and other hexagonal symmetries of Nature. One may argue that the near-hexagonal grid underlying the Gosper curve suggests another such symmetry, but I tend to take it less seriously. Extensive experience with fractals drawn on square and triangular grids suggests that the former look unnatural because they are highly nonisotropic and the latter look natural because they seem very close to being isotropic. These topics are to be discussed in further detail in Mandelbrot 1977h.

HOW WIDE IS
THE MISSOURI RIVER?

In a second approximation, it is obvious that any "Peano" model of river networks would be much more realistic if rivers are given a positive width, gradually decreasing to zero as one proceeds upstream. For example, if one starts with a network built on the Gosper maze, one can cut off the zigzags up to some posi-

tive scale that would thereafter become visible in the form of *meanders*. The point of departure is a river draining a "tile" like on Plates 46 and 47. This river can be subdivided into successive links, a link having been defined in Plates 62 and 63 as a portion that borders upon subtiles of mth order but not of $(m-1)$ order. In order to give to this river a width of $(1/\sqrt{7})^{\sigma}$, it suffices to flood a judiciously chosen row of subtiles of order σ bordering said river – where σ is some function that increases with m. (The reader may check that such flooding need not disrupt the network's remainder.)

As soon as one begins to implement this broad idea, however, one faces the choice between two very different procedures. To make the meander cutoff self similar, one needs $\sigma = m + m_0$. This amounts to choosing for the link width and the cutoff scale some fixed small proportions of said link's length (measured as the crow flies). The result is seen on Plate 67.

The trouble is that the shore's dimension must drop below 2. A small decrease in D would not matter in itself, but it also follows that the total area covered by the rivers equals the total available area. When rivers are wide (that is, if the value of m_0 is small, equivalent to the bottom figure of Plate 67), then, as one includes ever finer tributaries, the total area of the rivers converges rapidly. When rivers are narrow (large values of m_0, equivalent to the top figure of Plate 67), the total area's convergence is slowed but not avoided.

Thus the self similar width model necessarily refers to a situation in which the land resources devoted to drainage are so extensive that finally there is almost no area to drain! Such a model is of course untenable. To save something of it, one must argue that any passage to the limit is nonphysical. It is furthermore possible to hold the rivers' area down by simply stopping the construction on this nearly plane-filling curve after a finite number of stages.

The second procedure is more gradual, and we shall see that it brings in a very interesting new theme. We shall lose self similarity but gain something in exchange. The procedure consists in assuming that as one proceeds upstream the branches' relative widths and meander sizes decrease *more rapidly* than suggested by self similarity. In this way, it is easy to insure that the areas of the rivers taken together converge rapidly to the desired sum. For example, to return to the Gosper maze, one can replace the quantity $\sigma = m + m_0$ by the quantity $\sigma = 2m$ or by the integer closest to some smaller multiple of m.

Following this second procedure, we achieve a network of river shores that no longer has a similarity dimension. However, since its area is positive, it necessarily has the Hausdorff dimension $D = 2$.

Frankly, it is unlikely that the decision between the above models of river could be based on evidence alone. For this reason the evidence which suggests that river widths are indeed nonself similar will not be discussed here. Instead attention will be directed toward a theme of greater intellectual interest concerning bona fide "simple" curves that are fractals of dimension precisely equal to $D = 2$.

FRACTAL CURVES WITH D=2

It is worth restating that Peano "curves" are not of fractal dimension $D = 2$ because this is the value given by $D = \log N / \log(1/r)$, but because they are plane sets of topological dimension $D_T = 2$. Hence the inequalities $D \leq 2$ and $D \geq 2$ which imply $D = 2$. The basic dimensions being equal, these sets are not fractals after all! Nevertheless, Lebesgue 1903 and Osgood 1903 did construct bona fide simple plane curves that satisfy $D_T = 1$ and $D = 2$ so they are topological curves with a positive superficial measure. See also Gelbaum & Olmsted 1964, p. 136. Of course, the Lebesgue-Osgood *simple curves* are viewed as possibly even more monstrous than Peano curves, but in fact I find them to be the right tool in hosts of practical applications.

To construct such a curve, one follows roughly the same path as for a Peano curve, but one takes care to avoid points of self contact. It is convenient to classify said points into two kinds. In our island illustrations, the first (second) kind of self contacts occurs when black (white) points facing each other across a white (black) strip converge.

In the river networks interpretation, self contacts of the first kind occur when rivers narrow down to zero, and the preceding section showed how this feature can be eliminated without forcing the dimension below 2. Double points of the second kind are relative to the drainage divides, and there is no sensible way of eliminating them because they do express a correct observation, that two drops of water falling very close to each other may well go into different river basins. The resulting asymmetry is indeed an important characteristic of the water cycle. Only its first half consists in a river network that concentrates water from its sources to the main river's mouth; the second half lies in the atmosphere, and although we shall see in Chapter VI that it is also fractal, it is very different in its geometry.

VASCULAR GEOMETRY

It is desirable therefore to interrupt the study of rivers and to study instead some other natural phenomenon wherein the "up" and "down" half cycles are, so to speak, symmetric of each other.

Such an example is available. To quote from Harvey 1628, "The blood's motion we may be allowed to call circular, in the same way as Aristotle says that the air and the rain emulate the circular motion of the superior bodies; for the moist earth, warmed by the sun, evaporates. ... And similarly does it come to pass in the body, through the motion of the blood, that the various parts are nourished, cherished, quickened by the warmer, more perfect, vaporous, spirituous, and, as I may say, alimentive blood; which, on the other hand, owing to its contact with these parts, becomes cooled, coagulated, and so to speak effete." The example of the vascular system is so suitable to our present purpose that it would be too painful for us to be prevented from exploring it here, even though the fractal aspects of anatomy are in principle reserved for another occasion and also despite the fact that this chapter is devoted to curves in the plane, so that we shall have to deal, so to speak, with flattened blood vessels.

Our goal will be to establish that blood vessels crisscross organs so tightly that tissue is a spatial counterpart of the Lebesgue-Osgood monster.

The first underlying factor is that the arterial and venous systems involve no self contact.

The second factor is as follows. It is characteristic of the Harvey view of the circulation of blood that both an artery and a vein are found within a small distance of nearly every point of the body. (See also *The Merchant of Venice.*) This view is crude on the topic of capillaries, but in order to understand any view fully, it is best to begin by taking it textually. In the present case, interpolation demands that there should be both an artery and a vein *infinitely near every* point – except of course that points that lie *within* an artery or a vein are prevented from being very close to a vessel of the other kind.

Stated differently (but somehow this restatement makes the result sound much odder!): Every point in what we shall call tissue should lie on the boundary between the two blood networks.

Here is the third design factor. The volume of all the arteries and veins must be only a small percentage of the body volume, leaving the bulk to tissue.

Superficially, we seem to have boxed ourselves into an exquisite anomaly. A shape that is topologically two-dimensional, because it forms the boundary between two shapes that are topologically three-dimensional, is required to have a volume that is not only nonnegligible compared to the volume of the shape it bounds but much larger!

A virtue of the fractal approach to anatomy is that it shows that the above requirements are by no means impossible to fulfill. To give a proof in a two-dimensional reduction, the first step is to relabel everything on Plate 63: the gray lines will be viewed as veins and the black lines as arteries. (The fact that blood enters by a point and exits by a ring is surely not universal, but it is readily avoided in three dimensions, so it need not stop us here.) The second step is to attach to each vessel a positive thickness. We know that self similar vessels would not do. But we also know that width can be made to decrease faster than ordered by self similarity. (The construction is long to describe but easy to perform.) The result will fulfill all the requirements we chose to impose upon the design of a vascular system. One more detail is given in Chapter XII, at the end of PARADIMENSION.

To repeat: In the present planar reduction veins and arteries both have interior points, and small circles can be drawn entirely within them. On the other hand, the vessels occupy only a small percent of the overall area. The tissue is very different; it contains no piece, however small, that is not crisscrossed by both artery and vein. Such a tissue is a bona fide fractal curve: topological dimension of 1 and fractal dimension of 2.

Back to space, where the present structure belongs, tissue is a bona fide fractal surface: a topological dimension of 2 and a fractal dimension of 3.

Lebesgue-Osgood monsters are the very substance of our flesh!

The topic of fractal facets of anatomy will warrant more detailed attention on a better occasion. Then we shall also explore the notion that the lung exhibits yet another construct reputed monstrous, since *three* sets – arteries, veins and bronchioles – have a common boundary.

◁ *Postscript.* A last-minute check of Osgood 1903 reveals that its construct was explicitly described as the common shore of branching channels and dikes, in other words, through a somewhat forced interpretation of the hydrological problem which was alluded to but not developed in this section. Needless to say, I could find no trace of follow-up on this intepretation. ▶

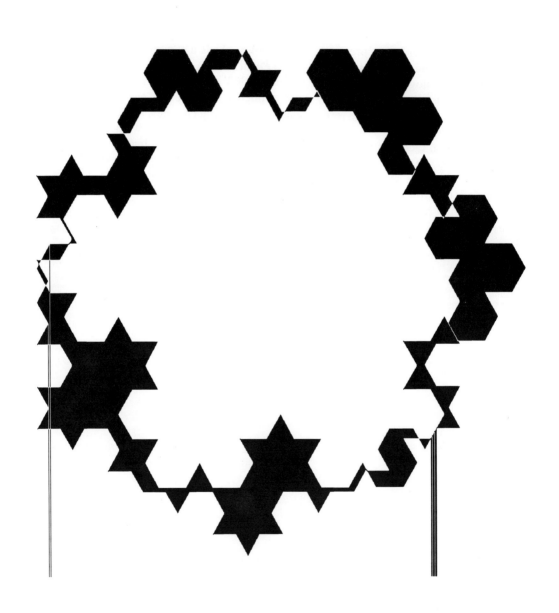

CHAPTER III

Uses of Nonconstrained Chance

This chapter will inject an essential new theme referred to in this Essay's subtitle. The simplest models of chance are related to the classical Brownian motion. Their principal characteristic is that the chance they involve can be viewed as totally nonconstrained. As a result, they are particularly easy to simulate, and their fractal properties are simple. Unfortunately, we shall find that their other properties make them unsuitable as models of geomorphology. On the other hand, the chapters that follow will show that nonconstrained chance deserves a careful study because many basic natural phenomena respond to it very well.

RANDOM COASTLINES

Much as it reminds us of real maps, the Koch curve has two major defects we shall encounter, almost unchanged in spirit, in the early models of every other phenomenon studied in this Essay. Its parts are identical to each other, and the self similarity ratio r cannot be chosen at will but must be part of a strict scale, namely, $1/3$, $(1/3)^2$, and so on.

One might try to improve the model by making the algorithm more complicated, while preserving its deterministic character. However, this approach would be not only tedious but ill-inspired. It is indeed obvious that each coastline has been molded throughout the ages by multiple influences that have not been recorded and cannot be reconstituted in any detail. The traditional goal of achieving a full description is hopeless and need not even be entertained.

We shall see very shortly how the theory of Brownian motion deals with this difficulty as it presents itself in the case of colloid particles bombarded by immense numbers of molecules in a fluid. The key lies in statistics. In geomorphology, the situation is even more complicated. Indeed, on the level that may be called local, fluid molecules are governed by the laws of mechanics, which are viewed as known. The open problems concern exclusively the precise nature of interactions on a global level. In geomorphology, however, even the underlying laws are uncertain. Hence we have an even more compelling reason to seek salvation in models that imply the renunciation of any search for a precise description of reality, in other words, in statistics. In other fields with which we shall deal, the current knowledge of local interactions is of a quality that lies somewhere between those characteristic of physics and geomorphology.

The fascinating question of the relationship between statistical unpredictability and local determinism continues to arise, but we shall have little to say about it beyond an allusion at the end of this chapter. In other words, while we recognize that the notion of chance evokes all kinds of quasi-metaphysical anxieties, we are determined not to worry about them. A fortiori we shall not be concerned whether or not, in Einstein's words, "the Lord plays with dice."

One might say that we shall systematically attempt to make the expression "at random" revert to the old intuitive connotations it must have had at the time when medieval English had just borrowed it from French, a time when the phrase "a horse at random" (*un cheval à randon*) is reputed to have been unconcerned with the horse's psychology and merely denoted irregular motion that one cannot *predict*.

More specifically, without worrying about philosophical problems of foundations, this Essay will interpret chance as the phenomenon that is described by the theory of probability. This theory is the only mathematical model available to the scientist who seeks to map the unknown and the uncontrollable; fortunately, it is at the same time extraordinarily powerful and convenient.

EMPTY INVOCATION OF CHANCE VERSUS ACTUAL DESCRIPTION

This model is also very tricky. To keep to the example treated in Chapter II, one is very much tempted to say that in a search for a model of coastlines, it is easy to eliminate the defects of the Koch curve while preserving its desirable characteristics. It suffices to *invoke* chance by saying, "just beat the sequence of the parts while varying their sizes."

This argument is frequently encountered (although ordinarily hidden under lengthy calculations), and in some cases it does work. In other cases, it is grossly insufficient. While an empty invocation of chance is all too easy, an actual description of the rules that generate acceptable random curves is often very hard indeed. The basic difficulty is that the curves (and other sets) with which we are dealing in this Essay are all imbedded in a surrounding space. By merely varying the sizes and beating the order of a coastline's parts, one is often left with pieces that will not fit together.

NONCONSTRAINED VERSUS SELF CONSTRAINED CHANCE

And we hit immediately upon a dichotomy that is somewhat informal but of great practical impact. In certain instances, it is quite permissible to let chance affect the parts of a set independently of each other, because there is no fear of a resulting mismatch. We shall describe such models as involving a nonconstrained form of chance. An example is presented in Plate 85. In other instances, such an approach is not permissible, and chance is *self constrained*.

To exemplify the contrast, we may note that any problem involving the total number of 2n-sided polygons on a lattice, regardless of whether or not they self intersect, is related to nonconstrained chance. By contrast, the so-called self avoiding polygons (like the polygonal approximations of coastlines) are related to self constrained chance.

The distinction is unfortunately of great importance, because counting all 2n-sided polygons turns out to be easy, but counting self avoiding 2n-sided polygons has thus far eluded the best minds. Therefore, the problems involving self constrained chance will be by far the harder ones, and they will tend to be postponed to the end of this Essay.

Our studies will deal exclusively with phenomena that relate to shapes embedded in E-dimensional Euclidean space, or *E-space*. We shall sometimes say they "sit" in such a space. In all cases, and especially when chance is self constrained, everything will very much depend on the value of this corresponding Euclidean dimension E.

SEARCH FOR THE RIGHT AMOUNT OF IRREGULARITY

When we get down to business, the following question arises: Granting that chance may bring about irregularity, is it able to bring about a degree of irregularity as strong as that encountered in the coastlines we are presently seeking to model? It so happens that not only is chance able to do so but that it is most difficult in many cases to keep it from going *beyond* the desired goal. In other words, the power of chance is widely underestimated. The reason would seem to lie with the fact that the physicist's concept of randomness was shaped by quantum theory and thermodynamics, two theories in which chance is essential at the microscopic level, while at the macroscopic level it is insignificant. Quite to the contrary, in the case of the natural objects that concern us, the importance of chance tends to remain constant at all scales. In particular, these objects remain very erratic even at the macroscopic level. As a result, while irregularity had to be artificially introduced in nonrandom constructs like the Koch curve, it becomes hard to contain when randomness is let in.

BROWNIAN MOTION

Our observation concerning the difficulty of preventing chance from going overboard had already been made by J. Perrin in 1906 when (as we saw in Chapter I) he compared the motions of Brownian particles with the continuous nondifferentiable curves of the mathematician. Perrin's illustration has already been reproduced in our Plate 11, and we have noted that it helped inspire

the young Norbert Wiener to a full mathematical study of the process, around 1920. Since then, the Brownian motion has starred with special frequency in the works of Paul Lévy. The precursor was Louis Bachelier, who is featured in Chapter XI.

Basically, the mathematical Brownian motion $P(t)$ in time (t) of a point P is a series of very small movements, mutually independent and isotropic. *Isotropy* means that all directions of motion are equally probable; in the plane, one can select the direction by throwing a point at random on a circle graduated in degrees and reading off an angle. *Independence* expresses that past positions of the motion do not affect its future evolution. This explains the term *nonconstrained*.

Brownian motion had been first defined by Bachelier as the limit of a simple random walk when the steps are made infinitely small. A simple random walk on the line is illustrated in Plate 87.

Similarly, a point $P(t)$ in the plane is said to perform a simple random walk if, at successive instants of time separated by the interval Δt, it moves by steps of fixed length $|\Delta P|$ in randomly selected directions. Usually the direction either is restricted to a lattice or is isotropic.

However, the most direct characterization of Brownian motion proceeds through a more general random walk.

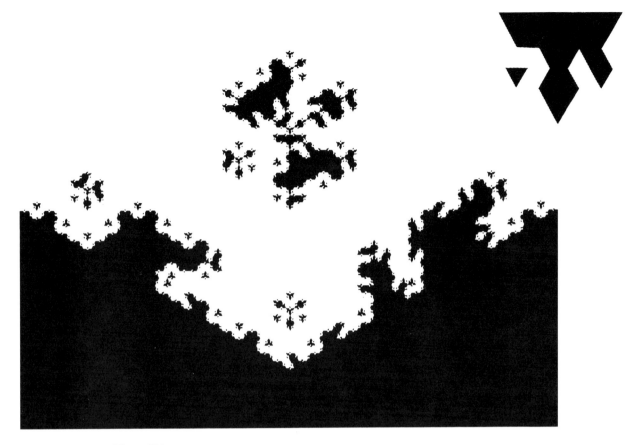

Plate 85

RANDOM KOCH COASTLINE
(DIMENSION D=1.4800)

In many instances, a Koch curve with prescribed D and no self contact can be achieved in several different ways by using the same overall grid and the same initial polygon. Suppose in addition that at least two different standard polygons (roughly speaking) fit within the same overall outline. Then it is easy to randomize the construction by selecting among said polygons by chance. For example, one can alternate between the standard polygon used on Plate 51 and a variant on the polygon used to the right on Plate 50. The result is shown to the left on Plate 50.

The present Plate presents another example. The standard polygons, shown in the top right corner, are drawn on the two bottom sides of an equilateral triangle. The desirable dimension D=1.5

with N=8 and r=1/4 involves unavoidable self contact. Hence, we had to settle on D just below 1.5 (the standard polygons' two end sides were made longer than their six other sides). The overall form of the random Koch island constructed in this fashion is very dependent on the initial shape. In particular, all the initial symmetries remain visible throughout. On the other hand, small scale representations of the same construction performed with an equilateral triangle as the initial shape can be seen in the three largest islands offshore.

Because of the constraints listed at the beginning of this caption and of the strong continuing effect of the initial polygon, random shuffling of the parts of a Koch curve is a method of limited scope. The fact that bolder shuffling would lead to innumerable self intersections is at the core of the distinction between nonconstrained and self constrained chance.

Plate 87

A RANDOM WALK APPROXIMATING
A BROWNIAN LINE-TO-LINE
FUNCTION, AND ITS ZEROSET

The longest running (and least skilled) of all games of chance has pitted Peter against Paul since around 1700, when the Bernoulli family was ruling over probability theory. The coin started and remains eternally fair. Each time it falls on heads, Peter wins a penny; otherwise Paul wins. Some time ago, William Feller came by to observe this game, and he reported Peter's cumulative wins on the upper Figure of this Plate, which is from Feller 1950, the celebrated textbook probabilists tend to simply call *Volume I*. (Reproduced from *An Introduction to Probability Theory and Its Applications, Volume I* by William Feller, by kind permission of the publishers, J. Wiley and Sons, copyright 1950.)

Likewise, the middle and bottom Figures represent Peter's cumulative winnings in a greater total number of coin tosses. For the sake of clarity, data were entered only at intervals of 20 tosses.

If data had been taken over a much larger period and squeezed in ever further before being reported, one would asymptotically obtain a sample of values of a Brownian line-to-line function.

Feller has confided that these Figures are by no means "typical" and were selected in preference to several others that looked too wild to be believed by the student. Be that as it may, seemingly endless contemplation of these Figures played a decisive part in elaborating the theories described in this Essay.

Mandelbrot 1963e observed that the whole graph's shape was reminiscent of a mountain's silhouette or of a vertical section of Earth's relief. Through several generalizations, this observation led to the successive models described in Chapter IX.

Now let us limit our attention to this graph's zeroset, that is, those moments when Peter's and Paul's fortunes chance to be back where they were when we started looking. By construction, the time *intervals* between the zeros are mutually independent. However, it is clear that the *positions* of the zeros are in no way independent, rather *very distinctly clustered*. For example, if the second curve were examined in the same detail as the first curve, most zeros would be replaced by a whole cluster of points. If we were dealing with mathematical Brownian motion, we could sub-

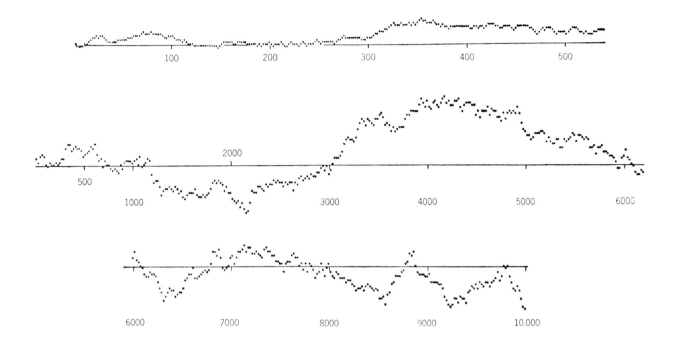

divide these clusters in a hierarchical manner, ad infinitum.

By good chance, I thought of Feller's diagram when asked to model the distribution of telephone errors. The diagram suggested that although such errors were known to be grouped in bursts (this being the gist of the practical problem they raised), it was useful to inquire whether the intervals between the errors might not be independent. A detailed empirical study did confirm this conjecture and led to the models discussed in Chapter IV.

◁ We shall see that the zeroset of the Brownian motion, approximated by the discrete zeroset of the above random walk, constitutes the simplest Lévy set, namely, a random Cantor set of dimension $D = 1/2$. Any other D between 0 and 1 may likewise be obtained through the zeros of other random functions. Through this model it is possible to define the fractal dimension of a telephone channel. Actual estimates are quite variable and depend upon the precise characteristics of the underlying physical process. ▶

One assumes that over a succession of equal time increments Δt, the corresponding vectors ΔP(t) are independent, isotropic, and random, with a probability distribution such that the projection of ΔP(t) on each axis is a Gaussian scalar random variable, with zero mean and a variance equal to $|\Delta t|$. Hence the root mean square is $\sqrt{|\Delta t|}$.

◁ The usual definition given in Chapter XII takes still another alternative path and dispenses with the division of time into equal steps. It requires isotropy for the motions between any pair of instants t and $t_0 > t$. It requires independence of future motion with respect to the past position. Finally, it requires that the vector going from P(t) to P(t') be such that its projection on any axis is a Gaussian scalar random variable of zero mean and variance $|t_0 - t|$. Hence, the vector divided by $\sqrt{|t_0 - t|}$ has the same probability density for all values of t and t_0. ▶

THE EXPECTATION ⟨X⟩

Each discipline seems to denote the expectation of X differently. The physicists use the notation ⟨X⟩, which has the virtue of including its own portative parentheses. It will be adopted in this Essay so that, for example, the above-listed conditions on the Brownian ΔP(t) will be written

$$\langle \Delta P(t) \rangle = 0 \text{ and } \langle [\Delta P(t)]^2 \rangle = |\Delta t|.$$

BROWNIAN FUNCTION AND TRAIL

Before going any further, one should stress that the term *Brownian motion* as defined here is geometrically ambiguous. It can, first of all, designate the ordinary graph of P(t) viewed as a *function* of t. When P(t) is the ordinate of a point in the plane, the graph is a plane curve like on Plate 87. When P(t) is a point in E-space, said graph is a curve in a (1+E)-space (the time coordinate being added to the E coordinates of the space of the P's).

In many instances, however, one is only interested in the *trail* left behind by the motion. This is a curve in E-space. It takes account of the order in which points were visited but not the timing of the visits. In the case of a random walk, this fine point need not be raised, since the trail has bends at uniformly spaced instants of time, so that the function and the trail are easily deducible from each other. However, in a continuous Brownian motion, the two aspects are not equivalent.

Hence, the custom which designates them by the same term, *motion,* is very confusing. Whenever ambiguity threatens, we shall use instead either *Brownian function* or *Brownian trail.*

In addition, we shall eventually need "time" to be made multi-dimensional. (Once one has let dimension be a fraction, anything seems to go!) For example, one of the models of Earth's relief will assume that the altitude is a Brownian function of latitude and longitude. This definition will be postponed until needed, but it is clear that further specification of the terminology is required. When necessary, we shall speak of Brownian line-to-E-space functions or of Brownian line-to-E-space trails, shortened to *line-to-E trails*.

STATISTICAL SELF SIMILARITY; FOR BROWNIAN TRAILS, D=2

Let a Brownian line-to-plane motion start at the point $P(0)=0$. Its trail until time 4 divides into $N=4$ partial trails: between times 0 and 1, 1 and 2, 2 and 3, and 3 and 4, all of which are statistically identical except for translation. It suffices, therefore, to consider the first partial trail. One can show that it is deduced from the whole by a similarity of ratio $r=1/2$.

◁ Since the root mean square of $P(4)$ is double that of $P(1)$, the value of $P(4)/2$ has the same distribution as that of $P(1)$. And more generally, for every fixed h satisfying $0<h<1$, the distributions of $P(4h)/2$ and $P(h)$ turn out to be identical. It is easy to show that this identity in statistical distribution extends to the trails defined by $P(4h)/2$ and $P(h)$. ▶

Applying the definition of D with the values $D=4$ and $r=1/2$, we obtain

$$D=\log4/\log2=2.$$

The preceding "proof" is actually no more than an heuristic one, because it slurs over the possible role of double points, of which Brownian motion has an infinity. The very notion of similarity dimension is therefore shaky in this instance, and one must turn to the Hausdorff Besicovitch dimension. Luckily, it confirms that $D=2$.

Self similarity as applied to random sets is less demanding than the notion introduced in Chapter II, since the parts need no longer be precisely similar to the whole. It suffices that the parts and the whole reduced by similarity should have identical distributions.

Consider now a Brownian line-to-line function. It can be deduced from the line-to-plane function by being squeezed into a line instead of having the full plane to roam through. Two line-to-line Brownian motions (statistically independent of each other) can be obtained simply by projection, that is, by taking the coordinates $X(t)$ and $Y(t)$ of the above planar motion $P(t)$. Euclid had taught us that the projection of the whole plane upon a line can have no more than the dimension of said line, namely 1, despite the fact that it covers this line superabundantly. And now we see that just the same will hold for the graphs of either $X(t)$ or $Y(t)$.

Of course, Brownian motion has no necessary link to the plane. As a matter of fact, Perrin's experimental motions proceeded in 3-space and were drawn on the plane as watched through their projections on the microscope's focal plane. It is obvious, however, that our derivation of the similarity dimension not only is not dependent upon $E = 2$ but becomes increasingly convincing as the number of double points decreases, which requires an increase in E.

The Hausdorff Besicovitch dimension confirms that the value of D equals 2 for every Brownian line-to-E trail. When $E > 2$, the inequality $E > D$ expresses numerically that Brownian motion fills space very scantily.

Despite its double points, a Brownian trail can be shown to be sufficiently curvelike to have a topological dimension of $D_T = 1$, and the fact that $D = 2 > D_T = 1$ confirms that Brownian trails are fractal sets according to the definition in Chapter I. And of course (this is a new subsidiary theme) $D = 2$ does not by any means force a set to be a surface.

DEPENDENCE OF MASS ON RADIUS

Scaling by $\sqrt{t}$ is characteristic of most aspects of Brownian motion. For exam-
ple, the distance from $P(0) = 0$ to the farthest point reached in time t is a random multiple of $\sqrt{t}$. Also, the total time spent in a circle of radius R about $P(0) = 0$ is a random multiple of R^2. If one succeeds somehow in weighting the different pieces of the Brownian trail by "masses" proportional to the time it takes to run through them, one finds that the total mass in a circle of radius R is proportional to R^2.

Formally, this relationship is precisely the same as in the case of the Koch curve examined in Chapter II. It is a fortiori the same as in the classical cases of a segment, disc, or sphere of uniform density, for which the total mass satisfies the familiar relationship mass$\sim R^D$. Later chapters will show the relationship to be of wide generality.

BROWNIAN FUNCTIONS AND ZEROSETS

The Brownian functions, such as the graphs of a two-dimensional $P(t)$ or of its coordinate functions $X(t)$ and $Y(t)$, are also of great practical importance. Of special simplicity are their zerosets, that is, the values of t for which $X(t) = 0$. The zerosets can be seen to be themselves self similar, and the obvious fact that they are an extremely sparse set is expressed by their having the fractal dimension $D = 1/2$.

Let us stop to emphasize this result. It is the first time that we encounter a nonintegral D in a stochastic context.

The graphs of the function $X(t)$ and $P(t)$, however, are *not* self similar, merely *self affine*. This last term expresses a property we have already used indirectly when deriving the dimension D. The whole curve from $t=0$ to $t=4$ can be paved by $N=4$ portions obtained if the space coordinate(s) continue to be reduced in the ratio $r = 1/2$ while the time coordinate is reduced in the different ratio $r^2 = 1/N$. Hence, similarity dimension is not defined for either $X(t)$ or $P(t)$. On the other hand, the Hausdorff Besicovitch definition does extend to $X(t)$, and it happens to yield the value $D = 3/2$.

This last example agrees with the assertion in Chapter II that Hausdorff Besicovitch dimension is the most general way of catching the intuitive content of fractal dimension. This is perhaps why it is also the most unwieldy!

THE FRACTAL DIMENSION
OF LINEAR SECTIONS

The following is a well-known property of Euclidean plane shapes. Knowing that the shape's dimension D satisfies $D \geq 1$, its section by a line, if nonempty, is "typically" of dimension equal to $D-1$. For example, a nonempty linear section of a square $(D = 2)$ is of dimension $1 = D-1$, while the linear section of a line $(D = 1)$ is in almost every case of dimension $0 = 1-1$.

Fortunately, fractal dimension preserves this important theme. Indeed, the zeroset of the Brownian line-to-line function is the section of a set of $D = 3/2$ by a horizontal line, and its dimension $1/2$ is, as expected, equal to $3/2 - 1$.

Similarly, the rule says that a linear section of a Brownian line-to-plane trail should have the dimension $2-1 = 1$, and such is indeed the case.

This rule suffers exceptions, however, that are especially conspicuous in the case of fractals that are not isotropic. For example, the section of the Brownian line-to-line function by a vertical line is simply a point. However, the rule, properly generalized to E-space, does play a very important role in model making (see Chapter XII).

BROWNIAN MOTION IS NOT A
SUITABLE MODEL OF A COASTLINE

The fact that our Brownian line-to-plane trail is of dimension $D = 2$ implies, most regrettably, that it is not at all suitable as an image of a coastline. Its irregularity is excessive. It is bad enough that it should include innumerable multiple points, but it goes even farther and (almost like the Peano curve) fills almost the whole plane!

The Brownian line-to-line function seems to correct all three defects. It is a nice one-to-one function, without double point, and while its dimension $D = 3/2$ is rather large in view of Richardson's data, it is not wildly inappropriate, and it does not imply that the curve fills the plane. What a pity, therefore, that the Brownian motion should achieve these aims by being overwhelmingly nonisotropic, that is, that it should require peninsulas and bays to jut out or in along a well-defined direction, along which they narrow down monotonically. I cannot think of any island having such a shore; the model is plainly absurd. And yet it gives an impression of progress. Without invoking chance emptily, we have at last a somewhat coastlike curve that goes beyond the geometric shapes in Euclid.

A BROWNIAN FUNCTION MODEL OF A RIVER'S COURSE

Now we turn to river courses. A model based on the Brownian line-to-plane trail is, again, absurd. But a model using a Brownian line-to-line function, while awkward, is worth a moment's

thought. With some good will, it helps rescue for a moment an acknowledgedly crude model due to Leopold & Langbein 1962. The idea is to imagine an incline that never goes up as one proceeds east and to assume that rivers could only flow east, north, or south at random.

Though Leopold & Langbein have not done so to my recollection, it is useful to examine the model's bearing upon the Hack relationship, which (as seen in Chapter II) links stream length and basin area. If the model were self similar, we know (again from Chapter II) that Hack's exponent would be linked to a dimension, but the present model is *not* self similar. In the case of a river without a tributary, the basin area has a rough counterpart in the product of the east-west and north-south spans of the river. If the former is denoted by L, the latter will be proportional to $\sqrt{L}$, so the area will be proportional to $L^{3/2}$ and the quantity $(\text{area})^{1/2}$ will be proportional to $L^{3/4}$. As to human-length, it is readily checked that in this case it must be proportional to L. Finally, we obtain

$$(\text{human-length})^{3/4} = (\text{area})^{1/2}.$$

In conclusion, the exponent in Hack's relationship takes the value $4/3$. Again, unfortunately this $4/3$ is *not* the dimension of anything, but the argument may have some weak explanatory value.

◁ The fact that the exponent $4/3$ is

not the dimension of the underlying Brownian function has the virtue of confirming that the definition of dimension through the relationship $(mass)^{1/D} \sim (area)^{1/2}$ is very much related to self similarity. In other cases, it readily leads to absurdity. ▶

A TENTATIVE CONCLUSION ABOUT NONCONSTRAINED CHANCE

The preceding sequence of successes and failures is unfortunately typical. It seems to be the fate of Brownian motion that for many problems, it should be of help in roughing out some aspects of the issues but must soon be discarded as inadequate.

The ordinary response to this failure is to introduce modifications that seem local but in fact tend to destroy the self similarity of Brownian motion and often its being a fractal. Our procedure will by and large be the opposite, to try and save fractal self similarity even at the cost of having to accept much deeper global modifications of Brownian motion.

PSEUDO RANDOMNESS AND PRIMARY CHANCES

The question of why deterministic transformations very often mimic a random set as described by the theory of probability must be faced separately in different fields. It already arises in particularly exemplary fashion in the context of pseudo random operations simulated on the computer. Thus this volume is full of diagrams that will be described as due to chance but that were really constructed in a well-determined fashion. The procedure starts with a sequence of numbers one views as having been obtained by throwing a ten-sided die (one that yields integers from 0 to 9). Yet this sequence is in fact nonrandom; the numbers in question are always formed in a way that is fully known and that can be repeated. They are the product of a "canned" generating program, starting with some arbitrarily chosen number called *seed*.

This last image is expressive and can no longer be changed, but it conveys rather poorly the intention of one who seeks to simulate chance. Every gardener hopes that his harvest will depend above all on the seed he sows. By contrast, I hope that my seed will have no significant effect and will only trigger a computer program.

Such a computer program is the hinge point in any simulation. Upstream, one performs operations that are universal in nature, and involve in every case the same interface between number theory and probability theory. The downstream portion of the argument, on the contrary, varies according to the program's ultimate goal.

However, there is no reason to demand a single universal hinge point in all cases. It is better to prepare in advance a kit in which one has stored a number of alternative "primary chances" and to select in each case the most convenient one. The hinge used most often is the computer die. However, there are many others, such as points that fall on the circumference of a circle with a uniform probability distribution. In cases where the problem involves not just one but two or several or even an infinity of variables, as with a whole continuous curve, the primary chance typically assumes them to be independent. Such has been the case for the steps of a Brownian motion.

Again, primary chance intervenes as the point of separation between two stages of a theory – an upstream, of which almost nothing will be said in this Essay, and a downstream, which requires a very different turn of mind. It consists in each instance in replacing unexpected forms of chance by forms to which we have already become accustomed, and which we have included in our kit.

CHAPTER IV

Fractal Events and Noises

This chapter's principal goal will be to acquaint the reader concretely and painlessly with another mathematical object ordinarily viewed as pathological – the Cantor set. More precisely, a family of sets having dimensions between 0 and 1 shall be introduced. They are formed by points on a straight line, which makes them easy to study. In addition, they involve in its simplest form a concept that is central to fractals, the "cutouts."

However, these sets are so "thin," seeming evanescent, that it is exceedingly difficult to drawn them and to otherwise achieve an intuitive grasp of their nature. Also, the only concrete problem I could find as a support to the discussion is somewhat esoteric.

DESCRIPTION OF A CATEGORY OF DATA-TRANSMISSION LINES

Every transmission line is a physical system capable of transmitting electricity. However, electric current is subject to numerous spontaneous fluctuations which, even when one does not try to transform them into sounds, are known as "noises." The quality of transmission depends on the likelihood of error or distortion due to noise This likelihood, in turn, is highly dependent on the ratio of signal to noise.

This chapter is concerned with lines used to transmit computer data. In the hope of obtaining a communication that is as near perfect as possible, one tends to use very strong signals, and errors are indeed greatly thinned out – but they are never completely eliminated. This means that from time to time noise is sufficient to cause the disturbed signal to be incorrectly read by the receiver.

An interesting fact is that while the distribution of errors reflects the distribution of noise, it simplifies it, so to speak, down to the bone. A function having several possible values is replaced by a function that has only two possible values. For example, it may be set to zero in the event that there is no error at time t and to one if there is an error.

Physicists understand very well the structure of the noises that predominate in the case of weak signals. The main one is thermal noise. In the problem just described, however, the signal is so strong that the classical noises are relatively speaking negligible, so that the nonnegligible noises are not classical.

The nonnegligible noises are called *excess noises*. They are difficult and fascinating because little is known about them. This chapter will study an example of excess noise that involves fractals.

◁ Its degree of generality is not yet entirely clear, but there is no question that around 1962 it was viewed as being of great practical importance by many electrical engineers and diverse talents were called to investigate it. As it happens, my own contribution to this effort, to be examined later in this chapter, was the first concrete problem in which I experienced the need to use fractals. It was the first of many stages that eventually led to the present Essay. No one remotely imagined at that time that a careful study of this unyielding but apparently modest engineering difficulty would lead to theoretical considerations of such a far-reaching nature. ▶

INTERMISSIONS

In order to characterize the fractal excess noises, let us investigate the resulting errors by means of an increasingly refined analysis. A rough analysis mostly reveals the presence of long periods during which no error is encountered. Let these periods be called "rank 0" intermissions if their duration exceeds one hour. By contrast, any time interval flanked by error-free intermissions is singled out as being a "burst of errors of rank 0." As the analysis is made ten times more accurate, it reveals that the original burst is itself "intermittent." It includes in its midst a number of shorter intermissions lasting six minutes or more, which separate correspondingly shorter bursts "of rank 1." Likewise, each of the latter contains several intermissions lasting 36 seconds, separating bursts "of rank 2," and so forth and so on, with each stage based on intermissions that are ten times shorter than the previous ones.

If each level of analysis were only three times as accurate as the preceding one, the preceding description could be illustrated very roughly by Plate 99. (One should not pay attention to the caption until the next section.)

The preceding description suggests something about the relative positions of the bursts of rank k within a burst of rank k−1. The probability distribution of these relative positions seems independent of k. This form of invariance is obviously an example of self similarity, and fractal dimension cannot be far behind, but I shall not rush into defining it. This Essay discusses diverse subjects not only for their own sake but also with the hope of using them to elicit new themes or to refine old ones. In this perspective, I shall, as in Chapter II, reverse the historical order. Instead of starting out with a model I recommend, I shall start by another model, which is very rough and nonrandom but much simpler and otherwise important and instructive.

A ROUGH MODEL OF ERROR BURSTS: THE CANTOR SET

The preceding section constructed the set of errors by starting with a straight line, the time axis, and cutting out shorter and shorter error-free intermissions. This procedure may be uncommon in Natural Science, but it has been of use in pure mathematics since G. Cantor and probably even before (see T. Hawkins 1970, especially p. 58). Cantor 1883 made use of cutouts to construct a certain fundamental set called either dyadic or triadic (or ternary); the first term is due to the fact that, as will be seen, $N=2$

and the other terms to the fact that $1/r=3$. It is also called *Cantor space* or *Cantor discontinuum*. In comparison to the real line, this set is so peculiar as to be pathological. It is taken for granted that it cannot conceivably be of use in modeling Nature. However, we are in the process of being led straight to the Cantor set by Nature's own peculiarities!

The triadic set's construction involves interpolation, then extrapolation. The interpolation (illustrated in Plate 99) proceeds by successive stages. The point of departure is the closed interval [0,1]. (The term "closed" and the direction of the brackets indicate that the extreme points are included; this notation was already used in Chapter II, but there was no need until now to make it explicit.) The first construction stage consists in removing the middle open third, designated]1/3, 2/3[(here, the term "open" and the reversion of the direction of the brackets indicate that the extreme points are excluded). Next one removes the middle of each of the two remaining thirds. And so on to infinity.

Since a point on the time axis marks an "event," the Cantor set is a fractal sequence of events.

Using terms borrowed from the study of turbulence, one can view the Cantor set as the outcome of a cascade. "Stuff" that was uniformly distributed over an initial interval [0.1] is first subjected to a centrifugal eddy of length 1, which sweeps it into the interval's extreme thirds. While a "first-order cutout" is thus emptied, the total stuff is conserved and shall be assumed to be redistributed over the outer thirds with uniform density. Then two centrifugal eddies come in and repeat the same operation, starting with the two intervals [0,1/3] and [2/3,1]. The process continues as a Richardsonian cascade.

As a prelude to the extrapolation, let us recall a point of history. At the time when he introduced the triadic set, Cantor had barely left the field in which he had started his career, the study of trigonometric series. Since such series are concerned with periodic functions, the only extrapolation needed consists of endless repetition. Using the self-explanatory terminology of inner and outer scales, which is used systematically in the study of turbulence, Cantor restricted himself to the outer scale $L=1$. To achieve any other L (satisfying $0<L<\infty$), it suffices of course to enlarge the original set in the ratio L before repeating it.

However, repetition inevitably destroys self similarity, which this Essay views as valuable. Fortunately, it is readily saved by using a different method of extrapolation. In the first stage, the set interpolated from [0,1] is replicated on segment [2,3] in such a manner as to yield the original set enlarged by a ratio of 3. Then two replicas are placed on [6,7] and [8,9], thus arriving at the original set made larger by a ratio of 9. After that, four replicas are placed on

$$[2\times9,2\times9+1], [2\times9+3,2\times9+4],$$
$$[3\times9-4,3\times9-3] \text{ and } [3\times9-1,3\times9].$$

This operation leads to the initial set enlarged by a ratio of 27, and so on.

DIMENSIONS BETWEEN 0 AND 1

It is easy to see that the set yielded by infinite interpolation and extrapolation is self similar and that its dimension is

$$D=\log2/\log(1/3)=\log2/\log3,$$

which is a fraction between 0 and 1.

In addition, the "dissection rule" can be changed. First of all, the middle segment to be removed can be of length $1-2r$ other than 1/3. In this fashion, one reaches other values for the dimension, but all are of the form $\log2/\log(1/r)$, that is, are smaller than 1.

Further, the value of N need not be 2. Here are a few examples. The set obtained by cutting out the second and fourth fifths of [0,1] yields

$$D=\log3/\log5=0.6826.$$

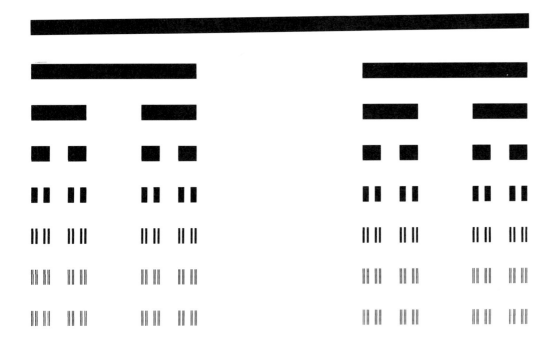

Plate 99

A CANTORIAN TRIADIC BAR
DIMENSION D=log2/log3=0.6309

The original Cantor set is extraordinarily difficult to illustrate because it is thin and spare to the point of being invisible. To give an idea of its form, it is best to thicken it through replacement by what may be called a Cantor bar. The construction begins with a round bar (seen in projection as a rectangle). One hammers matter out of its middle third into the side thirds, so that the positions of the latter remain unchanged. Then one hammers matter out of the middle third of each side third into its side thirds, and so on ad infinitum until one is left with an infinitely large number of infinitely thin slugs of infinitely high density. Moreover, they remain spaced along the line in the very specific fashion induced by the generating process. In this illustration, this process of hammering is carried on until the last step seems to lead to a gray slug rather than two black slugs parallel to each other.

◁ In technical terms, the diagram as drawn attempts to illustrate a geometric shape that is the Cartesian product of a Cantor set of length 1 by an interval of length 0.1. ▶

Plate 101

CANTOR'S DEVIL'S STAIRCASE

The cumulative distribution function of mass along the Cantor bar of Plate 99 is odd indeed. Set the bar's length and mass as both equal to 1, and define the (cumulative) distribution function for the abscissa R as being the mass contained between 0 and R. Since there is no mass in the intermissions, the distribution function remains constant along almost the whole length of the bar. However, since hammering does not affect the total mass, the distribution must manage to increase *somewhere* from the point of coordinates (0,0) to the point of coordinates (1,1). It increases over infinitely many, infinitely small, highly clustered jumps corresponding to the slugs.

The cumulative sums of the widths and of the heights of the steps both equal 1, and one finds in addition that this curve has a well-defined length equal to 2. A curve of finite length is called rectifiable and is of dimension $D=1$. This example has the virtue of demonstrating that sharp irregularities do not necessarily prevent a curve from being of dimension $D=1$, as long as they remain sufficiently few and scattered. The properties of this union are described fully in Hille & Tamarkin 1929.

The Devil's staircase accomplishes the feat on mapping the drastic nonuniformity of the Cantor bar into something uniform and homogeneous. Starting with two different intervals of the same length on the vertical scale, the inverse function of the Cantor staircase yields two collections of slugs that contain the same mass but ordinarily look very different.

Fractal homogeneity. It is convenient to describe the mass in the Cantor bar as fractally homogeneous in its distribution.

Generalization. Since science thrives on uniformity, some regularizing transformation of the general type of the Cantor staircase is invaluable if one wants to make fractal irregularity accessible to analysis. In Chapter V, the regularizing role will be played by ordinary clock time through Brownian or Lévy motion.

See also, in Chapter XII, FRACTAL TIMES, INTRINSIC AND LOCAL.

The set obtained by cutting out the second and third fourths yields

$$D = \log 2 / \log 4 = 1/2.$$

The set obtained by cutting out the second, third, fourth, sixth, seventh, and eight ninths yields the same value

$$D = \log 3 / \log 9 = 1/2.$$

Clearly, there is at least one Cantor set for every $D < 1$. Two conclusions can be drawn from these examples. One agrees with old intuition: the "thinner" a Cantor set, the smaller its dimension. The other calls for new intuition: two sets having the same dimension may "look" very different.

It is also possible to prove that in the Cantor set the number of intermissions of length U above u is roughly u^{-D}. More precisely, there is a constant σ such that said number is represented by a stairlike curve that passes constantly from one side to the other of the graph of σu^{-D}. Here comes dimension again.

Being well aware that a physicist is expected to be automatically "turned off" by a mention of Cantor sets, I had postponed discussing them. As is clear by now, I do not claim that these sets are so nice that they *must* be good for something. My claim is that the very same properties that cause Cantor discontinua to be viewed as pathological turn out to be indispensable in a realistic model of intermittency.

AVERAGE NUMBERS OF ERRORS IN THE CANTORIAN MODEL

As in the case of a coastline, a rough idea of the sequence of errors can be obtained if Cantor's iteration is interrupted as soon as it reaches segments equal to a small but positive inner scale η. For example, η may be the duration of a communication symbol. And similarly, extrapolation is to be stopped as soon as it reaches a large but finite outer scale L, leaving the outcome to be repeated periodically.

Assuming one has performed such double truncation, let us derive the number $M(R)$ of errors contained in a sample of increasing length R. This $M(R)$ constitutes an alternative method of keeping time, an example of *fractal time* (see Chapter XII).

First method: If the sample begins at $t = 0$, the formal derivation of $M(R)$ proceeds precisely as in the case of the Koch curve. It is easy to see that as long as R remains smaller than L, the number of errors doubles each time R is multiplied by 3. As a result, the total number of errors increases roughly like R^D.

This expression is of course identical

to the classical expression for the mass of a disc or ball of radius R in D- dimensional Euclidean space. It is also identical to two expressions obtained in earlier Chapters: in Chapter II for the Koch curve and in Chapter III for the Brownian trail.

As a corollary, the average number of errors will vary roughly like R^{D-1}; for medium-sized values of R it will exhibit an overall tendency to decrease.

When L is finite, the decrease in the average number of errors continues to the final value of L^{D-1}, which is reached with R = L. Then the density remains more or less constant. If L is infinite, the average number of errors behaves more simply; it continues to decrease to zero. Finally, the data may very well suggest that L is finite and very large but fail to determine its value with any accuracy. If this is the case, the average number of errors has a lower limit that does not vanish but is so ill-determined as to be devoid of practical use.

The second method of taking the average: L being finite, let us assume that the sample begins at a randomly selected origin. This origin is very likely to fall in the middle of an intermission, and there are no errors in the sample for as long as this intermission lasts. However, when R reaches the value R = L, the average number of errors must by necessity reach the same final value as in the first method, L^{D-1}. The larger the value of L, the smaller the final average.

In addition, the expected duration of the error-free initial period turns out to be, on the average, proportional to L. Furthermore, given a finite duration R, the probability that the randomly placed sample from t to t+R will be error-free increases with L. ◁ As L→∞, this last probability tends toward certitude. All manner of problems arise, which led to the notion of conditional stationarity (Mandelbrot 1965c and Chapter V) and to sporadic processes (Mandelbrot 1967b). See also, in Chapter XII, two subentries under STATIONARITY AND KIN.

CONDITIONALLY STATIONARY ERROR PATTERNS

It is good that the Cantor set should turn out to be a reasonable base approximation, but one must hasten to improve on it because its inadequacies as a model of any concrete phenomenon are evident. It is excessively regular and cannot be superimposed upon itself through translation. In particular, its origin plays a privileged role which lacks concrete justification. Irregularity is, however, easy to inject. As stated in Chapter II, the tool of choice is randomness. As to invariance translation, our hoped-for substitute for the Cantor set will only be required to be superimposed upon itself from the statistical point of view. In probabilistic terminology, this means that it has to be stationary.

A simple means of accomplishing part of this goal might have been found in Selety 1922, which subjects the whole set to a random translation along the time axis. A more thorough randomization is proposed in Berger & Mandelbrot 1963. The key is as follows: start out from a finite approximation of the Cantor set, with scales satisfying $\eta > 0$ and $L < \infty$, and shuffle its intermissions at random so as to render them statistically independent of one another. To improve the model further, it is good to assume that the number of intermissions of length exceeding u is precisely a hyperbolic, rather than the near hyperbolic stairlike function applicable to the Cantor set. In summary, Berger & Mandelbrot 1963 assume that the successive intermissions are integers, that they are statistically independent, and that the distribution of their lengths satisfies

$$\Pr(U > u) = u^{-D}.$$

$\Pr$ is the abbreviation for *probability* that will be used throughout the Essay. The model's fit is surprisingly good, the value of D being ~ 0.3. Other authors study different channels in the same per-

spective and find D's ranging from 0.2 to nearly 1.

In the Berger & Mandelbrot model, the intermissions are independent, so that fractal errors constitute what probabilists call a "renewal" or "recurrent" process. This concept implies that if the present is set at a point of recurrence, the past and the future are statistically independent. If the origin is arbitrarily chosen, however, the past and the future are not independent.

CLUSTERING

It is obvious by construction that in the Cantor model errors present themselves in hierarchical bursts. That is, they are "clustered," the intensity of clustering being measured by the exponent D. This property is preserved when intermissions are shuffled at random.

◁ To establish this result, choose a "threshold" u_0. Define a "u_0-burst" as a group of errors contained between two intermissions of length exceeding u_0. Classify the intermissions as either "external $> u_0$" (separating bursts) or "internal $< u_0$." Define the relative durations of these intermissions by dividing the durations by u_0. When D is small, the relative durations of the external $> u_0$ intermissions are mostly *very much greater than* 1. For example, knowing that $U > u_0$, the conditional probability that

Plate 105

RANDOM PATTERN OF STREETS

The black street-like stripes are placed at random, and in particular their directions are isotropic. Their widths follow a hyperbolic distribution and rapidly become so thin that they cannot be drawn. Asymptotically, the white remainder set (the "blocks of houses") is of zero area and of dimension D less than 2. The intersection of this white set by a straight line (say, by one of the edges) is almost surely empty when $D < 1$, and is almost surely of dimension $D - 1$ when $D > 1$.

$U>5u_0$ is 5^{-D}, so it is highly nonnegligible. On the contrary, the relative durations of the internal $< u_0$ intermissions are, for the most part, very much less than 1. This behavior of the probabilities can be shown to lead to the conclusion that the u_0-bursts are really well separated, and fully justifies the term "burst." Moreover, the same result holds for every u_0. Consequently the bursts are hierarchical. For larger values of D, the separation between bursts becomes less accentuated. ►

NUMBER OF ERRORS IN
THE RANDOMIZED MODEL

Various calculations relative to the Cantor set not only remain valid after randomization but both the arguments and the conclusions are considerably simplified, particularly if $L=\infty$.

◁ For example, let us consider the average number of errors in an interval from t to $t+R$, where R is much greater than the internal scale η and much smaller than the external scale L. It is best to begin by assuming $L<\infty$ and to calculate the average in two stages.

◁ The first stage is relative to the case where one knows that there is an error at the instant t, or, more generally, at least one error between the instants t and $t+R$. Because of these added requirements, the values calculated in ei-

ther way are not absolute but are *conditional*. One finds that the conditional expectation of $M(R)$ is proportional to R^D, and, in particular, that it is independent of L. Also, assuming $t<\theta<\theta+d$ $<t+R$, one finds that the conditional expectation of $M(\theta+d) - M(\theta)$ is proportional to d^D. The preceding results are mere corollaries of self similarity. Furthermore, the ratio $M(R)/\langle M(R)\rangle$ is a random variable independent of R and of L and every conditional distribution is independent of the position of t in relation to the origin. A final corollary is that

$$Pr[M(\theta+d)-M(d)>0\mid M(R)>0]=(d/R)^D,$$

where $Pr(A\mid B)$ is the abbreviation for "Probability of A, given B."

◁ By contrast, the absolute probability of the conditioning event depends strongly on L. In particular, if the truncation to $L<\infty$ is done properly, one finds that

$$Pr[M(R)>0]=(R/L)^D.$$

This last expression results from the final expression in the preceding paragraph by replacing R by L and d by R. In other words, the event "$M(R)>0$ knowing that $L<\infty$" can be treated like the event "$M(R)>0$ knowing that $M(L)>0$." In the limit case where L tends toward infinity, the probability that the interval from t to $t+R$ falls completely within a very long intermission converges to 1, so that the probability of observing an error becomes infinitely small. But the previously derived conditional probability of the number of errors is unaffected.

◁ The preceding discussion should be further clarified in Chapter V by the discussion of the "conditional cosmological principle".

SETS ON THE LINE
OBTAINED BY RANDOM CUTOUTS

The set obtained by shuffling the intermissions of the truncated Cantor set is surprisingly close to reality but is defective in several ways: (a) the fit of the formula to the data remains imperfect in its details, (b) the restriction to $\eta > 0$ is annoying from the esthetic point of view, (c) the construction itself is arbitrary, and (d) it is too far removed in spirit from the Cantor construction. Mandelbrot 1965c suggested an alternative that has proved to be better in every way: the Lévy set. Once the dimension is prescribed, the Lévy set is the uniquely determined combination of two desirable properties. Like the randomized truncated Cantor set, it is a recurrent (renewal) random process. Like the Cantor set, it is a self similar fractal. Better than the Cantor set, this Lévy set is statistically identical to itself reduced in an *arbitrary* ratio r between 0 and 1.

Unfortunately – perfection is not of this world – its customary construction involves major complications from the outset, because this set's intermissions follow the distribution $\Pr(U > u) = u^{-D}$ extended down to $u = 0$; that is, instead of constraining u to be an integer ≥ 1, one lets it be a positive real number. Because $0^{-D} = \infty$, it follows that the total "probability" is infinite. Appearances notwithstanding, the result is not absurd, but its study involves great technical difficulties.

Fortunately, the present context does not require us to stop and explain these difficulties away, because they vanish in a different but more natural construction of the same set proposed in Mandelbrot 1972z.

As a preliminary, it is useful to give a different description of the Cantor set, by means of "virtual cutouts." Again, one starts from [0,1] and cuts out its middle third]1/3,2/3[. From then on, substance remains the same but its description changes. One makes believe that the second stage cuts out the middle thirds of *each* third of [0,1]. Since the middle third of [0,1] has already been cut out, cutting out the middle third of the middle third has no perceivable effect. Such "virtual cutouts" nevertheless prove to be very convenient. In the same way, one cuts out the middle third of *each* ninth of [0,1], of *each* 27th, and so on. Note that the distribution of the number of cutouts of length exceeding u is no longer of the form u^{-D}. One finds instead that this distribution is roughly proportional to $1/u$. The same dependence upon u continues to hold with different rules of dissection (triadic and other), the sole difference being that the factor of proportionality takes on different values.

Now, let us randomize the construction of the Cantor set by choosing the lengths and positions of the cutouts at random, independently of each other and in such a way that among the cutouts centered on a point in $[0,1]$, the expected number of those of length exceeding u is equal to $(1-D_*)/u$. The reason for the notation $1-D_*$ will soon become clear. Being independent, the cutouts must be allowed to overlap, and many among them are going to be virtual in the sense defined in the preceding paragraph. The essential result is as follows. When $D_* \leq 0$ and the cutting out stops at $\eta > 0$, there is little likelihood of anything being left over, except perhaps one single small interval. Thereafter, when $\eta \to 0$, it becomes almost certain (the probability tends to 1) that no point is left uncovered. On the other hand, when $0 < D_* < 1$, the cutouts leave a thin uncovered set, which is precisely a Lévy set of dimension $D = D_*$.

RANDOM "STREET" AND DISC CUTOUTS

It is regrettable, as stated at the beginning of this chapter, that the results of the last paragraph should be so hard to illustrate. However, the Cantor set can be visualized indirectly as the intersection of the triadic Koch curve with its base, and in the same way the Lévy set can be imagined indirectly through the city with random streets represented on Plate 105. The construction consists in lengthening each cutout in a direction that was chosen at random. As long as the remaining "blocks of houses" have a dimension $D > 1$, their intersection by an arbitrary line is a Lévy set of dimension $D-1$. On the other hand, if $D < 1$, the intersection is almost certainly empty. This result is, however, not very apparent on Plate 105 because the construction could not be carried out sufficiently far. (This Plate is mostly decorative.)

Chapter VIII implicitly provides a better method of visualization of Cantor sets, one to be carried out sufficiently far to underline the "thinness" of these sets. The method consists in cutting random discs from the plane and is exemplified by Plates 178 to 181.

CHAPTER V

Fractal Clusters of Stellar Matter

The problem of the spatial distribution of celestial objects, such as stars, galaxies, clusters of galaxies, and so on, is fascinating to the amateur as well as the specialist and has been the topic of a considerable number of publications. Yet it remains clearly marginal in relation to astronomy and to astrophysics as a whole. The basic reason is that no one has yet explained why the distribution of celestial matter is, at least in part, organized in an irregular hierarchy. While there are allusions to clustering in most works on the subject, serious theoretical developments very quickly suppose that on scales beyond some large but unspecified scale threshold, stellar matter is uniformly distributed.

A less basic reason for the hesitation in dealing with the irregular is that one did not know how to describe it mathematically. As a last resort, statistics is called upon to decide between two assumptions, only one of which is thoroughly explored (asymptotic uniformity). Is it surprising that such tests are inconclusive?

Would it not be useful to attempt to *describe* clustering without waiting for an explanation? In this chapter, I shall try to demonstrate that an influential theory of the formation of stars and galaxies, due to Hoyle, the principal descriptive model of their distribution, due to Fournier d'Albe, and, most important, the empirical data can all be interpreted in fractal terms.

I shall argue that the distribution of stellar matter includes a zone of self similarity in which the fractal dimension satisfies $0 < D < 3$ and lies close to the value $D = 1$. The Fournier model must admit to very serious deficiencies, but the fractal viewpoint will suggest methods for improving it drastically. Theoretical reasons for expecting $D = 1$ will be sketched, and we shall raise the question of why the actually observed dimension is greater than predicted, namely, approximately equal to 1.3.

IS THERE AN UPPER LIMIT
TO THE SELF SIMILAR ZONE?

Without a doubt, the zone in which $0 < D < 3$ ends before one reaches "small" objects with well-defined edges such as the planets. An important issue, always controversial and now the subject of renewed activity, concerns the very large scales. Many authors either state or imply that the self similar zone ends with the clusters of galaxies. Other authors disagree, notably deVaucouleurs 1970, who asserts that "clustering of galaxies, and presumably of all forms of matter, is the dominant characteristic of the structure of the universe on all observable scales with no indication of an approach to uniformity; the average density of matter decreases steadily as even larger volumes of space are considered, and there is no observational basis for the assumption that this trend does not continue out to much greater distances and lower densities." See also deVaucouleurs 1956, 1971; Wilson 1969.

The debate between these two schools of thought is interesting and important to cosmology but not for the purposes of this Essay. Even if the zone in which $0 < D < 3$ is bounded above, it is sufficiently important in itself to warrant a careful study.

In either case, the Universe – just like the ball of thread discussed in Chapter I – appears to involve several different physical dimensions. Starting with scales of the order of Earth's radius, one first encounters the dimension 3 (stellar matter being viewed as including solid bodies with sharp edges). Then the dimension jumps to 0 (stellar matter being viewed as a collection of points). Then one has the zone of interest to us, a zone ruled by some nontrivial dimension satisfying $0 < D < 3$. If self similar clustering continues ad infinitum, so does the applicability of this last value of D. If, on the contrary, there is an upper bound to self similar clustering, a fourth zone is added on top, in which points lose their identity in a uniform fluid, and one can argue that the dimension is again equal to 3.

On the other hand, the most naive idea regarding the stars views them as distributed near uniformly throughout the Universe. Under this untenable assumption, the nontrivial D is not encountered, and one has the sequence of dimension: 3, then 0, and again 3.

◁ The general theory of relativity asserts that the Universe is globally flat and Euclidean with the presence of matter making it locally Riemannian. Here we could speak of a globally flat Universe of dimension 3 with local disturbances where $D < 3$. This type of disturbance is considered in Selety 1924, which obviously makes no mention of fractal dimension, but includes as its Figure 1 (p. 312) an example of the Koch construction we studied in Chapter II. ▶

WHAT IS THE GLOBAL DENSITY OF MATTER?

Let us begin with a close examination of the concept of global density of matter. Like the concept of the length of a coastline, density does not a priori seem to pose any problem. In fact, however, things go awry very quickly and most interestingly. Of the many possible procedures for defining and measuring such density, the most direct consists of measuring the total mass $M(R)$ contained in a sphere of radius R centered on Earth. Then the approximate density,

$$M(R)/[(4/3)\pi R^3],$$

is evaluated. After that, the value of R is made to tend toward infinity, and the global density is the limit toward which the approximate density converges.

But does the global density really converge to a positive and finite limit? Perhaps so, but the rapidity of its convergence leaves a great deal to be desired. Furthermore, it has behaved very oddly in the past. As the depth of the

world perceived by telescopes has increased, so the approximate density of matter has steadily diminished. It has varied in a very regular manner and, according to deVaucouleurs 1970, it has remained roughly proportional to an expression of the form R^{D-3}. The observed exponent D is positive but very much smaller than 3, its order of magnitude being $D = 1$ and a more accurate estimate being $D \sim 1.3$.

The thesis of deVaucouleurs is that the behavior of the approximate density reflects reality, meaning that the quantity of matter $M(R)$ increases roughly like R^D. This formula recalls the classical result that a ball of radius R in a Euclidean space of dimension E has a volume proportional to R^E. In Chapter II we had encountered the same formula for the Koch curve, with the major difference that the exponent was not the Euclidean dimension $E = 2$ but a fraction-valued fractal dimension D. In Chapter III we examined Brownian trails in E space and found that when $E \geq 2$, one has $M(R) \sim R^D$ with an exponent independent of E and equal to the fractal dimension $D = 2$. In Chapter IV we derived $M(R) \sim R^D$ for the Cantor set used to model clustering on the time axis (for which $E = 1$).

All these precedents suggest very strongly that the deVaucouleurs exponent D is also a fractal dimension. It remains to construct a fractal that behaves in this manner and in addition agrees with other accepted viewpoints concerning the Universe.

◁ A different version of the relationship $M(R) \sim R$, valid in the interior of galaxy clusters, is reported in Wallenquist 1957 and elsewhere. ▶

SUMMARY OF THIS CHAPTER

The first model in which the approximate density of matter converges to a zero global density is due to E. E. Fournier d'Albe (see Chapter XI). Fournier 1907 is largely a work of "science fiction" disguised as science, but it also contains genuinely interesting considerations. Some had the good fortune of attracting the attention of C. V. L. Charlier, and Charlier 1908 and 1922 described Fournier's model and made it a bit more general, but in the process neglected Fournier's reasons for expecting $D = 1$. This approach was much discussed at that time, in particular in Selety 1922, 1923a, 1923b, 1924. Furthermore, the comments in Borel 1922, while dry, were very perceptive. Then, aside from fitful revivals, the Fournier-Charlier model fell into complete neglect (for not very convincing reasons noted in North 1965, pp. 20-22 and 408-409), but it refused to die. The basic idea was independently reinvented many times, notably by Lévy 1930 (who told an interesting story about this work; see Chapter XI). More important, the Fournier universe is implicit in the considerations about turbulence and galaxies described by von Weizsäcker 1950 (see Chapter VI), and in the model of the genesis of the galaxies due to Hoyle 1953. Lévy, like Fournier and Charlier but unlike Hoyle, was driven by a desire to avoid the paradox of the Flaming Sky, or *Olbers paradox,* which will be discussed.

Fournier's construct will be seen to share every one of the characteristic defects of first fractal models. To begin with, just like the Koch curves of Chapter II and the Cantor set of Chapter IV (of which it is nothing but a spatial generalization), the Fournier model is so regular as to be grotesque, but this is not really its greatest failing.

The main drawback is that it is in direct conflict with the cosmological principle, which also will be discussed soon. A first specific ground for conflict is that this model's properties depend too much on a specific origin. Even more fundamentally, the cosmological principle in the ordinary strong form is incompatible with the very idea that the approximate density in a sphere of radius R tends toward 0 when R tends toward infinity.

The customary response is that the overall density must be positive, but the fractal viewpoint suggests an entirely different way out. I shall argue that the cosmological principle as ordinarily stated goes beyond what is reasonable and desirable. It demands too much and I will suggest that it ought to be replaced by a weaker form, to be called *conditional,* which will not refer to *all* but only to *material* observers. There is no doubt that the majority of astronomers will find this weaker form acceptable. They would have studied it long ago had they felt it had the slightest interest. And indeed I shall show that the conditional form generalizes the customary cosmological principle on an essential point: it implies no assumption concerning the global density and allows the average density to be proportional to R^{D-3}.

The demonstration of this last possibility will proceed through two explicit constructions, both very much in the spirit of earlier chapters in this Essay. One is based on Brownian motion and the other is based on a generalized Brownian motion, Lévy flight.

In a certain technical sense, a Lévy flight is equivalent to the unjustified replacement of an unsolvable N-body problem by a manageable combination of many two-body problems. Mandelbrot 1975u shows how one can use Lévy flight to derive two- and three-point correlations on the celestial sphere. These correlations turn out to be identical to those which Peebles 1975 and Peebles & Groth 1975 obtain by curve-fitting. ►

THE FOURNIER UNIVERSE

Let us consider five points on a plane, placed like the configuration near the center of Plate 114. Roughly (details will be given momentarily) they are the four corners of a square and its center. Two points are added above and below our plane, on the perpendicular drawn from the center of the square, and at the same distance from this center as the corners of the initial square. The seven points obtained in this manner form a centered regular octahedron.

To be precise, each point stands for the equatorial intercept of a sphere of radius 1. Furthermore, the distance between the centers of the middle and corner spheres is set to 6. In this way, the smallest sphere including the basic seven points is of radius 7.

If each little sphere is interpreted as a basic celestial object, call it a "stellar aggregate of order 0," the above collection of seven spheres is an "aggregate of order 1." The build-up continues in the following manner. An aggregate of order 2 is achieved by enlarging an aggre-

Plate 114

DIAGRAM OF THE MULTIUNIVERSE
OF E.E. FOURNIER D'ALBE

This Plate represents to scale the "equatorial" section of the construction described in the text.

To freely translate the caption in Fournier 1907: "a multiuniverse constructed upon a cruciform or octahedral principle is not the plan of the world but is useful in showing that an infinite series of similar successive universes may exist without producing a 'blazing sky.' The 'world ratio' in this case is 7 instead of 10^{22}, as in reality. *The matter in each world sphere is proportional to its radius.* This is the condition required for fulfilling the laws of gravitation and radiation. In some directions the sky would appear quite black, although there is an infinite succession of universes."

Plate 115

A FLAT FOURNIER UNIVERSE
WITH D=1

Plate 114, because it has been drawn to exact scale, is potentially quite misleading. Indeed, if taken by itself, it is of dimension $\log 5/\log 7 \sim .8270 < 1$, while the whole universe is of dimension $D = 1$. Therefore, in order to avoid implanting the wrong intuition, we hasten to exhibit a regular Fournier-like planar pattern of dimension $D = 1$. The construction could be carried one step further than had been possible on Plate 114.

gate of order 1 in the ratio $1/r=7$ and by replacing each of the resulting spheres of radius 7 by a replica of the aggregate of order 1. In the same way, an aggregate of order 3 is achieved by enlarging an aggregate of order 2 in the ratio $1/r=7$ and by replacing each of the resulting 49 spheres by a replica of the aggregate of order 1. And so on.

In sum, between any order of aggregation and the next, the number of points and the radius are enlarged in the ratio $1/r=7$. Consequently, the function $M_0(R)$ expressing the number of points contained in a sphere of radius R is $M_0(R)=R$ whenever R is the radius of some aggregate. For intermediate values of R, $M_0(R)$ is smaller (reaching down to $R/7$), but broadly speaking $M_0(R)$ is proportional to R.

Starting from aggregates of order 0, it is also possible, by successive stages, to interpolate without end to obtain aggregate of orders −1, −2, and so on. The first stage replaces each aggregate of order 0 with an image of the aggregate of order 1, reduced in the ratio $1/7$, and so forth. If one does so, the validity of the relationship $M_0(R) \sim R$ is extended to ever smaller values of R. In addition, after being infinitely interpolated and extrapolated, our set is made self similar. Thus it becomes possible to define a similarity dimension for it, and we find the value $D = \log 7 / \log 7 = 1$.

◁ We may also note incidentally the new form taken by a theme encountered for Brownian motion, which also has an integer fractal dimension. When the dimension of an object in 3-space is 1, that object need not be a straight line or any other rectifiable curve. It need not even be topologically connected. The same similarity dimension is compatible with different topological dimensions (see DIMENSION (TOPOLOGICAL) in Chapter XII). In particular, the doubly infinite Fourier inverse being totally disconnected, its topological dimension is 0. ▶

DISTRIBUTION OF MASS; FRACTAL HOMOGENEITY

The step from geometry to the distribution of mass is obvious. Between the extrapolation to the infinitely large and the interpolation to the infinitely small, one can view each stellar aggregate of order 0 as having uniform density and unit mass. Then the total mass M(R) within a sphere of radius R is simply identical to $M_0(R)$ and hence roughly proportional to R. Furthermore, the stage from aggregates of order 0 to order −1 amounts to breaking up a ball that had been viewed as uniform and finding it to be made of seven smaller balls (one around the center and the others sticking to the old ball's boundary). This stage, therefore, makes it possible to interpolate $M(R) = R$ down to radii below $R = 1$.

When viewed over the whole 3-space, the resulting mass distribution is grossly inhomogeneous, but over the Fournier fractal it is as homogeneous as can be. (Recall Plate 101.) In particular, any two geometrically identical portions of the Fournier universe carry identical masses. Such a distribution of mass will be called fractally homogeneous.

◁ The preceding definition is phrased in terms of self similar fractals, but the concept of fractal homogeneity is more general. It applies to any fractal such that the Hausdorff measure corresponding to some gauge function $h(\rho)$ is positive and finite (see Chapter XII). Fractal homogeneity requires the mass carried by a set to be proportional to said Hausdorff measure. ▶

GENERALIZED FOURNIER UNIVERSES

Later we will examine how Fournier justifies the value $D = 1$ starting from basic physical phenomena, and how Hoyle obtains this same value by an ostensibly very different argument. It must be noted, however, that $D = 1$ is in no way inevitable from the point of view of the spirit of the geometric construction given above. Even if we keep to the octahe-dron and the value $N = 7$, we may assign to $1/r$ a value other than 7, thus obtaining $M(R) \sim R^D$ with $D = \log 7 / \log(1/r)$ different from 1. Any value between 3 and infinity is acceptable for $1/r$; therefore D may assume any value between 0 and $\log 7 / \log 3 = 1.7712$.

Further, Fournier assumes $N = 7$ to help his draftsman produce a legible drawing, while claiming that the "true" value is $N = 10^{22}$. On the other hand, Hoyle argues that $N = 5$. Whatever the case, given a D satisfying $D < 3$, it is easy to construct variants of Fournier's model having this fractal dimension.

THE CHARLIER UNIVERSE

One of the innumerable defects of Fournier's model is that it is too regular. Charlier proposed to improve upon this feature by letting N and r vary from one hierarchical level to another, taking on the values N_m and r_m. Of course, the object he obtained in this fashion is no longer strictly self similar nor does it possess a true dimension. To put it more precisely, the quantity $\log N_m / \log(1/r_m)$ attached to stage m may vary with m. Charlier made it stay between two bounds, $D_{min} > 0$ and $D_{max} < 3$, thus introducing yet another theme. The physical dimension may very well have no single exact value but only upper and lower limits. This theme, however, need not be developed here. At any rate, by imposing the condition $D_{max} < 2$ (which Fournier had amply satisfied by assuming that $D = 1$), Charlier sidesteps the Olbers paradox, which we now discuss.

THE OLBERS PARADOX

Fournier introduced his model of the Universe as a way out of the so-called *Olbers paradox.* This motivation has ceased to be of direct interest, insofar as relativity theory and the theory of the expansion of the Universe provide *sufficient* ways out. Nevertheless, we must dwell on the paradox for a moment. It resides in the fact that under the assumption of uniformity in the distribution of celestial bodies ($D = 3$ for all scales), the background of the night sky would not be black. In fact there would be no night as opposed to day, because at any moment and in any direction, the sky would be lit uniformly to the brightness of the solar disc.

This inference, which will be justified momentarily, appears to have been first drawn by de Chéseaux 1744, 223-229, and by Lambert, but it is ordinarily credited to Olbers 1823. For an historical discussion, reference should be made to Gamow 1954, Munitz 1957, North 1965, Wilson 1965, Jaki 1969, or Clayton 1975. (Chapter IV of Clayton gives star billing to Jean Philippe Loys de Chéseaux and quotes him).

The argument credited to Olbers is simplicity itself. Assuming that the light emitted by a star is proportional to its surface area, the amount of light reaching one observer at a distance of R is reduced in the ratio of $1/R^2$, but the star's apparent surface is itself proportional to $1/R^2$. Thus the apparent luminous intensity remains independent of distance. Also, assuming that the distribution of stars in the Universe is uniform, almost any direction traced in the sky sooner or later intersects with the apparent disc of some star. Therefore, the apparent density of brightness is the same throughout the sky.

On the other hand, the assumption that $M(R) \sim R^D$, with $D < 2$, suffices to resolve the paradox because it follows that a nonvanishing proportion of the directions are lost in infinity without encountering any star disc. Along these directions, the night sky is black.

◁ And the day sky will also be black, except in the direction of the Sun. A pitch-black day sky was indeed observed by the astronauts, but on Earth the day sky is blue because of scattering of sunlight by the atmosphere. Let us also comment on the case when clustering has an end and the zone in which $D < 3$ is followed at a great but finite distance by a zone in which $D = 3$. In this case, the sky's background is not black but slightly illuminated. ►

As indicated earlier, physicists have elaborated other methods of exorcizing the Olbers paradox. In my opinion, these new methods imply that clustering should be studied on their own merits. Historically, however, the new explanations have been used very differently, sometimes as an excuse for neglecting the study of clustering altogether, and sometimes even as an argument a priori for denying its reality. Both the excuse and the argument are, in my opinion, completely devoid of merit. In particular, as will be seen in a later section of this chapter, the expansion of the Universe is compatible not only with the usual but also with the fractal form of homogeneity.

FOURNIER'S REASONS
FOR EXPECTING D=1

We now return to the Fournier model in order to describe the further argument which leads Fournier 1907, p. 103, to conclude that D must be equal to 1. This argument is impressive. As a reason for not forgetting its author it is perhaps even more convincing than the hierarchical Universe itself, which somehow has long been "in the air."

Consider a stellar aggregate of arbitrary order with mass M and radius R.

Using without misgivings a formula that in principle is only applicable to objects with spherical symmetry, it can be assumed that the gravitational potential on the surface takes the form GM (G being the gravitational constant). As a result, a star falling on this universe impacts with the velocity $(2GM/R)^{1/2}$. Fournier postulates that this velocity is essentially the same for all such universes. (One wonders what the basis for such an an assertion could have been in 1907. Even today, it seems novel.)

To paraphrase Fournier: "An important conclusion may be drawn from the observation that no stellar velocity exceeds 1/300 of the velocity of light. It is that the mass comprised within a world sphere increases as its radius, and not as its volume, or in other words, that the density within a world sphere varies inversely as the surface of the sphere. By 'world sphere' I mean a sphere enclosing a 'visible universe' of any order. To make this clearer, the gravitational potential at the surface would be always the same, being proportional to the mass and inversely proportional to the distance. And as a consequence, stellar velocities approaching the velocity of light would not prevail in any part of the universe."

◄Fournier's argument is really an application of old-fashioned dimensional analysis of mechanics. The value D = 1 might not have been guessed, but the fact that the predicted D is an integer might have been expected, since old-fashioned dimensional analysis knows of no fractional dimension.

◁ For modern development on the theme of "the gravitational potential at the surface being always the same," see POTENTIALS ... in Chapter XII. ▶

HOYLE CURDLING; THE JEANS CRITERION ALSO YIELDS D=1

Let us now show that a theory advanced in Hoyle 1953 and sketched in Hoyle 1975, according to which galaxies and stars were formed by a cascade process starting with a uniform gas, also yields a hierarchical distribution of stellar mass. The main differences with the Fournier universe are that $N=5$ instead of $N=7$ and that the construction stops after a finite number of stages. Furthermore (this fact was not noticed as such by Hoyle), one must have $D=1$. Although the argument is controversial, it does take into account a certain facet of physical reality, since it associates the value $D=1$ with the classical criterion for the equilibrium of gaseous masses due to Jeans.

Consider a gas cloud of temperature T and mass M_0, distributed with a uniform density over a sphere of radius R.

Jeans had examined the problem of equilibrium of such spheres and found that a "critical" situation prevails when one has $M_0/R_0 = JkRT/G$. (Here, k is the Boltzmann constant, G the gravitational constant, and J a certain numerical coefficient.) In this critical case, the primordial gaseous cloud is unstable and must inevitably contract.

Hoyle postulates (a) that M_0/R_0 in fact takes on this critical value at some initial stage, (b) that the resulting contraction stops when its volume has diminished to one-fifth, and (c) that each cloud then splits into five clouds of equal size, all having a mass $M_1 = M_0/5$ and a radius $R_1 = R_0/5$.

For reasons that are suggested by other fields where the same cascade process is encountered (one is discussed in Chapter VI), these five clouds will be called *curds* and the cascade process will be called *curdling*.

In contrast to Fournier's $N=7$, which was injected merely to facilitate illustration, Hoyle claims his $N=5$ has a physical basis, but we need not dwell on this point. The main fact is that the Jeans stability criterion also provides a method for determining the dimension D. The values of N and r it suggests satisfy $r=1/N$, so that the final result is exactly the same as in the Fournier model: a dimension equal to 1 must be part of the design if curdling is to end as it began, in Jeans instability. The first stage is followed by a second stage of contraction and subdivision, then a third, and so on.

Further, if the duration of the first stage is taken to be 1, the duration of the contraction of order m is 5^{-m}. It follows that the same process could conceivably continue to infinity within a finite total duration of 1.2500. (In fact, curdling must stop as soon as clouds become so opaque that the heat due to gas collapse can no longer escape.)

It is obvious that Hoyle uses the same geometric device as Fournier. The difference is that the curds are not placed in a strict pattern but are allowed to scatter around the cloud from which they had condensed. To give an actual illustration of the curdling mechanism, it suffices to simplify it by assuming that the original cloud is a cube, while the curds are cubes of side 5 times smaller, their positions being selected at random among the $5^3 = 125$ possibilities. Two illustrations involving four curdling stages are shown in Plates 123 and 124 and the idea is amplified in Plate 125.

ASIDES ON THERMODYNAMICS AND RELATIVITY THEORY

◁ At the edge of an unstable gas cloud satisfying the Jeans criterion, GM/R is equal both to $V^2/2$ (Fournier) and to JkT (Jeans). Hence $V^2/2 = JkT$. Let us now be reminded that statistical thermodynamics shows the temperature of a gas to be proportional to the mean square velocity of its component molecules. Hence the combination of the Fournier and Jeans criteria suggests that at the edge of a cloud the velocity of the fall of a macroscopic object is proportional to the average velocity of its molecules. This conclusion seems eminently sensible. It must be related to the virial theorem, and it must be telling us that a more careful analysis of the role of temperature in the Jeans criterion would show the two criteria to be equivalent.

◁ Most likely, the analogy extends to the various deviations of the M(R)~R relationship within galaxies, as reported, for example, in Wallenquist 1957. ►

◁ Relativity theory also involves a ratio of the form

$$2GM/(a \text{ velocity squared}).$$

Indeed, to "quote" again in paraphrase from deVaucouleurs 1970: "Relativity theory led us to believe that to be optically observable, no stationary material sphere can have a radius R less than the Schwarzschild limit

$$R_M = 2GM/c^2,$$

where c is the velocity of light. In a plot

Plate 123

IMPLEMENTATION OF HOYLE'S MODEL (DIMENSION D=1) BY RANDOM CURDLING IN A GRID

Hoyle's model of the hierarchy of stellar clusters is based upon a cascade, whereby a very low-density gas cloud collapses repeatedly to form clusters of galaxies, galaxies, and so on. Hoyle's original description, however, was extremely schematic, and actual implementation requires a large number of specific geometric assumptions. This Plate shows a plane projection of the simplest implementation.

A big cube of side 1 is subdivided into $5^3 = 125$ subcubes of side 5^{-1}, and so on successively into 125^k subcubes of the kth order, each of side 5^{-k}. In the kth cascade stage, the matter contained in a (k−1)th order subcube collapses into 5 subcubes of the kth order, to be called k-curds. On this diagram, the first three stages are illustrated in superposition, using increasingly dark shades of gray to represent increasing gas density – except that the 4-curds obtained after the last collapse are shown in white.

One could oblige the positions of the k-curds to be aligned along the main diagonal of the (k−1)-curd, from which said k-curds collapsed. In this case, the cascade would concentrate mass from a cube into a straight line. Similarly, other systematic methods would concentrate mass into different classical lines. In all these cases, curdling would reduce the dimension from 3 down to 1, while remaining among the classical shapes of Euclid.

A random distribution of the curds, on the other hand, almost certainly breaks up mass into small bits and leads to an extraordinarily spare dust.

From the "qualitative" viewpoint of topology, this random cascade's outcome is entirely different from a classical line. But surely it must be possible to express the intuitive feeling that from other viewpoints the curds' positions do not matter, only the number of curds at each stage. Such a viewpoint had been suspected by Carathéodory and fully implemented by Hausdorff. I propose to call it the fractal viewpoint.

Since we present a plane projection, it is not rare that two subcurds should project on the same square. In the hypothetical limit, however, the projections of two points almost never coincide. There are so few limit points that they leave space essentially transparent.

Compared to the illustration in Hoyle 1975, p. 286, the present Plate may seem crude. But within the constraints of cubic eddies, it is carefully drawn to scale, and in all questions relative to dimension accuracy is very important. Furthermore, the process's outcome, to be exhibited alone on Plate 124, is less crude than the process itself.

123

Plate 124

FINAL OUTCOME
OF HOYLE'S MODEL

This figure is analogous to Plate 123, except that the first three stages of Hoyle curdling have been erased, and we only show the final outcome. It is especially interesting to note that on this pattern the underlying grid is no longer perceptible. Note also that due to different choices of pseudo random seed, Plates 123 and 124 differ in general appearance.

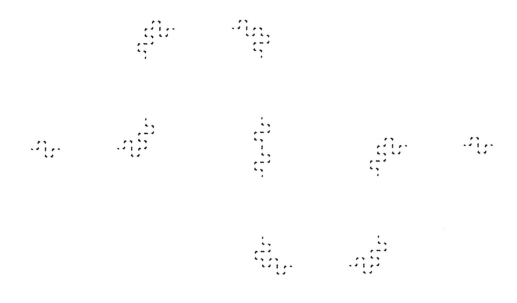

Plate 125

KOCH-LIKE POINT CLUSTERS
OF DIMENSION D=1

The caption of Plate 123 described two actual implementations of Hoyle curdling: systematic and wholly random. A further implementation exhibited on this Plate is meant less to model reality than to illustrate two mathematical points.

First of all, we see that the Koch construction can be modified so that it breaks a line systematically and leads to dust that has the same dimension ($D=1$) but entirely different topology and appearance.

Imagine that a rubber band laid along [0,1] is extended to follow the standard Koch polygon that had been used in Plate 49 to generate the Koch curve of dimension $3/2$. Then the corners are pinned down permanently, each of the 8 pieces of the band is cut in its middle, and the 16 pieces allowed to snap back to their original lengths 1/16. Their free ends are then pinned down and the process is repeated. The end result is a self

similar hierarchically clustered dust with $r=1/16$ and $N=16$, hence $D=1$.

This construction presents the subsidiary advantage of serving as transition to the construction on Plates 134 ff. Indeed, the points on this Plate are ordered intrinsically by the Koch curve from which the standard polygon was borrowed. Furthermore, it is easy to derive the frequency distribution of the snapback distances between successive pinned-down points. Roughly, the number of distances ≥u is proportional to $1/u$. Plates 134 ff. will use the same frequency distribution differently.

◁ In the terminology of generalized Koch curves and orientation indexes introduced in the caption of Plate 56, the construction of the present rubber band amounts to allowing said index to take a further value, say, X. To mark a side by X is to instruct it to be erased at the next stage of the Koch construction. ▶

of the correlation between the mean density ρ and the characteristic radius R of cosmical systems of various sizes, the line of equation

$$\pi_M = 3c^2/8\pi GR_M{}^2$$

defines an upper limit. The ratio ρ/ρ_M of the actual density to the limit value for a system of observed radius R may be called the Schwarzchild filling factor. For most common astronomical bodies (stars) or systems (galaxies), the filling factor is very small, on the order of 10^{-4} to 10^{-6}."

◁ This argument reminds us that the square of 1/300, which was the velocity ratio postulated by Fournier, equals about 10^{-5}, thus falling precisely in the range deVaucouleurs claims for the empirical values of the filling factor. ▶

THE COSMOGRAPHIC PRINCIPLE

The innumerable defects of the Fournier-Charlier model (including its excessive regularity) may be summed up by saying that its origin plays a privileged role. This model is resolutely geocentric and therefore anthropocentric. It is contrary to the "cosmological" principle, which postulates that our time and our position on Earth are in no way special or central, that the laws of Nature must everywhere and always be the same. This assertion is discussed at length in Bondi 1952.

To be exact, we are solely concerned with one special topic, the distribution of matter. Further, we deal less with theory ($\lambda o \gamma o \sigma$) than with description ($\gamma \rho \alpha \varphi \eta$). Finally, we shall propose a weakened version of Bondi's statements of this principle.

Therefore, we propose the expression *strong cosmographical principle* to designate the customary assertion that the distribution of matter follows precisely the same statistical laws regardless of the system of reference (origin and axes) used to examine it.

Unfortunately, it is either difficult or impossible to reconcile this principle with the notion that the actual distribution is extremely far from being uniform. If the global density of matter ρ in the Universe is to vanish, the strong cosmographic principle must be wrong. If, on the contrary, ρ is small but positive, then the strong cosmographic principle may well be true, but for the scales in which we are interested it is of no actual use. One may like to keep it in the background as being reassuring. One may prefer to avoid it as being potentially misleading. Finally, one may settle on replacing it with a statement that is meaningful for all scales, is independent of whether $\rho = 0$ or $\rho > 0$, and is in keeping with a vision of the world containing fractals. We will show that in order to do so, it is good to divide the strong cosmographic principle into two parts.

THE CONDITIONAL COSMOGRAPHIC PRINCIPLE

Let the spatial distribution of mass in the Universe be described in a frame of reference that satisfies the condition that its origin is itself a material point. The resulting probability distribution of mass is called conditional.

Assumption. The condition distribution of mass is the same for all conditioned frames of reference. In particular, the mass $M(R)$ contained within a sphere of radius R is a random variable independent of the frame of reference.

THE ADDITIONAL ASSUMPTION OF POSITIVE OVERALL DENSITY

The limits of $M(R)R^{-3}$ and $\langle M(R)\rangle R^{-3}$ for $R \to \infty$ are almost certainly equal, as well as positive and finite.

THE CLASSICAL CASE

The statistical laws of distribution of matter can be viewed in two ways. One can use the absolute probability distribution, which is in effect relative to an arbitrary frame of reference. Alternatively, one can use the conditional probability distribution relative to a frame centered on a material point. Under the above additional assumption, these alternative probability distributions are both nontrivial and are deducible from each other. The latter may be deduced from the former by the usual rules for calculating a conditional probability. The former may be deduced from the latter by taking the average relative to origins that are uniformly distributed over space.

◁ Parenthetically, the uniform distribution of origins integrated over the whole space results in an infinite mass. Consequently, it is not evident that the nonconditional distribution may be renormalized to add up to 1. Indeed, in order to do so, it is necessary and sufficient that the global density be positive. See Mandelbrot 1967b. ▶

THE NONCLASSICAL CASE

Suppose on the contrary that the additional assumption is wrong, and that $\lim M(R)R^{-3}$ vanishes. In this case the absolute probability distribution is degenerate. It simply states that a sphere with a finite radius R, if chosen at random, is almost certain to be empty. Hence, if one peeks around from a point selected at random, one is almost sure to be unable to see any star. Clearly, this result is of no interest whatsoever. Man is only interested in the probability distribution of mass in cases where mass does not vanish, and he finds no help at all to be told that such cases almost never occur. The very fact that the nonconditional distribution automatically disregards such cases implies it is grossly inadequate in the instances when $\rho = 0$.

This result faces the physicist with an almost sure event which can be disregarded, together with an event of zero probability which not only cannot be disregarded but must be analyzed into finer subevents. This contrast is precisely inverse of the one to which one is accustomed by statistical mechanics.

Indeed, every desirable thermodynamic conclusion (such as the principle of the increase of entropy) can be trusted to hold almost surely. The opposite (undesirable) conclusion has a vanishing probability and is *therefore* negligible. Similarly, the average number of heads in an increasing sequence of tosses of a fair coin may fail to converge to $1/2$, but the cases of nonconvergence are of zero probability and *therefore* devoid of interest. The point of the present discussion

is that the *therefore* in the preceding two sentences is the precise opposite of what is required in cosmography.

◁ Mathematicians also used to hold to the opinion, in the words of Frostman 1935, that "sets of measure 0 enter most often as exceptional sets which one must exclude in the statement of theorems of the highest importance." Today, however, they find the various "exceptional sets" to be of intrinsic interest. ►

The statement of the conditional cosmographic principle involves precisely the same words regardless of whether $\rho > 0$ or $\rho = 0$. This is esthetically pleasing and has the philosophical advantage of satisfying the spirit of contemporary physics. By subdividing the strong cosmographic principle into two parts, we highlight a statement which is relative to distributions that are observable, at least in principle, and it downgrades another statement that constitutes either an act of faith or a working hypothesis.

In fact, as previously indicated, it is highly probable that most astronomers will not object a priori to the idea of conditioning, and that this idea would have become commonplace long ago if it were acknowledged as having consequences worthy of attention, that is, recognized as constituting not a formal refinement but an authentic generalization.

In the next few sections we shall

prove that such is the case. We shall describe explicit constructions which induce a zero global density, satisfy the conditional cosmographic principle, and fail to satisfy the strong cosmographic principle. The model of the distribution of stellar mass that is incidentally involved in this proof is no more than an approximation but is instructive.

◁ The trivial absolute probability distribution is not only compatible with mass carried by any fractal satisfying $D<3$ but with any other mass distribution for which $\rho=0$. In other words, it tells absolutely nothing beyond $\rho=0$. The conditional probability distribution, on the contrary, distinguishes between fractals having different dimensions, between fractals that are or are not self similar, and between other alternative assumptions. ▶

RAYLEIGH FLIGHT STOPOVERS

In order to postpone technical difficulties, we shall begin with a very artificial example that has the virtues of being familiar to the reader from Chapter III and of being old (it goes back to Rayleigh 1880). This example will begin to respond to deVaucouleurs 1970, which called for "the extension of Charlier's work to quasi-continuous models of density fluctuations that would replace the original oversimplified discrete hierarchical model." On the other hand, our first example's fatal defect is that it has neither the dimension nor the degree of topological connectedness required by the facts. This model will be followed by modifications that are less familiar but less unrealistic.

Our description of Rayleigh flight overlaps with the discussion of random walk in Chapter III. Assume that a rocket, starting from a point $\Pi(0)$ in space, goes off in an isotropic and random direction. Assume also that the duration of each jump is $\Delta t=1$, and that the distance between $\Pi(0)$ and the next stopover $\Pi(1)$ is also random with its distribution prescribed in advance. It may be imagined that the length of this jump is bounded or constant, and the most convenient assumption (as we shall see) is that it follows the exponential distribution. The essential requirement is that large jumps are very rare, in the sense that the mathematical expectation $\langle[\Pi(1)-\Pi(0)]^2\rangle$ is finite. The rocket then goes off again toward its next stopover $\Pi(2)$, which is such that the vectors $\Pi(1)-\Pi(0)$ and $\Pi(2)-\Pi(1)$ are independent and identically distributed. And it continues performing its independently distributed jumps ad infinitum.

Further, we include its previous sites $\Pi(-1)$, $\Pi(-2)$, and so on. In theory, they are obtained by means of the same mechanism applied in the opposite direction. However, the mechanism does not involve the direction of time, so it is sufficient to draw two independent trajectories starting from $\Pi(0)$.

Having done this, the rectilinear trail that the two rockets leave behind is erased, and the set of all stopovers is examined without taking into account the order in which they were reached.

It is clear, by construction, that this set of stopovers follows exactly the same distribution when examined from any of the points $\Pi(t)$. Therefore, they satisfy the statistical version of the conditional cosmographic principle.

THE DIMENSION D = 2

Furthermore, the sites of a Rayleigh flight do not satisfy the strong cosmographic principle because they take almost no room in 3-space. A sphere with radius R and center $\Pi(t)$ would have room for roughly R^3 uniformly spaced points, but the number of stopovers is only of the order of magnitude of $M(R) \sim R^2$.

The fact that the exponent is independent of the distribution of the jumps $\Pi(t)-\Pi(t-1)$ is a rather direct consequence of the classical form of the central limit theorem. This theorem asserts that when $\langle[\Pi(t)-\Pi(t-1)]^2\rangle < \infty$, then, after a large number of stages k, the distribution of the distance $\Pi(t)-\Pi(0)$ is asymptotically Gaussian, regardless of the exact distribution of the individual jumps $\Pi(t)-\Pi(t-1)$. Thus the mean site density is proportional to R^{-1} and tends toward 0 when $R\to\infty$. Furthermore, if the origin of the reference is chosen with uniform probability in space, it can be shown that a sphere with an arbitrary but finite radius R contains no site $\Pi(t)$. Seen from an arbitrary origin, the distribution of stopovers is degenerate, except in cases of zero total probability.

In sum, the basic idea of the cosmographic principle applies to stopovers, but only in a sense that is both statistical and conditional.

Furthermore, the fact that M(R) increases like R^2 conforms to the idea that the dimension of the whole system of sites $\Pi(t)$ is equal to 2.

However, the above-mentioned flight being discrete, the rigorous value of its Hausdorff Besicovitch dimension of the set of stopovers is 0. Also, this set is not self similar, so it does not have a similarity dimension. To achieve self similarity, it is necessary to make t continuous and to interpolate the flight's progress. This is easy to arrange if the jumps are exponentially distributed; the interpolate is an isotropic Brownian motion.

We are back in well-known territory, but there is no harm in running through it again to pick up fresh details. The interpolation of a Rayleigh flight with exponentially distributed jumps may proceed in stages, as in the Koch construction, but at random – as indicated in the quote from Perrin in the caption of Plate 11. First, the positions are established for integer values of t. Then follows interpolation for values of t that are multiples, say, of 1/3, yielding a lengthened trajectory, and so on to infinity. At the limit, the elementary jump between t and t+dt is an isotropic Gaussian vector of mean zero and a variance equal to dt. We know that Brownian motion is self similar with an exponent equal to 2.

This same D holds both on a plane and in space. Hence, if the Rayleigh flight had been confined to the plane, the present discussion would have yielded an entirely different result. The sites taken together would have satisfied the cosmographic principle in its usual strong form. Indeed, plane Brownian motion is "recurrent," meaning that the stopovers are dense everywhere (there will be one arbitrarily near every point prescribed in advance). Therefore, the distribution of the cloud of points Π(t), as seen from any point of a plane, has the same statistical distribution.

On the contrary, conditioning is necessary when E > 2 because the corresponding Brownian motion is transient.

The necessity of interpolating brings us back to a question already asked in the case of mature but finite approximations of the coast of Britain. When t is left discrete, does the concept of dimension remain the least bit useful? My own reply, related to the nature of physical dimension in the presence of cutoff, is affirmative, and there is nothing new I can add concerning this matter.

Nevertheless, the Brownian model exhibits two characteristics that are grossly unacceptable in cosmography. It is a continuous curve – yet no continuous curve can be seen in stellar distributions. Its dimension D = 2 is larger than suggested by the evidence. Therefore, in order to save the principle used to construct this curve, we shall soon have to modify it on one essential point.

A GENERALIZED DENSITY AND EXPANSION OF THE UNIVERSE

When mass is assigned at random to each of the stopovers of a Rayleigh flight, conditional stationarity may extend from position considered alone to position considered together with mass, as long as the various masses are identically distributed and statistically independent. If a uniform distribution is desired, the masses have to be equal. Likewise, it is convenient to think of the Brownian trajectory as having a uniform density δ, if it is true that the mass between the points $\Pi(t_0)$ and $\Pi(t)$ is proportional to the time increment t_0-t and equal to $\delta[t_0-t]$. As a corollary, $M(R)$ becomes the time spent in the sphere of radius R, multiplied by δ.

Let us consider this Brownian distribution from the point of view of uniform expansion of our Universe. Usually, such expansion starts out from a uniform density δ – an idea that is made explicit by the Poisson probability distribution in space. If the Universe expands, the value of δ is progressively modified, but uniformity is never destroyed. On the other hand, it is generally believed that all other distributions are changed by expansion. In fact, such need not be the case. Indeed, for the above Brownian distribution, the value of δ changes with expansion but remains uniform.

Therefore, despite the defects they exhibit, Rayleigh stopovers are eminently compatible with the expansion of the Universe, more specifically, neutral with respect to the question of whether the Universe does or does not expand. This is a property that must and will be preserved in the generalization of other values of D, to which we now proceed.

STOPOVERS OF A CAUCHY FLIGHT AND THE DIMENSION D = 1

In a first stage, we shall attempt to generate a dimension equal to D = 1 by starting with the Brownian trail interpolation of the Rayleigh flight and seeking to decrease D by unity. In the case of classical shapes from Euclid, such a decrease is easy to achieve. If one starts with shapes in the plane, it suffices to take the section by a line; in 3-space, it suffices to take the section by a plane; and in 4-space, to take a section by a 3-space. We also saw in Chapter III that this same rule carries on to fractals, since the Brownian line-to-line function has the dimension 3/2, while sections that are not perpendicular to the t-axis have the dimension 1/2.

Extended by formal analogy, this method for subtracting 1 from D leads us to suspect that appropriately selected sections of a Brownian trail are typically of dimension 2−1 = 1. The result should apply to plane sections of a trail in the ordinary 3-space. More important, it should apply to 3-spatial sections of a trail in the 4-space, in which the coordinates are x, y, z, and a fourth one, which we find it convenient to call *humor*.

Starting from a line-to-4-space Brownian trail, consider the "humorless" sites where humor = 0. It is clear that they can be generated in sequence, in the order in which they are visited in the underlying Brownian trail, and that the distances between such visits are independent and isotropic. As a result, the humorless sites can themselves be viewed as being the stopovers of some modified Rayleigh flight. It turns out to be the Cauchy flight, in which the probability distribution of the distance from $\Pi(0)$ to $\Pi(t)$ is some numerical multiple (we need not worry about its value) of

$$t^{-E}[1+|\Pi(t)-\Pi(0)|^2 t^{-2}]^{-E/2}.$$

Moreover, the formal hunch that the corresponding dimension is $D=1$ is correct, as has been shown by S. J. Taylor 1966, 1967. The Cauchy flight is illustrated on Plates 134 and 135 and also on one of the views of Plates 140 and 141.

STOPOVERS OF A LEVY FLIGHT; NONINTEGER DIMENSIONS < 2

One can show that the deep difference between the dimensions of the Cauchy flight and the Rayleigh flight is due to the fact that the former assigns an enormously higher probability to very large values of the distance U from $\Pi(t)$ to $\Pi(t+1)$. Indeed, the Cauchy distribution satisfies $P(U>u)\sim u^{-1}$, and it follows in particular that the mathematical expectation $\langle U^2(t)\rangle$ is infinite.

More generally, in order to preserve asymptotic self similarity with a more general D, it is sufficient to suppose that U is asymptotically a hyperbolic random variable, meaning that $Pr(U>u)\sim u^{-D}$. It is easy to see that the condition $\langle U^2\rangle = \infty$ requires that $0<D<2$.

The resulting degree of clustering is illustrated in Plates 137 through 141. Plate 137 is a single view relative to $D=1.5000$, Plate 138-139 is a multiple view relative to $D=1.3000$, and finally Plate 140-141 interpolates from $D=2$ down to $D=1$ and below.

Visually, the degree of clustering relative to $D=1$ is unquestionably excessive and the degree of clustering relative to $D=1.5000$ is inadequate. The value $D=1.3000$ is much better visually, so it is good to know that it also conforms to the estimates of deVaucouleurs. Finally, it conforms to the estimates resulting from a combination of Peebles 1975 and Mandelbrot 1975u.

CONCLUSION

The disagreement between the empirical dimension $D=1.3$ and the Fournier and Hoyle theoretical value $D=1$ can hardly be attributed to statistical sample variability. It raises an important issue, perhaps to be tackled by relativity theory (Peebles 1974).

Nevertheless, the preceding several sections show that thanks to the possibility of relating the value of D to the dis-

A

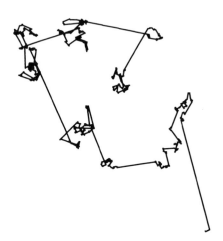

B

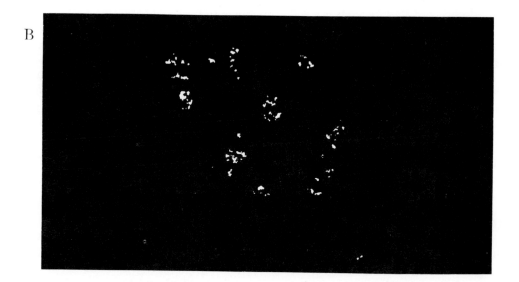

Plates 134 and 135

CLUSTERS OF DIMENSION D = 1:
CAUCHY FLIGHT AND STOPOVERS

A Cauchy flight is roughly a sequence
of jumps separated by stopovers. Only
the latter are of direct interest in this
chapter, but jumps are necessary to construct the process. Therefore, the flight
is illustrated both with and without the
jumps.

In each case we represent a graph

A′

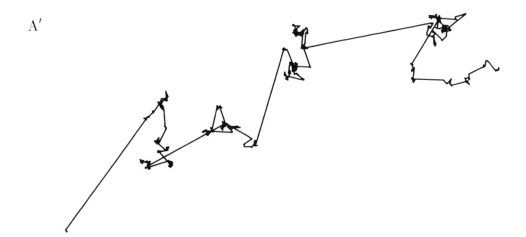

B′

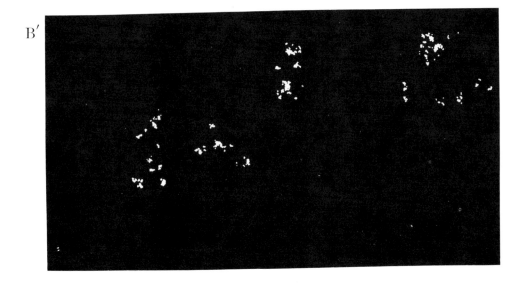

drawn in space through its projections on two perpendicular planes, which can be visualized by holding the book half open.

To proceed to the white-on-black portions, one performs a photographic inversion after wiping away the above segments. Each of the flight's stopovers is taken to be a "star" or, more precisely, to represent a blob of matter.

Additional comments on these two illustrations are found in the caption for Plate 137.

Plate 137

CLUSTERS OF DIMENSION D=1.5000: LEVY FLIGHT AND STOPOVERS

The black-on-white portions of this Plate and of Plates 134 and 135 are made of jumps that are straight segments with the following characteristics. Their direction in space is isotropic (that is, parallel to the vector joining the origin of space to a point chosen at random on a sphere). The segments are statistically independent, and the length follows the probability distribution $\Pr(U>u) = u^{-D}$, except that $P(U>u) = 1$ when $u<1$. In Plates 134 and 135, we set $D=1$. In the present Plate, we set $D=1.5000$, and only one projection is shown.

The drawings are on a scale in which the overwhelming majority of the segments are by far too small to be perceived. In fact we lined the plane with a uniform grid and marked all those cells containing one or more stopovers. In other words, each of the points as drawn actually stands for a whole minicluster.

In addition, regardless of D, the miniclusters are themselves seen to be clustered. In fact, they seem to exhibit such clear-cut hierarchical levels that it is hard to believe that the model as carried out had involved no explicit hierarchy, only a built-in self similarity. Also hard to believe but true is the fact that whether the set of stopovers is viewed from a cluster's center or periphery, its statistical structure is precisely the same.

On the other hand, this invariance becomes obvious when one looks at the black-on-white Figures here and in Plates 134 and 135. Let us elaborate by mentioning that all the Plates in the present portfolio represent the beginning of two distinct flights, forward and reverse, and such flights are nothing but two statistically independent replicas of the same process. Clearly, the same continues to hold true if the origin is displaced to some other stopover. Hence, every stopover has precisely the same claim to be called the Center of the World. This feature constitutes the essence of the *conditional cosmographic principle* propounded in this Essay.

While there is solid reason to believe that the present generative method accounts only roughly for galactic reality, it suffices to bring home one of my principal themes, namely, that the conditional cosmographic principle is compatible with multilevel clustering. A great variety of configurations may be present in a shape into which none has been inserted "to measure."

Examining the various stopovers more attentively, one is tempted to say that compared with the stars, the *degree of clustering* is unrealistically high in the Cauchy $D=1$ case and unrealistically low in the Lévy $D=1.5000$ case. This impression is confirmed by the fact that the observed value of D is between 1 and 1.5000. It may be useful to consider that the degree of clustering that corresponds to $D=1$ is of "average" intensity. In this sense, the degrees of clustering corresponding to $D<1$, respectively to $D>1$, would be considered as being above and below average.

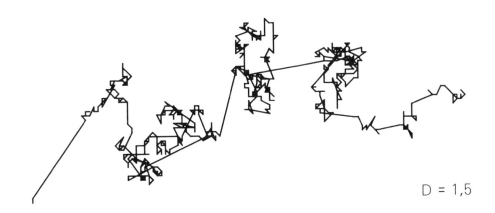

D = 1,5

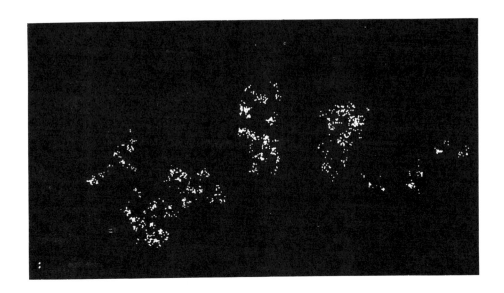

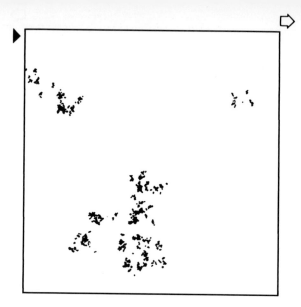

Plate 138-139

CIRCUMNAVIGATION OF LEVY CLUSTERS OF DIMENSION D=1.3000

The shape of clusters generated as sites of a Lévy flight in the plane is highly sample dependent, meaning that if one simulates clusters again and again while keeping the same dimension, one must expect to obtain a great variety of different shapes.

The same is true of a small isolated spatial Lévy cluster when viewed from many different directions – as seen by following the present "strip" clockwise from the top of Plate 138.

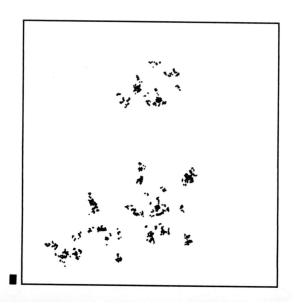

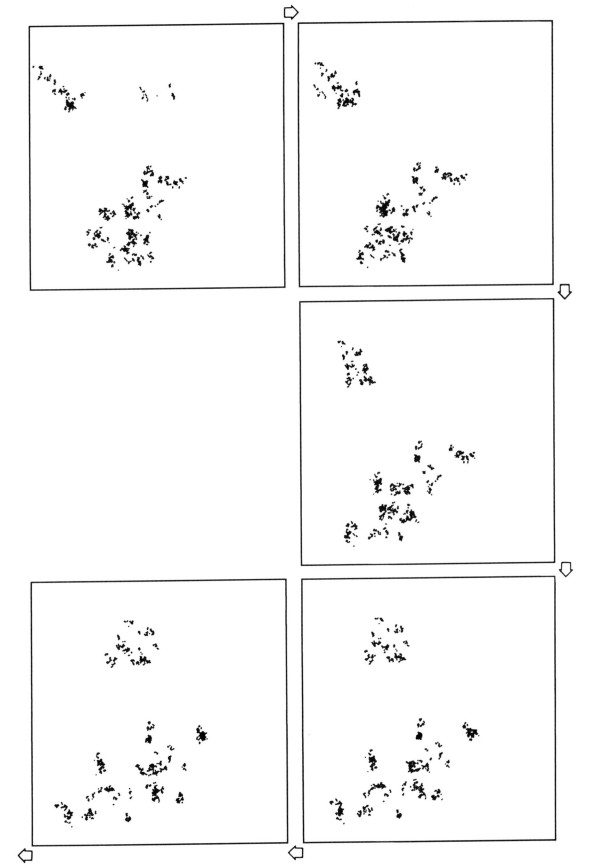

139

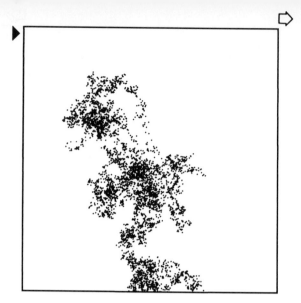

Plate 140-141

AS D DECREASES, LEVY CLUSTERS BECOME INCREASINGLY SEPARATE

In a Lévy flight, the degree of clustering of the sites depends upon the dimension. This Plate illustrates this phenomenon by keeping the seed and the direction of viewing fixed, while D is decreased starting from the Brownian value of D=2. Decrease of D means that the long jumps are made longer, while the short jumps are made shorter. However, the whole is continually being rescaled to fit in a graph of invariable size, so that the first visible effect of the decrease in D is that clusters become identifiable. Later each cluster is made to appear increasingly tightly packed and hence increasingly separate from neighboring clusters. Furthermore, the number of apparent hierarchic levels seems to increase.

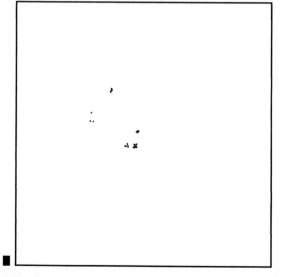

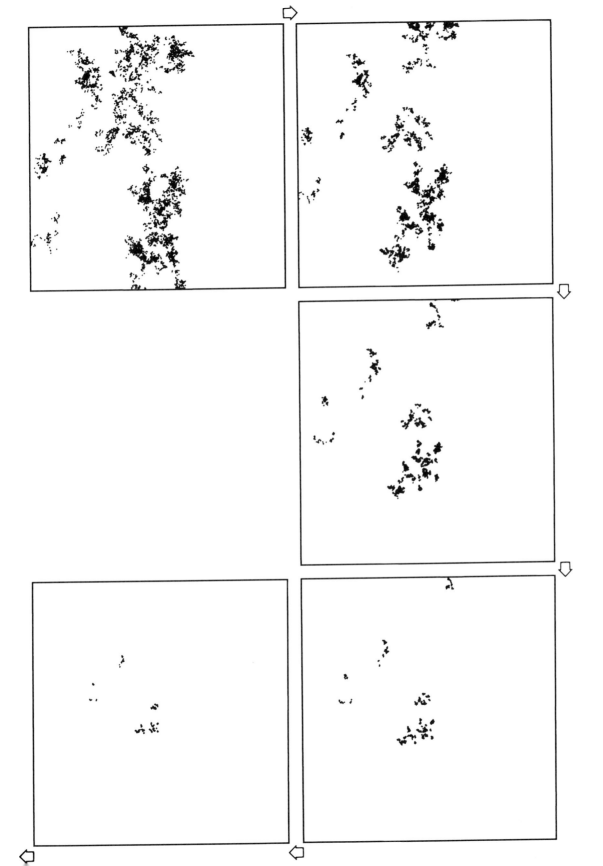

tribution of the jumps of a generalized Rayleigh flight, any dimension likely to be suggested by the facts and/or by theory is compatible with the conditional cosmographic principle.

ASIDE ON LEVY STABILITY

◁ The major consequence of the choice of a hyperbolic distribution for $Pr(U>u)$ is as follows. Whether for a plane or for space (and even, when $D<1$, for a straight line), the quantity $\langle M(R)\rangle$ becomes proportional to R^D, the ratio $M(R)/\langle M(R)\rangle$ being a random variable independent of R. Contrary to what has been noted for Rayleigh flight, this exponent is no longer universal but depends upon the distribution of the jumps.

◁ This dependence is a direct consequence of the fact that when $\langle U^2\rangle=\infty$, the classical form of the central limit theorem ceases to be valid, and a special central limit theorem must be used instead. This special theorem depends upon the distribution of the jumps, and its limit constitutes the tridimensional version of a "stable" random variable, in the sense of Paul Lévy (its definition is discussed under STABILITY ... in Chapter XII).

◁ The stable distribution corresponding to $D=1$ is the Cauchy distribution. The mathematics of unidimensional stable variables is discussed in many textbooks, such as Feller 1971. (These variables were thought to be devoid of applications until, starting in 1960, I discovered their role in economics, to be discussed in Chapter X. It is hardly necessary, however, to state that no acquaintance with economics is required in the present application.)

◁ As to tridimensional stable variables, the special case $D=3/2$ occurs in physics in the problem of Holtsmark, which is also discussed in Feller 1971, as well as in Chandrasekhar 1943. ►

COMPARISON WITH FRACTAL ERRORS

◁ If constrained to remain on a line, the Lévy flight becomes very similar to the construction of fractal errors as described in Chapter IV. There is one key difference, however. In the Berger & Mandelbrot construction, the intermissions follow each other from left to right. In the Lévy flight on the line, they go either way with equal probabilities. The isotropic flight is meaningful in an arbitrary Euclidean space and for all $D\le2$, while the method going from right to left is restricted to the line because neither plane nor space can be oriented, and also in practice is restricted to $D<1$ because a flight with $D>1$ covers the line.

◁ Nevertheless, in the case of a line, which offers a choice between two methods, the isotropic construction is the less satisfactory one. First, let the "past" portion and the "future" portion be viewed from a point belonging to the set. They are statistically independent in the Berger & Mandelbrot construction, whereas in the Lévy flight they are not. Second, each jump in the oriented flight is a single intermission. On the contrary, isotropic jumps constantly fall back in the middle of earlier jumps. Thus almost every intermission is the intersection of several jumps. ▶

CAN A FRACTAL UNIVERSE BE OBTAINED BY AGGLUTINATION?

The constructions based on the various Rayleigh flights yield qualitatively sensible results but must not be taken too seriously. Their principal virtue is that they simply and constructively demonstrate an important fact which was not obvious a priori. The conditional generalization of the strong cosmographic principle is not vacuous.

Now we return to physics. Many authors think one may explain the genesis of stars and other celestial objects in a manner that is, in a way, diametrically opposed to Hoyle's explanation. They invoke an *ascending* cascade (i.e., the agglutination of greatly dispersed dust particles into increasingly bigger pieces) rather than a *descending* cascade (i.e., the fragmentation of very large and diffuse masses into smaller and smaller pieces).

An analogous alternative arises in connection with the cascades postulated in the theory of turbulence, a topic discussed in the next chapter. Richardson's classical thesis postulates a cascade that descends toward ever smaller eddies, but evidence suggests that some cascades ascend toward structures of increasing bulk. Thus it may be hoped that the interrelations between descending and ascending cascades will be clarified simultaneously in both fields and that the confused debate between the supporters of fragmentation and the supporters of agglutination can be resolved in the not too distant future.

CHAPTER VI

Turbulence, Intermittency, and Curdling

We now shift our attention to a grand chapter of physics, the study of turbulence in fluids. The geometric aspects of turbulence that should prove to involve fractals and fractal dimension are so numerous and diverse that it is best to take them up separately. Only two have been studied in detail and both are tackled in this Essay. The better known one is relative to homogeneous turbulence, but it is more conveniently postponed to Chapter IX. We shall begin in this chapter by tackling intermittency. This phenomenon is not yet very widely known but is increasingly recognized as important and happens to respond to a model that is nearly identical to the Hoyle model of stellar distribution studied in Chapter IV.

This chapter splits naturally into four parts. It begins with an informal and general plea for a full geometric study of turbulence and the use of fractals, with or without chance.

The second part lists some conjectures about the fractal character of the singularities of solution of various equations of physics. This is a new theme (even though it was implicit in Chapter V that a Fournier universe has something to do with the equations of gravitation).

The third part of the chapter adds detail to the study of curdling. Much of it involves themes discussed previously.

The fourth part discusses further the relation or absence of relation between the fractal and the topological properties of geometric shapes. The first instance of this theme was encountered in Chapter II when we noted that fractal dimension measures both the irregularity and the fragmentation of coastlines, which means that fractal properties may be insensitive to topology. Furthermore, $D \sim 1.2$ was first encountered in Chapter II on coastlines and then in Chapter V on stellar matter as a characteristic of a new guise of fragmentation, namely clustering. Thus we are already well aware that the same fractal dimension is compatible with very different topological structures. In the latter part of this chapter we shall obtain a result in the opposite direction, showing that under certain circumstances, the fractal properties do tend to affect topology.

ANALYSIS AND GEOMETRY OF TURBULENCE

Let us start with some general comments about the physical phenomenon of turbulence and the way it is ordinarily investigated. The shapes it involves are easily seen or made visible and therefore almost cry out for a proper description. In addition, they are practically important. But they have not received anything near the attention they deserve. Regular configurations ◁ such as Bénard cells and Kármán streets ► are exceptions but it is arguable as to which ones should be called turbulent. In the last decade, attention has been drawn to other turbulent configurations in fluids which combine irregular and systematic features, but they also have not yet given rise to any sophisticated geometric study. In other words, the main approaches to turbulence have been analytical.

In particular, the relationship of turbulence to probability theory has been anomalous. On the other hand, through Norbert Wiener, the theory of stochastic processes had been explicitly influenced, not only – as we already know – by Perrin's work on Brownian motion but also by G. I. Taylor's early paper on turbulence. In return, techniques such as spectral analysis have been invaluable in Taylor's later papers and ever since. On the other hand, probabilists fully grasp Perrin's observation that random lines almost invariably exhibit features that, if they were encountered in a nonstochastic context, would be viewed as "pathological," and various tools are available to describe this kind of irregularity. However, we have already noted that sophisticated stochastic geometry is not used in any science, and includes the specific random surfaces and lines of turbulence. It is a pity. I believe that more imaginative stochastic geometry would be helpful in describing the truly disordered aspects of turbulence. Stochastic fractal geometry should be of particular interest.

Furthermore, it is hardly necessary to remind the reader that fractals need not be random. Hence fractal geometry may also be of interest in describing the more "systematic" turbulent configurations.

Finally, better geometry may hasten the emergence of better physics.

One typical problem of the geometry of turbulence is to describe the trajectory of a fluid particle ◁ from the Langrangian viewpoint. ► In a crude but useful approximation, one may think of this particle as carried up vertically by an overall current of unit velocity, and also by a hierarchy of eddies, each of which is simply a circular motion in a horizontal plane. ◁ This picture is obviously inspired by the Ptolemaic model of planetary motion. ► The resulting functions, $x(t)=x(0)$ and $y(t)=y(0)$, are sums of cosines and of sines. When the high frequency terms are very weak, the trajectory is continuous and differentiable, hence it is rectifiable and $D=1$. When, however, the high frequency terms are strong and the inner scale is $\eta=0$, $D>1$ and the trajectory is a fractal. Assuming that eddies are self similar, said trajectory happens to be identical to a famous counterexample of analysis called (complex) *Weierstrass function*. (See Chapters I and XII.) The generalization to truly spatial eddies is easy and leads to the following question. Can the transition of all the fluid to being turbulent be associated with the transition of the trajectory to being a fractal?

Another very different and more "generic' problem of the geometry of turbulence is the shape of the portion of the flow in which some characteristic of the fluid is encountered. When this characteristic occupies a volume, one wants to know the shape of its boundary.

Turbulence may be restricted to a portion of an otherwise laminar fluid. First, consider the boundaries of wakes or jets. In a "global" approximation, each is like a rod, or half a rod (or perhaps like the lovely shape of an oil tanker's spill). If, however, the boundary is examined in detail, it reveals a highly visible and very complicated "local" structure, one that does not evoke a rod as much as a rope with many loose strings floating around. Its typical cross-section does not evoke a circle at all. When the jet is irregular but systematic, its sections may be like Koch curves. When the jet is random, its sections are like the most rugged among the coastlines studied in Chapters II and IX. When vortex rings are present, their topology is of interest but does not exhaust the topic. A proper description should help one to understand how the shape in question was generated.

Related but different problems arise when studying the Gulf Stream. It is believed by many scholars to be not a single well-defined sea current but rather one divided into multiple branches, and these branches may themselves subdivide. An overall specification of its propensity to branch out would be most useful. In the case of clear-air turbulence, a similar concern is the local structure of the surface separating the turbulent from the laminar region.

A second very broad subclass of geometric problems occurs when a medium is completely filled by turbulence but a portion is somehow "marked" as different from the rest. The best example is generated when turbulence disperses a passive characteristic of the fluid, such as color. Branches of all kinds shoot off in all directions, endlessly, but existing critical analyses are of little help in describing the resulting shapes. A geometric phenomenological approach might be effective. (See Plate 53 and Mandelbrot 1976c.)

Similarly, it is most interesting to study the shape of the surfaces of constant temperature or the isosurfaces of any other scalar characteristic of the flow. The former may be delineated by the surface of the volume occupied by proliferating plankton that can only live in water at T>45°F, and fills all the volume available to it. The boundary of such a blob is extremely involved; we shall study it in Chapter IX.

THE PROBLEM OF INTERMITTENCY

A last example of geometric problems of the second class constitutes the topic of this chapter. It concerns intermittency and is the fact that in natural turbulence the distribution of dissipation is very far from uniform. Some regions are marked by very high dissipation, while other regions seem by contrast nearly free of dissipation. The importance of intermittency was first stressed by Landau & Lifshitz 1953-1959. See also Batchelor & Townsend 1949, p. 253; Batchelor 1953, Section 8.3; and Monin & Yaglom 1963, 1971, 1975.

The regions in which dissipation concentrates will be conveniently described as "carrying" or "supporting" it. An understanding of the geometry of the support of dissipation may help in understanding intermittency.

As may be expected from the inclusion of this topic in this Essay, the basic models of intermittency all lead to the conclusion that the support of dissipation is an approximate fractal with a dimension between 2 and 3. It cannot be an exact fractal because the inner scale η of turbulence is positive. The basic models suggest that no point in the support of turbulence is farther than η from its boundary; in other words, that this support is a very diffuse shape devoid of significant inside points.

The simplest model of fractal intermittency, which also happens to be the most useful and the one to which this chapter will be restricted, involves the new notion of *fractally homogeneous turbulence.* The corresponding generating model involves a form of curdling essentially identical to that which Hoyle uses to construct the fractal support of stellar matter; see Chapter V. The principal difference – a very considerable one, however – is due to the difference between the dimensions. We know it to be about 1 in the case of the stars, but it will be seen to lie between 2 and 3 for turbulence.

Incidentally, the fact that this Essay brings together the intermittency of turbulence and the distribution of celestial objects is natural and not new. A while ago, von Weizsäcker 1950 (as well as other physicists whose motivation was entirely distinct from mine) had thought of explaining the genesis of the galaxies by turbulence on a colossal scale. The only model available at that time involved homogeneous turbulence as conceived in G. I. Taylor 1935, and von Weizsäcker did not fail to recognize that a homogeneous model cannot by itself account for stellar intermittency. He was therefore led to introduce amendments which in effect constitute a prefiguration of the theory of turbulent intermittency. (They are quite in the spirit of the Fournier-Charlier model examined in Chapter V.)

Since the 1950's the theory of turbulent intermittency has developed dramatically, so that if von Weizsäcker's unifying effort were to be taken up again, it might consist more reasonably in establishing a physical link between two kinds of intermittency and the corresponding self similar fractals.

Intermittency is particularly clear-cut when the outer scale L of turbulence much exceeds its inner scale η. Such is the case when the classic measure of hydrodynamic scale (Reynolds number) is very large. It is immense in the case of the stars and extremely large in the ocean and in the atmosphere.

◁ To Richardson 1926 it had seemed doubtful that L is actually finite. He wrote that for most sample means relative to turbulence, "observation shows that the numerical values would depend entirely upon how long a volume was included in the mean. Defant's researches show that no limit is attained within the atmosphere." When $L=\infty$, one needs a generalization of random processes such as the sporadic processes of Mandelbrot 1967b. ▶

Since the curdling model involved in this chapter involves little suspense, a leisurely progression toward fractal homogeneity may prove instructive.

ROLE OF SELF SIMILAR FRACTALS

The relatively few studies made of the various geometric aspects of turbulence that we have listed do not appear to have gone beyond the most elementary shapes, and vocabulary has tended to be pre-Euclidean. There is much use of terms such as *spotty* and *lumpy,* and Batchelor & Townsend 1949 envision "only four possible categories of shapes: blobs, rods, slabs, and ribbons." Some lecturers (but few writers) also use the terms *beans, spaghetti,* and *lettuce.*

Kuo & Corrsin 1972 study this choice further, declare it "primitive" and call for *in-between patterns.* The use of self similar fractals to approach both the stochastic and the nonstochastic geometric facets of turbulence may be viewed as responding to Kuo & Corrsin's call (though I was using fractals in this context well before 1972.)

As to the notion of self similarity, it is quite natural that it should enter here from the outset, for it was first conceived with a view toward a theory of turbulence. The pioneer was the Lewis Fry Richardson whom we have already encountered. Richardson 1926 was also the first paper to introduce the concept of a hierarchy of eddies linked by a cascade. (See the biographical sketch in Chapter XI.) It is in the context of turbulence, also, that the theory of cascades and of self similarity achieved its triumphs of prediction between 1941 and 1948. The main contributors were Kolmogorov, Obukhov, Onsager, and von

Weizsäcker, but tradition denotes all the developments of the period by one name, Kolmogorov's.

Given that the role of self similarity in turbulence was suggested by the consideration of readily perceived eddies, it is odd that all the actual applications of this notion to turbulence were purely analytic. Now fractals make it possible to apply the technique of self similarity to the *geometry* of turbulence.

The originality of this approach is related to the peculiar fact that while earlier analyses of turbulence assume self similarity, the shapes of the abovementioned "primitive" four way choice fail to be self similar. Possible ad hoc corrective measures come to mind. For example, one might consider splitting rods into ropes surrounded with loose strands (remember the analogous situation with wakes or jets) and slicing slabs into sheets surrounded with loose layers. Somehow those strands and layers might be made self similar.

However, an ad hoc injection of self similarity has never been actually implemented. Besides, as readers of this book are well aware, I find it unpromising and unpalatable, and I shall follow an entirely different tack. By allowing the details of strand and layer to be generated by the cascade itself, we shall encounter new configurations in great variety.

Their very wealth, however, will

make us fail in one related task, since it is hard to coin or find names for them all. For example, even in shapes that are most nearly ropelike, the "strands" can be made so heavy that the result is really "more" than ropelike. Similarly, near sheets are "more" than sheetlike. Also, it is possible to mix sheetlike and rope-like features at will.

Since the basic self similar fractals are devoid of privileged direction, our study is restricted to phenomena that manifest some homogeneity. We can go beyond the classical Taylor homogeneity, but the interesting geometric questions that combine turbulence with strong overall motion are by necessity left aside.

◁ The first explicit uses of self similarity in studies of intermittency are in Obukhov 1962 and Kolmogorov 1962. These papers, in terms of influence, nearly match their 1941 predecessors by the same authors, but there are reasons to doubt that their influence will be equally durable. See Mandelbrot 1972j, 1974f, 1976o; Kraichnan 1974. ▶

CONJECTURE: THE SINGULARITIES OF FLUID MOTION ARE FRACTALS

Historically, there has been almost no interaction between the Kolmogorov theory of turbulence and the attempts to explain turbulence from the Euler equations of nonviscous fluids and the Navier-Stokes equations, which apply when viscosity is present. That is, the equations have not helped us understand Kolmogorov, and Kolmogorov has not helped solve the equations.

The notions I propose – including the assertion that turbulent dissipation is not homogeneous over the whole space, only over a fractal subset – may at first sight seem to make the gap even greater, but I contend that the opposite is the case.

Let us indeed review the procedure that has allowed other basic equations of mathematical physics to be solved successfully. Typically, one starts by drawing up a list of elementary solutions, some of which are obtained by actually solving the equation under special conditions, while others are simply guessed on the basis of physical observation. Next, neglecting details of these solutions, one concentrates on their singularities, and one draws a list of elementary singularities characteristic of the problem. From then on, more complex instances of the basic equations can often be solved in the first approximation by identifying the appropriate singularities and stringing them as required.

In this perspective, the difficulties experienced in deriving turbulence from the Euler and Navier-Stokes equations seem due to the fact that the classical singularities known from other equations fail to account for what we perceive intuitively to be the characteristic features of turbulence.

It is my contention that the turbulent solutions of the basic equations involve singularities of an entirely new kind, fractals.

A first specific conjecture included in my contention is as follows. The solutions of the Navier-Stokes equation, if they exist, are actual limit fractals.

Another conjecture is that the singularities of the solutions of the Euler equations are actual limit fractals.

The preceding statements are phrased to take account of the uncertainty as to the existence of Navier-Stokes singularities. If such singularities are indeed nonempty, dimension may help express the intuitive feeling that the Navier-Stokes solutions are smoother and hence less singular than the Euler solutions. The underlying conjecture is that the dimension is larger in the Euler case than in the Navier-Stokes case.

A final conjecture in implementation of my overall contention is that the peaks of dissipation involved in the notion of intermittency are Euler singularities smoothed out by viscosity.

The proof of either of these conjectures is beyond my analytic powers, but the first one has been fortunate in attracting the attention of V. Scheffer, who has studied a finite or infinite fluid subject to the Navier-Stokes equations with a finite kinetic energy at time $t=0$. If singularities are present, Scheffer 1976 shows they necessarily satisfy the following theorems. (a) Their projection over the time axis has at most the fractal dimension $1/2$. (b) Their projection on the space coordinates is at most a fractal of dimension equal to 1.

It turns out after the fact that the first of the above results is no more than a corollary of a theorem in Leray 1934, a paper well known among others for the result that the set of singularities is necessarily contained within a time span from $t=0$ to $t=t*<\infty$. This paper ends quite abruptly, shortly after the proof of a certain formal inequality, of which Theorem (a) is a corollary, in fact merely a restatement. But is it fair to say *merely*? The mere restatement of a result in more elegant terminology is (for sound reasons) rarely viewed as an advance, but I think that in the present instance the inequality in Leray's theorem was nearly useless until the Mandelbrot-Scheffer corollary had placed it in the proper perspective.

SINGULARITIES OF OTHER EQUATIONS OF PHYSICS

By discussing basic equations, the preceding section broke the custom of this Essay, which is to stress description. We may as well continue in the same vein for a few lines. The other phenomena which this Essay claims are fractal have nothing to do with either Euler or Navier. Therefore, their fractal character is likely to be related to some generic features shared by many different equations of mathematical physics. Can it be some very broad kind of nonlinearity?

CONCRETE CHARACTERIZATION OF THE SUPPORT OF INTERMITTENCY

While contending that the support of turbulent intermittency is an approximate fractal, we come close to claiming a definition of turbulence, but turbulence has never been defined properly. Part of the reason is that the same term has been applied to several noncoincident phenomena. Nevertheless the term ought not be used too loosely. The most regular configuration ◁ Kármán streets, Bénard cells, and the like ► are better left out, but other system configurations should be kept in, and it is not clear what else should be included.

This continuing lack of a definition becomes easy to understand if it is confirmed, as I claim, that a proper definition requires fractals.

The customary mental image of turbulence appears nearly "frozen" in the terms in which it was first isolated by Reynolds, about one hundred years ago, for fluid flow in a pipe: when the upstream pressure is sufficiently weak, the motion is regular and "laminar;" when the pressure is increased sufficiently, everything suddenly becomes irregular. In this prototype case, turbulence is either supported by a set that is "empty," meaning nonexistent, or it fills the entire tube. In either case there is not only no geometry to study but also no imperative reason to define turbulence.

In wakes, things become more complicated. There is a boundary between the turbulent zone and the surrounding sea, and one ought to study its geometry. However, this boundary is so clear-cut that, again, an "objective" criterion to define turbulence is not truly necessary.

In fully developed turbulence in a wind tunnel, matters are again simple, the whole appearing turbulent like the Reynolds pipe. Nevertheless, the procedures used to achieve this goal are sometimes curious, if we are to believe certain stubbornly held stories, which may not all be legends. It is said that wind tunnels when first blown are unfit for the study of turbulence. Far from filling up the volume offered to it, the turbulence itself seems "turbulent," presenting itself in irregular gusts. Only gradual efforts manage to stabilize the whole thing, after the fashion of the Reynolds pipe. Because of this fact, I am among those who wonder up to which point the nonintermittent "laboratory turbulence" in wind tunnels is the same physical phenomenon as the intermittent "natural turbulence" in the atmosphere. Hence we must define the terms.

We approach this task indirectly, starting from an ill-defined concept of what is turbulent and examining the one-dimensional records of the coordinates of velocity at a point. The motions of the center of gravity of a large airplane illustrate a rough analysis of such records. Every so often, the airplane is shaken about, which shows that certain regions of the atmosphere are strongly dissipative, the remainder seeming to be laminar. Now, we repeat the test with a smaller airplane. On the one hand, it "feels" turbulent gusts that had left the large airplane undisturbed. On the other hand, the small airplane experiences each shock received by the large airplane as a burst of weaker shocks. Thus when a strongly dissipative piece of the cross-section is examined in detail, laminar inserts become apparent, and further smaller inserts are seen when the analysis is refined further.

Each stage demands a redefinition of what is turbulent. The notion *turbulent minute of record* is meaningful if interpreted as "minute of record that is not completely free of turbulence." On the other hand, the more demanding notion of *solidly turbulent minute of record* seems devoid of actual significance. Proceeding to successive stages of analysis, turbulence becomes increasingly sharp over an increasingly short proportion of the total record length. The set that supports it seems to have less and less of an "inside." Our next task will be to model such a set.

◄ Incidentally, some aspects of turbulence can presently be visualized in space and followed in time. However, the older method based upon one-dimensional "probes" through a four-dimensional field is not exhausted. The same principle of investigation is furthermore encountered in exploratory geomorphology, when the subsoil is probed by means of "plugs." It is also encountered in computerized tomography of the interior of a living subject's organs. However, the analogy does not extend far because once a mine is opened, the limits of the geological layers can be explored, for they do not shift. And a cadaver can be mapped spatially in a computer's memory. For turbulence, this is inconceivable. ►

THE NOVIKOV-STEWART CASCADE, AN EXAMPLE OF CURDLING

The first explicit model of the intermittency of turbulence has been advanced in Novikov & Stewart 1964 and is illustrated in Plates 156 through 159. Though crude, it can be developed to lead to the main fractal conjectures. Therefore it deserves careful attention. Its most striking feature is that it unwittingly includes the principle of Cantorian "curdling," which we know from Chapter V to be a characteristic of the Hoyle model of stellar intermittency.

The image incorporated in the term *curdling* should not be taken too literally, but it is of further help in suggesting that the space outside of the curds be called *whey*. However, the set into which "stuff" is concentrated as a result of curdling will be called *concentrate*.

◄ It is hard to avoid thinking of curdled cheese as falling to the bottom of the whey, while one can think of concentrate as floating in space! Furthermore, the formation of real cheese may result from biochemical instability in the same way as Novikov & Stewart curdling is presumed to result from hydrodynamical instability. However, I could not put my hand on the data that are necessary to tell whether or not any real cheese is also a cheese in the fractal sense. ►

Since the principle of the Novikov & Stewart curdling is already familiar from Chapter V, it will suffice to sketch it while stopping every so often to settle details. The key assumption of Novikov & Stewart is that intermittency results from a cascade, each stage of which concentrates or "curdles" the dissipation from an eddy into N subeddies smaller in the ratio r than the eddies.

Like Hoyle, Novikov & Stewart postulate nothing of the spatial distribution of nonempty eddies. We interpret this lack of specification to imply total randomness. In any event, one item common to every method of arrangement is the fractal dimension

$$D = \log N / \log (l/r).$$

The difference between the species of curdling involved in astronomy and turbulence appears as soon as one examines the actual value of D. For stars, the exorcism of the Olbers paradox restricts D to satisfy $D < 2$. By contrast, and this will be one of the triumphs of the fractal vision of the intermittency of turbulence, it will be very easy to demonstrate that its dimension must satisfy $D > 2$. Another difference is that the value of D for turbulent intermittency was, until now, purely empirical. No theory claims to have deduced it from physical considerations (however controversial), while for galaxies $D = 1$ (however inadequate this value may be) can be derived theoretically.

Chapter V described how the Hoyle cascade, which involves randomness and some physics, was preceded by the Fournier model, which is nonrandom and pos-

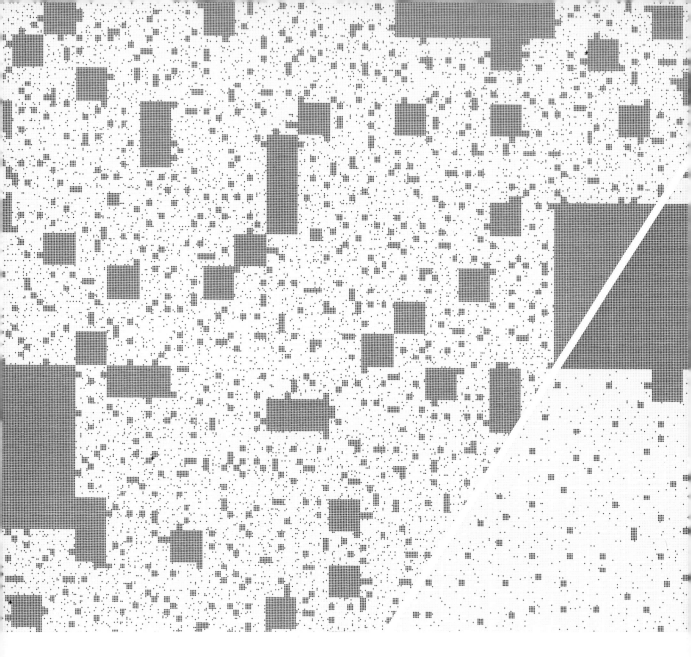

Plates 156 through 159

THE NOVIKOV-STEWART MODEL OF RANDOM CURDLING IN A PLANE GRID
(DIMENSIONS D=1.5936 TO D=1.9973) FOLLOWED BY PERCOLATION

The Novikov-Stewart cascade provides a useful general idea of how turbulent dissipation in a fluid concentrates (curdles) into a small relative volume. Conceptually, it is very similar to the Hoyle cascade illustrated on Plates 123 and 124, but the fractal dimension D of the concentrate is very much larger for turbulence than for

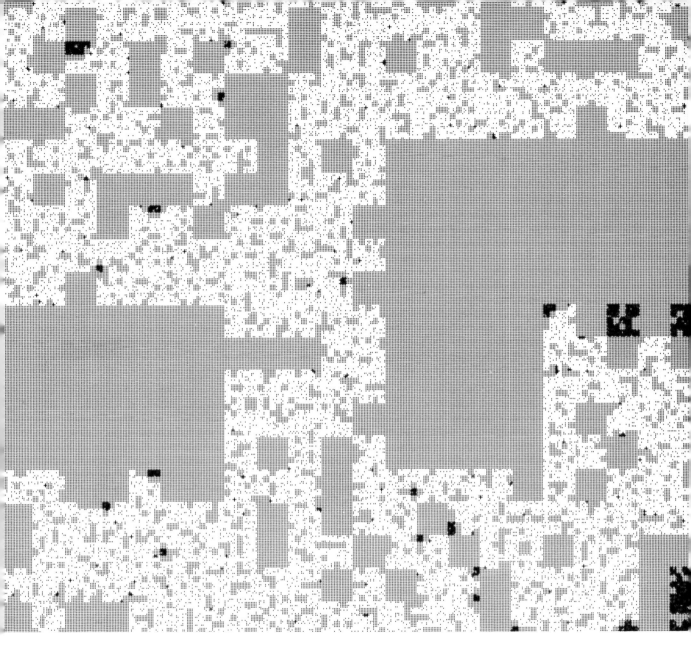

stellar matter. In cosmography D~1, while in turbulence D>2. In the present Plates, several different values of dimension are illustrated, because the actual value of D for intermittent turbulence is unknown and also for the sake of general understanding of the process of curdling. Throughout, $r = 1/5$, and the values selected for N are

$$N = 5 \times 24, \ N = 5 \times 22, \ N = 5 \times 19, \ N = 5 \times 16, \text{ and } N = 5 \times 13.$$

Hence the dimensions take the values

$$D = 1 + \log 24/\log 5 = 2.9973, \ D = 2.9426, \ D = 2.8505, \ D = 2.7227, \text{ and } D = 2.5936.$$

The whey is represented in gray, while the Novikov & Stewart concentrate is drawn in either black or white, for reasons that will only transpire late in Chapter VI.

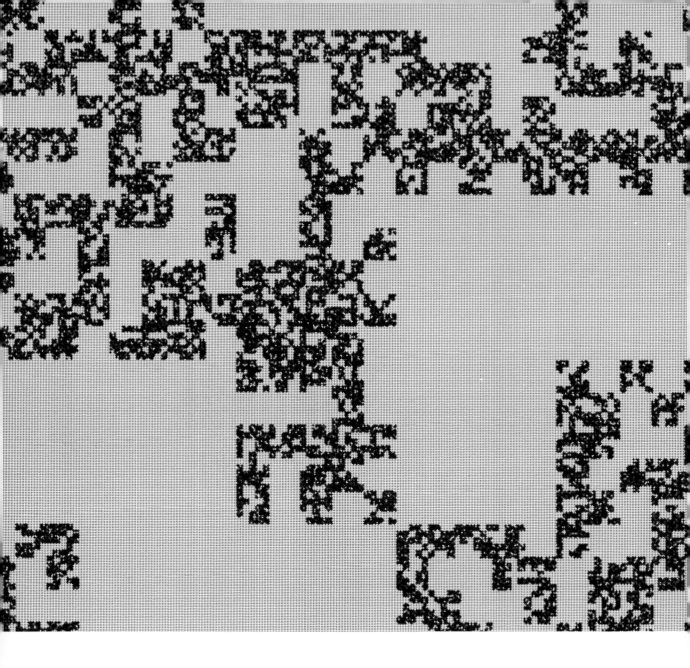

Because turbulence satisfies D > 2, Novikov & Stewart concentrates are essentially opaque, and (contrary to Hoyle concentrates) there is no point in attempting to draw them in projection. The process had to be tackled in cross-section, and the diagrams shown below are, respectively, of dimensions

$$D = 1.9973, \quad D = 1.9426, \quad D = 1.8505, \quad D = 1.7227, \quad \text{and} \quad D = 1.5936.$$

Although still different from a full square, the first Figure is nevertheless barren of

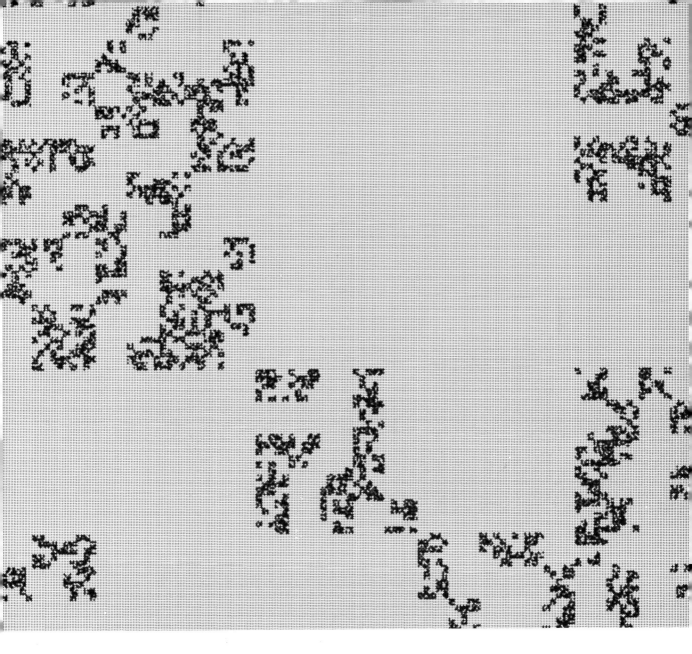

interesting detail, and the second Figure is not too exciting either. They have therefore been combined in Plate 156, the lower-right corner of which is relative to the highest value of D.

The generating program and the seed are the same throughout, and one can follow the progressive disappearance of the grays. One began by stacking at random the 25 subeddies of each eddy. Then for successive integer values of $5^D = N$, the top 25−N subeddies in the stack were "grayed away."

◁ For the lower dimensions, there is no percolation. For N = 19, there is a bit of black and much of white. A few other seeds percolate already for N = 18. ▶

tulates essentially the same final result without attempting to explain it. The Novikov & Stewart cascade, which involves the semblance of physics, failed to have such a nonrandom predecessor. In some way, this is a pity, and to make certain points conveniently we shall describe later in this chapter some shapes that *might* have come before Novikov & Stewart curds. The same approach was used in Chapter IV where the description of the random Cantorial structures of Berger & Mandelbrot 1963 and Mandelbrot 1965c was preceded by a plain Cantorian approximation.

FRACTALLY HOMOGENEOUS TURBULENCE

After a finite number m of stages of a Novikov & Stewart cascade, dissipation is distributed uniformly over γ^{mD} out of γ^{3m} mth order subeddies. A cascade continued without end predicts that the limit distribution of dissipation is spread uniformly over a fractal concentrate of dimension $D<3$. Further, while the Novikov & Stewart model is very special and crude, the resulting property of homogeneity restricted to sets of dimension $D<3$ is also encountered in other models, either exactly or approximately. The corresponding form of turbulence is therefore important; I propose to call it *fractally* homogeneous turbulence.

The added term 'fractally' is of course meant to contrast this turbulence with G. I. Taylor's homogeneous turbulence, which can be viewed as the limit case for $D\rightarrow3$. The salient fact is that curdling allows $D-3$ to be negative.

We shall see that the same value of D can be encountered in sets that differ from each other from many other viewpoints, for example, from the view of topological connectedness. The topological dimension yields a lower bound to the fractal dimension, but this bound is frequently exceeded by such a wide margin as to be of no use. A shape having a given fractal dimension D between 2 and 3 may be either "sheetlike," "linelike," or "dustlike," and either connected or disconnected. Intuitively, one might have hoped that some closer relationship should exist between fractal dimension and degree of connectedness, but mathematicians lost this hope between 1875 and 1922. We shall return to such problems late in the chapter, but it may be said that even to the pure mathematician, the study of the actual loose relationship between these structures remains essentially virgin territory.

It is fortunate, therefore, that many practically useful characteristics of fractally homogeneous turbulence turn out to depend solely upon D.

◁ Two examples are examined in Mandelbrot 1976o. The first relates to the spectrum of turbulent velocity. I showed that the classical Kolmogorov exponent 5/3 must be replaced by 5/3+B, where B=(3−D)/3. This result (announced in Mandelbrot 1967t, with θ=D−2) has been rederived in a different manner by Sulem & Frisch 1975.

◁ Another example is the kurtosis, a measure of the degree of intermittency. Specifically, the illustrative models of Corrsin 1962, Tennekes 1968, and Saffman 1968 all turn out to generate diverse forms of fractally homogeneous turbulence. Ostensibly, the models deal with shapes that share the topological dimension of either the plane (sheets) or the straight line (rods). However, this feature turns out to be almost completely irrelevant, since (a) each model also involves additional assumptions; (b) one can only test one aspect of the result, the value of of the exponent in the relationship between the kurtosis and a Reynolds number; and (c) this exponent can be shown to depend solely on the fractal dimension D of the shape generated by the model. A model should of course lead to the actually observed value, which is D=2.6 (and which agrees with other estimates of D). In fact, the Corrsin model predicts a value of D equal to the topological dimension it postulates, $D_T=2$. Hence this model does not involve a fractal. (This may be another a priori reason why it does not fit the evi-

dence.) On the other hand, the Tennekes model does involve an approximate fractal, since the predicted D greatly exceeds the postulated topological dimension $D_T=1$. Therefore, the inferences the above authors have attempted, from the kurtosis to a combination of intuitive "shape" and dimension, were unwarranted. ▶

CONSEQUENCES OF THE CURDS BEING "IN-BETWEEN" SHAPES

As seen in Chapter II, Koch curves have zero area and infinite length, which means they are *in-between shapes*. This last term also applies to the limit fractals obtained by curdling.

First, we know all classical shapes such that D<3 (points, lines, and surfaces) have a vanishing volume. The same is true for the limit curdling fractals.

The area also behaves very simply. When D>2, it is infinite. When D<2, it vanishes. When D=2, curdling leaves it essentially constant.

Similarly, when m → ∞, the cumulative length of the edges of approximate curdling fractals tends to infinity when D>1 and to zero when D<1.

These volume and area properties confirm that limit curdling fractals with a fractal dimension satisfying 2<D<3 lie somewhere between an ordinary surface and a volume.

◁ *Proofs*. The volume of the mth finite approximation is $L^3 r^{3m} N^m = L^3 (r^{3-D})^m$, which tends to zero with the inner scale $\eta = r^m$.

◁ For the area, the case $D < 2$ is settled on the basis of an upper bound. The area of the mth approximation at most equals the sum of the areas of the curds, because the latter sum also includes subeddy sides that neutralize each other by being common to adjacent curds. The area of each curd of mth order being $6L^2 r^{2m}$, the total area is at most $6L^2 r^{2m} N^m = 6L^2 (r^{2-D})^m$. When $D < 2$, it tends to 0 with $m \to \infty$, which proves half of our assertion. The case $D > 2$ requires a lower bound. One is obtained by noting that the surface of the union of mth-stage curds contained in an $(m-1)$th-stage curd has at least one little square of side r^m and area r^{2m} that is wholly contained in said $(m-1)$th-stage curd, and hence cannot possibly be erased. Hence the total area is at least $L^2 r^{2m} N^{m-1} > (L^2/N)(r^{2-D})^m$, which tends to infinity if $m \to \infty$. Finally, when $D = 2$, both bounds are finite and positive. ▶

FRACTAL DIMENSION OF THE CURDS' SECTIONS

Several themes ago, in Chapter III, it was observed that the Brownian line-to-line function has the dimension $3/2$, and its zeroset has the dimension $1/2$. Since the zeroset is the intersection of the graph of the function by a straight line of dimension 1, the example suggested that intersection of a planar fractal by a line decreases its dimension by 1. This rule was again used in Chapter IV, and it was asserted that it is of very general applicability. It is easy to prove it for the limit curdling fractals.

Consider the traces (squares and intervals) that the eddies and subeddies of the curdling cascade leave upon either a face or an edge of the original eddy of side L. Each cascade stage replaces a curd by a certain number of subcurds. Not only are their positions random, but their numbers and the resulting distributions involve classic birth and death processes. At each stage, each curd edge can be viewed as having acquired a random "offspring" made of subcurd edges whose number N' has the expectation $\langle N_1 \rangle = Nr^2$. Classical results on birth and death processes can therefore be used (Harris 1963).

They show that the number $N_1(m)$ of offspring present after the mth generation is ruled by the following alternative. When $\langle N_1 \rangle \leq 1$, that is, $D \leq 2$, it is almost certain that the offspring will eventually die out, meaning that the edge will eventually become empty, hence of zero dimension. When $\langle N_1 \rangle > 1$, that is, $D > 2$, then, on the contrary, the offspring will have a probability < 1 of dying off and a nonzero probability of expanding in numbers forever.

Here, similarity dimension requires fresh thinking because the usual ratio $\log N_1(m)/\log(1/r)$ is random and can no longer be interpreted directly. On the other hand, the following relation holds almost surely:

$$\lim_{m\to\infty}\log N_1(m)/\log(1/r^m)$$
$$= \log\langle N_1 \rangle/\log(1/r)=D-2.$$

This asymptotic relationship suggests a generalized similarity dimension $D-2$.

Two-dimensional eddy traces obey the obvious modification of the same argument, after replacement of N_1 by a random N_2 such that $\langle N_2 \rangle = Nr$. When $\langle N_2 \rangle \leq 1$, that is, $D \leq 1$, each eddy face will eventually become empty. When $\langle N_2 \rangle > 1$, that is, $D > 1$, one has

$$\lim_{m\to\infty} \log N_2(m)/\log(1/r^m)$$
$$= \log\langle N_2 \rangle/\log(1/r)=D-1.$$

◁ As a further confirmation that fractal dimension behaves under intersection in the same way as Euclidean dimension, the intersection of several curdled fractals of respective dimensions D_m satisfies $E-D=\Sigma(E-D_m)$. ▶

DIRECT EXPERIMENTAL REASON FOR BELIEVING THAT INTERMITTENCY SATISFIES D>2

The above results concerning one-dimensional sections have the following corollary. Had the carrier of turbulence satisfied $D < 2$, nearly all experimental probes would slip between turbulent regions. The fact that such is not the case suggests that in reality $D > 2$. This inference is extraordinarily strong, because it is so elementary. It relies upon an experiment that is repeated constantly and for which the possible outcomes are reduced to an alternative between "never" and "often."

Naturally, $D > 2$ does not by any means exclude the classical value $D = 3$, but it allows the novel possibility $D < 3$. Unfortunately, I cannot even begin to tackle the tasks of proving this inequality from basic principles, and a fortiori of deriving D from basic physics.

To put the significance of $D > 2$ into a broader perspective, one may recall that Chapter V features $D < 2$. While these conclusions are the precise opposite of each other, both spring from the effect of the sign of $D-2$ on the typical section of a fractal. In the present chapter, the section has to be nonempty. In Chapter V, on the contrary, the exorcism of the Olbers paradox required it to be empty, to insure that the majority of rays originating from the earth *never* meet a star.

NONRANDOM CANTOR CONSTRUCTS IN 3-SPACE

Although such denials may be tiresome, the present D > 2 is not a topological but a fractal inequality. ◁ However, a tentative topological counterpart of D > 2 is discussed later in this chapter. ▶ The remainder of this chapter is concerned with the relationship between fractal dimension and topology.

In a first stage, we provide the Novikov & Stewart model with a nonrandom ancestry it failed to have. To do so, several classical nonrandom mathematical "monsters", and obvious variants, are introduced as nonrandom illustrations of certain features of turbulent cascades. Historically, of course, the mathematicians whose work we discuss had no inkling of any conceivable application.

In the last sections of this chapter, we turn to the problem of percolation of random fractals obtained by curdling.

THE SIERPINSKI SPONGE, A SELF SIMILAR KNOTTED ROPE

Successive sections below describe self similar fractals that most resemble, respectively, knotted ropes with strands and layered sheets, either alone or in mixture, that is, *vortex tubes* and *vortex sheets*. We stress fractal dimensions of the right magnitude for turbulent dissipation, namely, between 2 and 3.

The first shape to be examined, Plate 167, is circumscribed by a cube of side L = 1. Its traces on the cube's sides were first considered in Sierpiński 1916 and are called *Sierpiński carpets,* and the whole will be called a *Sierpiński sponge.* Each stage of the construction is meant to correspond to a stage in a cascade of physical "instabilities."

◁ They resemble the Taylor-Görtler instability for curved sheets or the Taylor-Couette instability for tubes, but we think in less specific terms. ▶

In the illustration r = 1/3; hence, like the basic Koch curve, this set is "triadic." The "core" that is cut out or swept clean in each stage is shaped like a cross with spikes on its front and on its back. The process leaves N = 20 curds connected to each other and including their boundary points. Among them, 12 form thick "rods" or ropes, and the remaining 8 can be viewed as knots, connectors, or ties. The second cascade stage cuts out 20 sets shaped as above but reduced in the ratio 1/3, and so on. The limit set has the dimension

$$D = \log N / \log (1/r)$$
$$= \log 20 / \log 3 \sim 2.7268.$$

The above curdling implies very specific cutouts. Purposes for which the cutouts' form makes no difference and all that counts is the final D are purely fractal. For other purposes the cutouts' shape is important. For example, the term *sponge* is appropriate because the present cutouts were deliberately chosen so that the cutouts and the ropes remain connected sets. Hence one can conceive of water flowing around the latter, between any two points of the former. After a loop has been strung through several holes, it becomes impossible to deform it continuously down to a point without break.

The cubes that remain in the triadic sponge seem a bit thick to qualify as "rods" or "ropes," but their thickness is adjustable at will in either direction. One first replaces $1/r$ by an integer >3. When $1/r$ is odd, a thick sponge is obtained by thinning the cutout so that its trace on the planes bounding the initial eddy is a square of side r. For an arbitrary $1/r$, a thin sponge is obtained by thickening the cutout so its trace is a square of side $1-2r$. For example, if $r=1/5$, one constructs a thick pentadic sponge by cutting out 13 subcubes (the center and two cubes strung along each of six directions). When $r=1/15$, one cuts out one center and seven cubes strung along each of six directions.

Each sponge can be deformed into any other without breaking, so they are topologically identical. And the fact that D takes different values for different sponges proves again that fractal dimension is not a matter of topological form. The sponges contain no square, however small, and they do contain straight segments (in this sense, they are more rope-like than sheetlike; in fact, they are curves of topological dimension 1). However, this feature is entirely compatible with the fact that for thick sponges

$$D=\log\,(1/r^3-3/r+2)/\log\,(1/r)$$

satisfies $D>2$, because $1/r>3$ implies $1/r^3-3/r+2>1/r^2$. On the other hand, thin sponges only demand

$$D=\log\,(12/r-16)/\log\,(1/r)>1.$$

The inequality $D>2$ is due to constraints specific to thick sponges, but the inequality $D>1$ holds because any shape that includes topological lines must have a fractal dimension $D>1$.

A DUSTLIKE FRACTAL FOR WHICH DIMENSION IS CLOSE TO 3

However, a set of dimension $D>1$ need not contain lines. The result has already been exemplified by some Fournier dusts of Chapter V. Another example is obtained as follows. Section the initial unit cube by narrow slits of width $1-2r$ perpendicular to each of the coordi-

Plate 167

THE SIERPINSKI SPONGE
(DIMENSION D=2.7268)
AND SIERPINSKI CARPET
(DIMENSION D=1.2618)

The principle of the construction of this shape is evident. After it has been continued without end, it leaves a remainder known as the Sierpiński sponge. Its volume vanishes, while the area lining its holes is infinite. (Reproduced from *Studies in Geometry*, by Leonard M. Blumenthal and Karl Menger, by kind permission of the publishers, W. H. Freeman and Company, copyright 1970.)

Each of its external faces is known as a Sierpiński carpet (Sierpiński 1916). Its area vanishes, while the total perimeter of its holes is infinite. The intersections of the sponge with medians or diagonals of the initial cube are triadic Cantor sets. The fact that the sponge's D exceeds the carpet's D by more 1 is discussed in the text.

If one takes it to be the trace of a Sierpiński sponge, a Sierpiński carpet is obtained directly by taking out a succession of square-shaped cutouts. It may also be obtained indirectly by yet another generalization of the Koch construction, wherein self overlap is allowed but overlapping portions are counted only once. The initial shape is the big cutout in the middle, and the standard polygon is made of the segment from $(0,0)$ to $(0,1)$ plus the two squares of side $1/3$ having the points $(-2/3, 1/3)$ and $(1/3, 1/3)$ as their lower-left corners. As a result of the rule we have adopted for the overlaps, the usual formula for D is not valid, and the fact that $N = 11 > 1/4^2 = 9$ does not mean that $D>2$ (a result

that would have sufficed to dismiss the model). In fact, two familiar phenomena are encountered again: each island's coastline is rectifiable and therefore of dimension 1, and the dimension of the carpet expresses the degree of fragmentation of land into islands rather than the degree of irregularity of the islands' coastlines.

However, the present standard polygon and the rule for overlaps lead to novel features, too. Indeed, the previously considered examples of the Koch construction with islands and no lakes (Chapter II) involve no self contact or self intersection, and as a result the sea is open in two distinct senses. First, it is connected, which seems to be a proper topological interpretation of nautical openness. Second, it is open in the set topological sense of not including its boundary. The novelty brought in by the Sierpiński carpet is that it is possible for the Koch islands to touch each other *asymptotically*, and in effect to end up by forming a solid superisland of area 9 times that of the square island with which one had started.

◁ The Sierpiński carpet is a plane universal curve and the Sierpiński sponge is a spatial universal curve. These terms are explained in Blumenthal & Menger 1970, pp. 433 and 501. They express that in a certain purely topological sense, our shapes are respectively the most complicated curve in the plane and the most complicated curve in any higher dimensional space. ▶

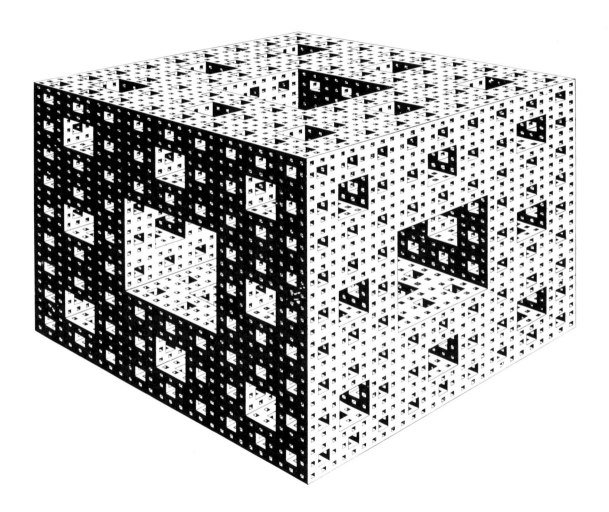

nate axes and co-centered with the cube. Then repeat the process ad infinitum. The resulting dimension is $\log 8/\log(1/r)$ and can be as close to 3 as desired, yet the set reduces to dust!

THE SIERPINSKI PASTRY SHELL, A SELF SIMILAR LAYERED SHEET

By changing the cascade without diminishing its connection with the ill-specified physical cascades of Novikov & Stewart and of Hoyle, one can reach a concentrate that is palpably different from the Sierpiński sponge.

As the "core" that is being swept out in each stage, select the middle cube of side $1/r$. In this way, no two cutouts intersect, and one is left with a kind of "pastry shell" structure made of highly involved sheets surrounding cubic holes (ready for stuffing).

The outer sides of the original cube are left unperturbed. Therefore, a view from the outside shows a blank cube, with no hint of the increasingly evanescent structure being constructed inside. If the shell is cut by a plane parallel to a side and passing through the center, the section is a Sierpiński carpet.

Again, if the cascade continues without end, the remainder set has zero volume but is nonempty. Its fractal dimension is $\log 26/\log 3 \sim 2.9656$, close to 3 and higher than the sponge's $D = 2.7268$.

Had we started by selecting for $1/r$ an integer other than 3, we could, as with the different sponges, proceed in either of two ways. We could either cut out one middle rth (this requires $1/r$ to be odd) or leave in many little cubes of side r "lining" the big one. ◁ We also get closer to Taylor-Görtler instability, with "cells" of size equal to the thickness of the original tube. ▶

For any Sierpiński pastry shell, D is above 2, which had to be the case for a shape that includes topological squares. It is striking, however, how slowly D converges to its lower limit 2, and how sensitive it is to detail that to the eye seems to be comparatively slight. In overall appearance when the holes are big and D is near 2, the Sierpiński pastry shell resembles an overly airy Emmenthaler. When D is near 3 and the holes are small, it resembles this other Swiss cheese delicacy, Appenzeller. Nevertheless, every hole is always entirely enclosed by an uninterrupted boundary split into infinitely many, infinitely thin layers of infinite density. In order to join two points situated in different holes, it is necessary to cross an infinite number of layers.

It is easy to obtain a mixture of ropes and sheets by modifying the method very slightly. For example, again starting with $r = 1/3$, it suffices to cut out a triadic cross continued by a single spike in front, so it is made up of the center cube and of only five out of its six continuations toward the sides. Then the finite mth approximation includes both slabs and rods, and the limit includes both sheets and ropes. By changing the direction of the slab every so often, one may choose to end up with punctured sheets.

"TYPICAL" SECTIONS TEND TO HAVE THE "WRONG" DIMENSION

Let us select at random a linear section of one of the above nonrandom and very regular shapes. One can show that if the intersection is nonempty, its dimension is D−2, just as in the case of random curdling fractals.

Unfortunately, this behavior is almost impossible to illustrate explicitly for either Sierpiński shape, because *every one* of the simple sections happens to belong to the exceptional set where the usual rules fail to apply.

For example, take the triadic shapes. The almost sure dimensions of their sections are conveniently written as

D = log 20/log 3−2 = log (20/9)/log 3
and   D = log (26/9)/log 3,

respectively. Nevertheless, sections by some lines parallel to the axes are straight segments, and their dimension is 1 = log 3/log 3, which is too large, while other parallel sections are Cantor sets

and their dimension log2/log3 is too small.

The variability of the value of the dimension is traceable to the excessive regularity of the original shape and is devoid of physical meaning.

CONNECTEDNESS/PERCOLATION IN RANDOM CURDLING

The remainder of this chapter concerns the topological connectedness (a) of the set in which "stuff" becomes concentrated as the result of random curdling, and (b) of its complement, which (it may be recalled) I propose to call *whey*. Connectedness is likely to be of importance in turbulence (see Mandelbrot 1976o), but following up this role would take us too far afield at this point. However, the mixture of topology and probability involved in curdling is of great intrinsic interest. Unfortunately it is a bit complex, and the reader who finds the going rough is advised to skip ahead.

We begin by rather obvious thresholds.

Let the two points P and P_0 be chosen at random in the primeval eddy of side L. Since D < E, both P and P_0 almost surely fall within the whey When D < E−1, so does the whole straight segment PP_0. A fortiori, any two points in the whey can almost surely be linked by a line entirely contained within the whey.

Secondly, let us divide the primeval eddy into cells, using a grid. When $E = 1$, we draw lines perpendicular to the coordinate axes such that their coordinates are of the form $x_\delta = a\gamma^{-b}$ (a and b being integers), and when $E > 1$, we draw analogous planes or hyperplanes. When $D < 1$, it is almost sure that every one of these hyperplanes will miss the concentrate. The latter will therefore be almost surely dustlike. A fortiori, it is almost sure that two points in the concentrate or two sides of the primeval eddy (say, top and bottom) cannot be joined continuously within this set.

The above threshold dimensions, $E - 1$ and 1, conform to the intuitive topologic aspects of dimension and are independent of γ and of any specific feature of curdling. On the contrary, the next obvious threshold to be described depends greatly upon γ and upon the fact that in the presently used definition of curdling N is nonrandom.

When $\gamma^E - N < \gamma$, the number of empty eddies left after each stage of curdling is not sufficient to build a column linking opposite sides of an earlier stage eddy. As a result, it is sure (not almost sure, but sure) that the concentrate is one piece surrounding separate alveoles or inclusions filled with whey. Further-

more, two points of the whey can be linked only when they are in the same alveole. In other words, the topology is as in the nonstochastic example of the Sierpiński pastry shell.

We now inject a finer analysis that exploits the structure of curdling to achieve thresholds that are tighter but of special validity. For two points P and P_0 in the whey to be almost surely continuously joinable within the whey, the previous condition, which we can write as $E - D > 1 = 2^{-0}$, can be replaced by $E - D > 2^{1-E}$. Similarly, for the concentrate to be almost surely dustlike, the previous condition, which can be written as $E - D > E - 1$, can be replaced by $E - D > (E-1)/2$. The expressions 2^{1-E} and $(E-1)/2$ coincide only for $E = 2$. For $E > 2$, one has $2^{1-E} < (E-1)/2$, which was to be expected since the concentrate cannot be dustlike without the whey being connected, but the contrary relationship is conceivable.

It may be observed that for $E = 3$ the above-mentioned limit implies that the concentrate is dustlike when $D < 2$, which includes the Fournier-Hoyle model and the models considered in Chapter IV. This feature explains why we did not raise the question of connectedness until now.

To establish the role of $E - D \sim 2^{1-E}$ as threshold, we construct a continuous link through the whey. We begin by observing that, the whey being an open set, P and P_0 are surrounded by closed subeddies entirely contained in it. Hence both P and P_0 can be joined within the whey to points situated in the interior of the primeval eddy of side L, and such that all their coordinates are of the form $x_\delta = a\gamma^{-b}$ (a and b are integers). These two points, in turn, can be linked to each other by a zigzag line that stays within the primeval eddy and follows eddy edges throughout. Because of E−D<1, the concentrate is almost surely hit by this line, but one can show that, because of $E−D>2^{E-1}$, one can modify this line slightly in such a way that the concentrate almost surely fails to intersect it.

◁ The argument is of course easiest to visualize when E=2. One can define the terms *edge* and *up* in such a manner that each edge lies between an "upper" and a "lower" (sub)eddy, which include respectively an upper and a lower concentrate. When neither intersects the edge, one can keep said edge as part of the continuous curve being designed to join P and P_0 within the whey. When only one intersects, the endpoints of the edge are almost surely in the whey, and it suffices to join them by a continuous line within the empty eddy. The harder case is when the upper and the lower concentrates both intersect the edge.

However, the condition $E−D>2^{E-1}$, which in the present case reads $2−D>2^{-1}$, insures that the mutual intersection of these two intersections is almost surely empty. Then a brief additional argument shows that the edge can be divided by a finite number of points into intervals that are, alternatively, exterior to the upper and to the lower concentrate. By making the link from P to P_0 proceed just above the former intervals and just below the latter, one can make it avoid the concentrate altogether. A proof along the same principle applies to E>2. ▶

When E−D>(E−1)/2, a proof of the breakup of the concentrate into dust can also be given by explicit construction. One begins with the grid of hyperplanes having coordinates of the form $x_\delta = a\gamma^{-b}$. To avoid the concentrate, one finds it is almost surely sufficient to modify these hyperplanes a bit, either up or down.

The above thresholds tell only a part of the story. Regarding the structure of the whey, the first open question concerns the first critical dimension, defined as the maximum value of the D such that P and P_0 can almost surely be linked within the whey. This maximum may well exceed $E−\gamma^{E-1}$, but at the present time I have no estimate for it.

Concerning the structure of the concentrate, the first question, again, concerns the second critical dimension, defined as the threshold value of D below which the set reduces to dust.

The next question is whether or not there exists an intermediate range of D within which the set is neither dustlike nor in one piece, but rather is made of disjointed pieces to be called *clusters*. I conjecture that such a range exists. By self similarity, the clusters must be almost surely infinite in number. Since the concentrate has no interior points, the clusters must take the form of infinitely branching curves.

In many ways, one should expect the concentrate to resemble those Koch curves constructed in Chapter II that decompose into an infinity of island coastlines. One difference is that coastlines are closed nonbranching curves, while clusters are branching (and may or may not include closed loops). Therefore, while the coastlines of individual islands were not self similar, the clusters are self similar. But of course they are expected to have a dimension below the concentrate's overall dimension D. If such is indeed the case, each cluster is negligible in comparison with the whole. Hence – if one gives a sensible meaning to "choosing two points P and P_0 at random in this set" – such points are expected to almost surely fall within different clusters. As a matter of fact, a point

P so chosen at random will almost surely fall in a cluster of diameter smaller than any prescribed threshold.

◁ This last proposition can be proven directly as follows. When

$$D < E - 1 + \log(\gamma - 1)/\log\gamma,$$

one can show that a point of concentration is almost certain to lie in an infinite number of mutually imbedded boxes of sizes tending to zero and such that their boundary lies entirely in the whey. It follows either that the concentrate is dustlike or that noninfinitesimal clusters exist but are such that the special Hausdorff measure which is positive and finite for the concentrate vanishes for each individual cluster. ▶

As a result of the decomposition of the concentrate into clusters, the method used to assess the degree of connectedness of the whey is inapplicable to the concentrate. To study the latter, it is best to inject the notion of percolation, which, like the present usage of the term cluster, first arose in a different problem of physics, to be examined in Chapter X. Curdling will be said to percolate if at least one cluster of the concentrate spans the primeval eddy of side L. Such a cluster will be called a *percolator,* and the degree of connectedness of the concentrate can be studied through the probability of percolation. Thus far, the best we could do in this direction has involved brute force computation, using the kind of simulation illustrated on Plates 156 through 159.

It may be observed that a percolator is a union of self similar clusters but is not itself self similar. Indeed, its first-order parts – defined as its intersections with the γ^E subeddies of the primeval eddy – are special percolators constrained to span more than a subeddy. The second-order parts spread even farther, and the very small parts are portions of an infinite percolator. As a consequence, the similarity dimension is not defined for the percolator. The Hausdorff dimension, on the other hand, being determined by local properties, is likely to be defined. In particular, it is defined if – as we shall suppose is the case – there is a positive probability for a group of clusters in the percolator to extend to infinity in all directions. The dimension would be the same for finite and for infinite percolators.

CANONICAL CURDLING

The definition of curdling used thus far follows Novikov & Stewart in assuming not only that $1/r$ is an integer but that the number of curds at each stage is a fixed integer N. Randomness enters only because the curds' positions are chosen uniformly from all the combinations of N out of $1/r^3$ objects.

Such curdling, which Mandelbrot 1974f calls microcanonical, is often inconvenient. For example, it binds D to a finite set of possible values. As N goes down through the integers from $1/r^3$ to $1/r^3-1$ and so on to 1, D goes through the discrete sequence

$$3, \log(1/r^3-1)/\log(1/r),$$

and so on.

An alternative procedure that, among others, frees D from this constraint consists in letting N itself become random. The simplest is to attach to each subeddy of any order a random number U between 0 and 1, then to select a probability threshold p. Whenever $U > p$, the subeddy is made to "die off" as whey, together with all the subsubeddies it includes. When $U < p$, the subeddy "survives" as curd. The resulting process is a familiar one, namely, a birth and death process (of the kind already used in this chapter to study the fractal dimensions of sections). Therefore, we may recall that when $p < 1/r^3$, it is almost sure that the whole process will die off. Hence $D = 0$. Otherwise, there is a nonvanishing probability for the process to converge to a fractal having a dimension given by the formula

$$D = \log (p/r^3)/\log (1/r)$$
$$= 3 - \log p/\log r.$$

This method was introduced in Mandelbrot 1974f, under the term *canonical curdling*. In the present context, its basic advantage is that it allows D to lie anywhere between 0 and 3, and that it makes it possible to follow the characterization of concentrate, whey, and all other sets of interest as continuously varying functions of dimension. It suffices to hold all the above-mentioned random numbers U fixed, while the threshold p decreases continuously from 1 to 0.

Canonical curdling involves two different critical dimensions. The first is the critical dimension $\geq E - 2^{-1}$ below which the probability of percolation is 0 and the concentrate is less than "linelike." The second is the critical dimension $\geq (E+1)/2$, below which the probability of junction of P and P_0 within the whey is 1 and the concentrate is less than "surfacelike."

The brute-force simulations referred to in the preceding section have actually been performed on canonical curdling.

CHAPTER VII

Meteorites, Moon Craters, and Soap

Chapter IV concluded with a construction in which an infinite number of random segments were cut out from the real line, and the remainder was studied. The logical development of this model now leads us to consider concrete problems that involve cutouts from the plane and from 3-space. In order to be isotropic, the generalized cutouts must take the form of circles and spheres. The practical problems we shall consider will take us all over the landscape of science, from meteorites to moon craters and on to the structure of "smectics A," a soap-like phase of liquid crystals.

However, circles and spheres are a bit difficult to manipulate. The main complication is an interesting one: the property of self similarity is further weakened in this case and becomes merely approximate, hence the corresponding path toward the concept of dimension becomes

impassable. One is reduced to the all-weather (unfortunately ill-lit) approach due to Hausdorff. No such problem arises when the planar generalization of cutouts involves triangles, which is why triangles sneak into the discussion. Similarly, no problem arises with squares and cubes, especially when they are constrained – as in the original Cantor cascade – to belong to a hierarchy of mutually embedded eddies.

Therefore, before we examine circles, it is useful to generalize the virtual cutouts that (in Chapter IV) had constituted the last stage before the introduction of an easily manageable form of randomness. In the planar version of the process of curdling within a grid, which is studied in Chapters V and VI, the first cascade stage had consisted in marking N out of $\gamma = 1/r^2$ squares and keeping them as curds. Alternatively, one may say that the first stage cut out $1/r^2 - N$ squares. The next stage cut out second-order squares numbering

$$(1/r^2)(1/r^2 - N),$$

it being understood that this last number includes (a) $N(1/r^2 - N)$ real new cutouts and (b) $(1/r^2 - N)^2$ virtual cutouts, each of which attempts to eliminate again something that had already been eliminated in the first stage, and so on.

Counting both real and virtual cutouts, we find that the number of those having an area in excess of s is proportional to $1/s$. The corresponding result relative to curdling in 3-space is that the number of cutouts of a volume in excess of v is proportional to $1/v$. Incidentally, looking with half-closed eyes, the cutouts of Plates 156 through 159 begin to look roundish.

Now we proceed to a direct study of round cutouts. To be concrete, we shall start with Moon craters, jump to Earth-impacting meteorites, and go on to liquid crystals.

LUNAR CRATERS

The origin of the craters of Earth's Moon, of Mars, and of other planets or satellites is controversial. The term itself implies a volcanic origin, but an alterna-tive theory that traces them to the impact of meteorites is more widely held.

It views the Moon as having swept, over large periods of time, through the same meteorite cloud as Earth. The larger the meteorite, the larger and deeper the hole resulting from its impact. However, a later, heavier impact may wipe out the trace of several previous meteorite craters, while a small meteorite may "dent" the rim of a large older crater. As for the sizes, there is solid empirical evidence that at the moment of meteorite impact the areas follow a hyperbolic distribution with an exponent γ near 1, hence the diameters are hyperbolic with an exponent 2γ near 2. This evidence is discussed in Marcus 1964, which has prompted the present application, and in Arthur 1954. A convenient survey, with references, is Hartmann 1977.

To simplify we shall not reason in terms of the surface of a sphere but shall approximate lunar craters by circular cutouts in the plane. Two related problems arise. One must determine whether or not any part of the Moon remains perpetually untouched by craters. If the reply is affirmative, one must characterize the remainder's geometrical structure.

If the Moon went on perpetually scooping up meteorites from a statistically invariant environment, the lunar surface would be covered again and again ad infinitum. (This is obvious but will be proved in a moment anyhow.) However, craters are wiped clean every so often, say by volcanic lava, so the set they fail to cover may be nontrivial. To study it, we shall assume that after a time the number of craters with an area exceeding s km^2 and such that their centers are located within a square of 1 km^2 takes the form V s$^{-\gamma}$. Thus V measures the time since the last attrition of the craters on an intrinsic clock. Doubling V also doubles the number of craters for each area.

If theorems and conjectures are combined, the answers to the two problems raised above are as follows.

First, there are two cases that happen not to be encountered in reality. When the exponent γ of the law of crater areas is larger than 1, the Moon is almost certainly oversaturated with small craters. The result of meteorite bombardment is at least one crater for every point of the Moon's surface – regardless of the craters' lifespan.

When the exponent is smaller than 1, the probability for each square of the Moon's surface to remain craterless is positive. The resulting pocked surface resembles a slice of Swiss cheese. The song that tells English children the Moon is made of green cheese errs in regard to color. The greater the value of γ, the smaller the number of small holes, and the more "chunky" the resulting cheese.

However, regardless of the value of γ, the cheese is of positive area, hence it is a (nonself similar) set of dimension 2. On the other hand, I have no doubt that its topological dimension is 1, hence it is a fractal, a variant of the networks examined in Chapter II.

Let us now consider the case of interest to both the mathematician and the selenologist, that which corresponds to $\gamma = 1$. When the lifespan V of the craters exceeds a certain constant V_o, we have the same result as when $\gamma > 1$. The surface is oversaturated, every point being almost certainly covered by at least one crater, ordinarily a small one. In particular this would be the case if the Moon's surface were never wiped clean and were to continue scooping up meteorites endlessly. When $V < V_o$, the uncovered set is not empty but contains no square, no matter how small. It is a fractal of area equal to zero, and its fractal dimension D lies between 0 and 2. As V increases, the surface becomes increasingly saturated and D decreases and reaches 0 for $V = V_0$.

When V is very small, we deal with a slice of cheese that is almost entirely pierced by very small pin holes that overlap hardly at all. The lover of Swiss cheese may join me in recognizing here a wildly extreme extrapolation of the structure of Appenzeller. When V in-

Plates 178 and 179

SMALLISH ROUND CUTOUTS, IN WHITE, AND RANDOM SLICES
OF "SWISS CHEESE" (DIMENSIONS D=1.9900 and D=1.9000)

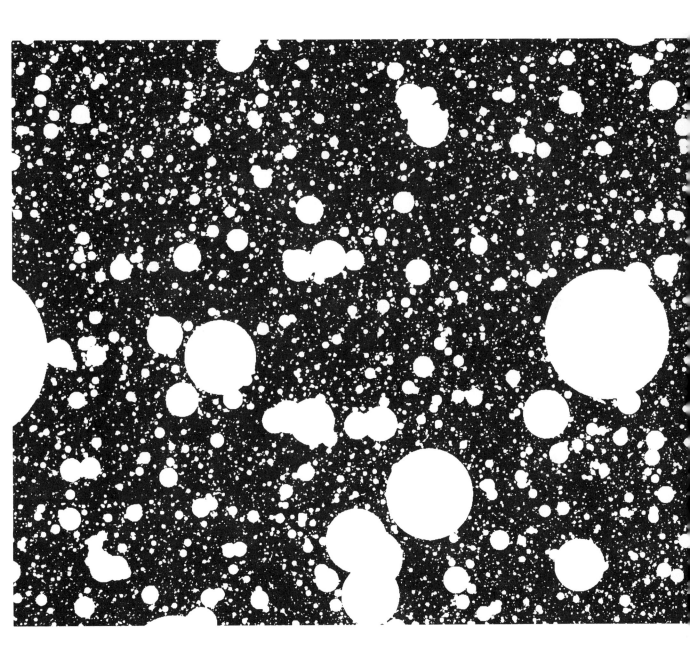

The holes or cutouts are black circular discs. Their centers are distributed at random on the plane. For the disc of rank ρ by decreasing size, the area is $K(2-D)/\rho$, with K a numerical constant. Plate 178 shows a sort of Appenzeller of dimension $D = 1.9900$ and Plate 179 a sort of Emmenthaler of dimension $D = 1.9000$.

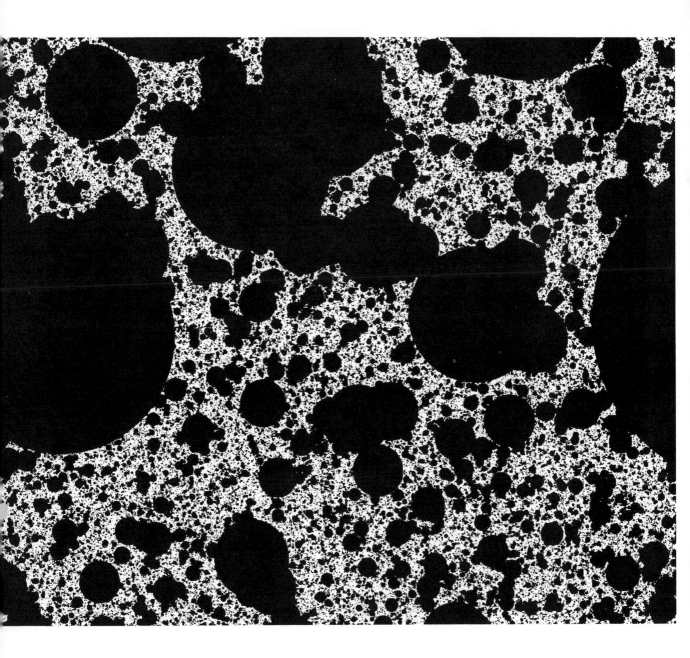

Plates 180 and 181

LARGER ROUND CUTOUTS, IN BLACK, AND RANDOM FORKED
WHITE THREADS (DIMENSIONS D=1.7500 AND D=1.5000)

The construction proceeds as in Plates 178 and 179, but the discs are bigger so hardly anything is left out.

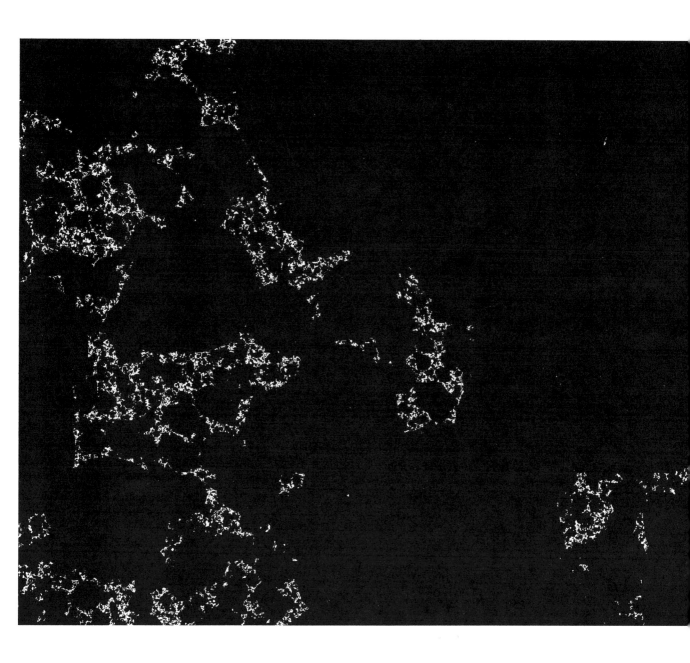

The scales of Plates 178, 179, and 181 differ a bit from each other, but the growth of the cutouts' sizes is easy to follow anyhow. The Pattern shown as the Frontispiece is related to this drift in scale. It resulted from an accidental superposition of Plates 178 through 181 and of other patterns in the same sequence.

creases, we turn progressively to a wildly extrapolated Emmenthaler that also becomes evanescent because of the large holes that often have parts in common.

The dependence of D upon V is illustrated by Plates 178 through 181.

THE PROBLEM OF CONNECTEDNESS

It is inevitable that one should inquire about the degree of connectedness of the set that round random cutouts leave uncovered. The discussion begins, as it did in Chapter VI where we dealt with the concentrate of curdling, with two easy thresholds. When D < 1, every line grid drawn in advance is almost surely entirely covered by cutouts, and a fortiori the uncovered set is dustlike. When D < E−1, the uncovered set is almost surely not intersected by the line joining any two points P and P', hence the union of all the cutouts (the *whey*) is almost surely connected.

On the other hand, the other threshold dimensions applicable to curdling in square grids have no counterpart in the present problem. Since these thresholds were intimately linked to specific features of curdling in grids and to its nonisotropy, their failure to have a counterpart shows the cutouts model to be more intrinsic than curdling. However, the cutouts model also involves two critical dimensions relative respectively to the connectedness of the whey and to percolation of the uncovered set. Thus far all the information I have about them is due to brute-force simulation, and this is not the place to dwell on it.

METEORITES

The distribution of meteorite masses has been studied carefully, for example in Hawkins 1964. Over successive ranges of sizes, it is ruled by hyperbolic distributions with varying exponents. In the mid-range of sizes, meteorites are made of stone, and 1 km^3 in space contains roughly $P(v) = \mu v^{-\gamma} = 10^{-25}/v$ meteorites of volume exceeding v km^3.

◁ This claim is ordinarily expressed differently, using the following very mixed units. During each year, each km^2 of Earth's surface is on the average host to $0.186/m$ meteorites of mass above m grams. Their average density being 3.4 g cm^{-3}, this relation boils down, in more consistent units, to $5.4 \cdot 10^{-17}/v$ meteorites of volume exceeding v km^3. Moreover, Earth moves on by roughly 1 km during 10^{-9} years – the inverse of the order of magnitude of Earth's trajectory around the Sun in km. Hence, using completely consistent units, and keeping to orders of magnitude so that 5.4 becomes 10, we find that while Earth moves on by 1 km in space, each km^2 of Earth's surface is host to $10^{-25}/v$ meteorites of volume exceeding v km^3. Assuming that the meteorites impacting Earth as it sweeps through space are a representative sample of the meteorites' distribution in space, we obtain the result that has been asserted.

◁ This $10^{-25}/v$ law is reminiscent of the Vs^{-1} law encountered in the preceding section, but in fact they refer to different ranges. ▶

Meteorite sizes follow a different law when v is very small. Nevertheless there is something to be gained in investigating what would happen if $P(v) = 10^{-25}/v$ held down to v = 0 and if meteorites could overlap. Adding the assumption – this one is quite innocuous – that meteorites are spherical, the remainder set is easy to investigate. The sections of meteorites made by straight lines randomly thrown in space are rectilinear cutouts, and it can be shown that the number of such segments centered within 1 km and of length exceeding u km is $10^{-25}/u$. (There is also a numerical factor of the order of magnitude of 1, but it is unimportant in this context.) Hence a result in Chapter IV shows the dimension of the remainder to be $1 - 10^{-25}$. Adding 2 when we go back from the linear sections to the full shape, we find $3 - D = 10^{-25}$.

This result is obviously inane, since it implies in particular that meteorites nearly fill space, even after one has allowed for overlap. The key is of course that the very small meteorites obey a different distribution. Nevertheless, the codimension $3 - D = 10^{-25}$ deserves just another glance. Let us assume in a first approximation that the $10^{-25}/v$ relationship holds down to a positive cutoff $\eta > 0$ and that there is no meteorite of smaller size. The argument we have sketched asserts that if one could actually perform the pass to the limit $\eta \to 0$, the set outside of all meteorites would converge to a limit set of dimension $D = 3 - 10^{-25}$.

However, this limit set would be attained so extraordinarily slowly as to have no practical importance whatsoever. Similarly, a theory that must involve an interrupted passage to an impossible limit is most inelegant. A different way out, which also happens to be more realistic, consists in observing that the distribution of meteorite volumes is better fitted by a hyperbolic law $\mu v^{-\gamma}$, with γ slightly below 1. Thus the integral of $P(v)$ near v = 0 is convergent, meaning that the total meteorite volume is finite and that adjustment of μ can make it as large or as small as desired. The remainder set is no longer self similar. Its fractal dimension is $D = 3$ and its topological dimension is doubtless 2. Furthermore, a random placement of the meteorites produces only a limited overlap, and we are led to wonder which values of γ and μ are compatible with a total absence of overlap. This question is tackled in the next section.

NONOVERLAPPING
SPHERICAL CUTOUTS IN SPACE

We shall describe two very different problems that concern packing of E-space by nonrandom balls that cannot overlap. In the first problem, which was posed and solved in Lieb & Lebowitz 1972, packing is slow and inefficient, and the dimension of the remainder set outside of the balls is close to E. In the other problem, which refers to liquid crystals (it was first described to me by P. G. deGennes), packing is optimally fast and efficient, and the dimension of the remainder is much smaller. The examples involve no randomness.

SLOW PACKING OF E-SPACE
BY NONOVERLAPPING SPHERES

Cantor's original method, described in Chapter IV, can be viewed as packing the complement of the Cantor set by self similar nonoverlapping intermissions. The following generalization, which is especially simple at the cost of being very far from optimal, is found in Lieb &

Lebowitz 1972. ◁ The remainder set had no interest for these authors, hence their contribution to fractal geometry was unwitting and incidental. ▶

Since we expect self similar packing of E-space to be possible, we expect from self similarity that the radii of the packing spheres must be of the form $\rho_k = \rho_0 r^k$, with $r < 1$, and that the per-unit-volume number of spheres of radius ρ_K must be of the form $n_k = n_0 \nu^k$, with ν an integer. It is shown by Lieb & Lebowitz that a first constraint on ν and $\vee$ is that $\nu = (1-r)r^{-E}$. The remainder set is not quite self similar, but an argument we shall skip shows the dimension to be

$$D = \log\nu/\log(1/r) = E - \log(1-r)/\log r.$$

Further, Lieb & Lebowitz derived the following nonobvious upper bounds on r:
for $E = 1$, $r \leq 1/3$; for $E = 2$, $r \leq 1/10$;
for $E = 3$, $r \leq 1/27$; as $E \to \infty$, $r \to 0$.
Hence the codimension has the following upper bounds:
for $E = 1$, $E - D \leq 1 - \log 2/\log 3$;
for $E = 2$, $E - D \leq 1 - \log 9/\log 10$;
for $E = 3$, $E - D \leq 1 - \log 26/\log 27$;

for $E \to \infty$, the bound tends toward E, and the covering can be said to tend toward vanishing efficiency.

The above results do not imply in the least that nonoverlapping packing could not proceed more rapidly. They only describe one set of conditions under which a primitive and mechanical method of packing is not penalized by a resulting risk of overlap. The procedure is to spread around the biggest spheres of radius ρ_1 on a lattice of side $2\rho_1$; the vertices of a lattice of side $2\rho_2$ that lie outside of the biggest spheres turns out to be numerous enough to serve as centers for the next smaller spheres, and so on.

In the case of the line $(E=1)$, we saw that the maximum r is $1/3$. When one places the centers of kth-order spheres, which in this case are line segments, in the middle of $(k-1)$st-order intermissions, one is simply brought back to the original triadic set of Cantor! The nontriadic Cantor sets demonstrate that one dimensional covering can be tighter, leaving a remainder of arbitrarily low dimension. However, the grid used in those sets has a richer structure than the grid used by Lieb & Lebowitz.

FASTEST PACKING OF THE PLANE BY NONOVERLAPPING CIRCLES

We now outline the part that packing by circles and the concept of fractal dimension play in the description of a category of "liquid crystals." Since they are relatively little known, we may stop to describe them by paraphrasing Bragg 1934. These are beautiful and mysterious substances that are liquid in their mobility and crystalline in their optical behavior. Their molecules are relatively complicated structures, lengthy and chainlike. Some liquid crystal phases are called *smectic* from the Greek word σμηγμα signifying soap, because they constitute a model of a certain kind of soaplike organic systems.

In discussing the structure of what is called smectic A, we encounter a new application of a very ancient problem of geometry, one that goes back to a Greek mathematician of the Alexandrine school circa 200 B.C. and close follower of Euclid, Apollonius of Perga. Because of these roots, the problem is easy to state, but it still remains open.

In addition, we shall cast a first glance toward more general perspectives relative to one of the most active areas of physics, the theory of *critical points*. The best-known among them is the "point" on a temperature-pressure diagram that describes the physical conditions under which solid, liquid, and gaseous phases can coexist at equilibrium in a single physical system. Physicists have recently established that if a physical system is in the neighborhood of a critical point, its behavior is governed by several "critical" exponents. By combining analysis with computer simulations, it is possible to obtain the numerical values of many such exponents.

Their conceptual character, however, remains obscure. The example of soaplike liquid crystals will show that at least one critical exponent is a fractal dimension. This interpretation suggests that the same holds true for other critical exponents. ◁ A further argument in this direction will be obtained in Chapter VIII, when we study Bernoulli percolation. ▶

PACKING A TRIANGLE WITH INVERTED TRIANGLES

As a preliminary, let us begin with a construction that is even closer to the spirit of those we encountered earlier in this Essay (Plate 57). We start with a closed equilateral triangle (including its boundary) such that the peak is pointed upward and the base has length 1 and is horizontal. We try to cover this triangle as best we can by means of open equilateral triangles for which the bases are also horizontal, while their peaks are pointed downward. It so happens that the optimal covering consists of filling the central quarter of the initial triangle with a triangle of side 0.5, and then doing the same for the three remaining quarters. The set of points that will never be covered is the Sierpiński gasket mentioned in the caption of Plate 57. Its area is zero and it is self similar with the fractal dimension $\log 3/\log 2$. The complete proof is given in Eggleston 1953.

"APOLLONIAN" MODEL OF SOAP

In its smectic, or soaplike phase, a liquid crystal is made of molecules that are arranged side by side like corn in a field, the thickness of the layer being the molecules' length. The resulting layers or sheets are very flexible and very strong and tend to straighten out when bent and then released. At low temperatures, they are piled regularly, like the leaves of a book, and form a solid crystal. When temperatures rise, however, the sheets cease to be bonded together, and they become able to slide easily on each other. Each layer constitutes a two-dimensional liquid.

Of special interest is the focal conics structure, one aspect of which is that a block of liquid crystal can be separated into two sets of pyramids, half of which have their bases on one of two opposite faces and vertices on the other. Within each pyramid, liquid crystal layers are folded to form very pointed cones. All the cones have the same peak and are approximately perpendicular to the plane. As a result, their bases are circles. Their minimum radius η is linked to the thickness of the above-mentioned liquid layers. Starting with a simple volume other than a cone, perhaps a square-based pyramid, let us fill it with soap in an optimal manner. To do so, the circles that constitute the bases of the cones

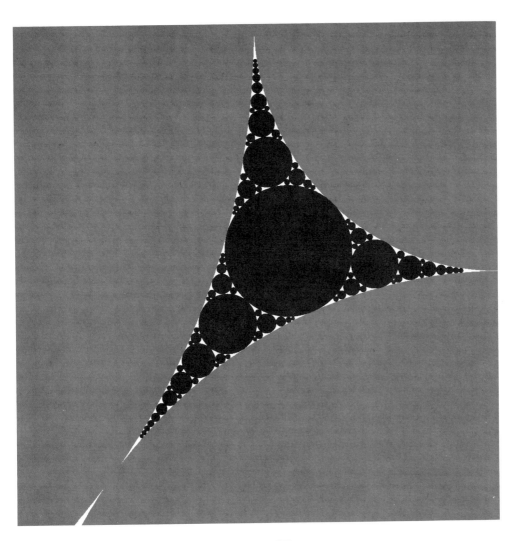

Plate 187

APOLLONIAN GASKET
(DIMENSION ABOUT D~1.306951)

To construct a circle tangent to three given circles constitutes one of the geometric problems that tradition attributes to Apollonius of Perga. Begin with three gray circles tangent two by two, forming a circular triangle, and let the above construction be iterated to infinity. The black Apollonian circles (less their circumferences) will "pack" our triangle, in the sense that almost every point of it will eventually be covered. The remainder will be called *Apollonian gasket*. Its surface measure vanishes, while its linear measure, defined as the sum of the circumferences of the packing circles, is infinite. Thus the shape of the Apollonian gasket lies somewhere between a line and a surface. It enters in the theory of Smectic A liquid crystals.

must be distributed all over the square that constitutes the pyramid's base.

It can be shown that in the distribution that corresponds to equilibrium, one begins by placing in the base a circle of maximum radius. Then another circle with as large a radius as possible is placed within each of the four remaining pieces (see Plate 187), and so on and so forth. If it were possible to proceed without end, we would achieve exact Apollonian packing. In addition, it is postulated that the circles are open, that is, do not include their perimeters. The points that do not fall within any of the circles form a set of vanishing area and infinite length, which may be called an Apollonian gasket.

Our construction bears a certain resemblance to the preliminary problem regarding the packing of triangles, but its order of difficulty is higher due to the fact that our remainder is no longer self similar. Still, the Hausdorff Besicovitch definition of D (as an exponent used to define measure) continues to apply to the present case. The precise derivation of the value of the D has proven surprisingly difficult, and has given rise to quite an extensive literature. Thus far (Boyd 1973), the best one can say is that

$$1.300197 < D < 1.314534 ,$$

but the numerical experiments in Melzak 1969 suggest the more precise value

$$D = 1.306951 .$$

In the present context, D is neither a measure of irregularity nor of fragmentation but is definitely close in spirit to the latter. When, for example, the Apollonian packing is "truncated" by prohibiting the use of circles of radius smaller than η, the remaining interstices have a perimeter proportional to η^{1-D} and a surface proportional to η^{2-D}.

To return to physics, the properties of a soap thus modeled do in fact depend upon the surface and upon the perimeter of the sum of its interstices, with the link being effected through the exponent D. In other words, the properties of the soap with which we are concerned are most easily expressed through a kind of photographic "negative," namely, the gasket that the molecules of that soap are unable to penetrate. Details of the physics are found in Bidaux et al. 1973.

◁ How sad that this critical exponent should fail to take a "simple" form, like 4/3. The facile assumption that critical exponents tend to be simple becomes hard to entertain. ▶

CHAPTER VIII

Uses of Self Constrained Chance

Chapters IV through VII attempted to demonstrate that certain phenomena respond well to models based on surprisingly unstructured random processes, in which the effects of elementary random events are statistically independent and are merely allowed to accumulate in an environment one may describe as inert. The basic trick or pattern has been pioneered by the mathematical Brownian model of physical Brownian motion. The resulting model is so effective that refinements (for example, the Gauss-Markov process of Uhlenbeck & Ornstein) did not become necessary until a full generation after the foundation of the theory. For seacoasts and rivers, on the contrary, an analogous approach could not survive for one minute.

Therefore, we should be grateful for any instance in which mere accumulation of chance effects suffices in a first approximation. In Chapters III, V, and VI, independent effects were accumulated to obtain Brownian motions, Rayleigh or Lévy flights, and Hoyle or Novikov & Stewart curds. In Chapters IV and VII, independent random cutouts were excised from the real line, the plane, or space. In each instance, the earlier members of the primary series of events only entered in a very global fashion or not at all. The underlying space did play a role, but only through the value of its Euclidean dimension E.

It is fortunate that this old technique could find new and worthwhile applications, because we noted in Chapter III that it fails to apply to our initial problem of describing the shapes of coastlines and of rivers, and because the failure is due to an elementary (that is, fundamental) *geometrical* reason: neither shape is allowed to cross itself. Consequently, each piece of a coastline may well affect other pieces, even distant ones. This is why the random Koch curves exemplified by Plates 50 and 85 require such a great deal of deliberate structure. While direct effects resulting from the constraint of self avoidance propagate from neighbor to neighbor, indirect effects may well propagate to infinity, and in the last analysis local constraints are capable of creating very global structures. In such a "live" space, in which local phenomena propagate far and wide, the importance of the actual value of the Euclidean E is enormous.

Models involving self constrained chance will be studied in this chapter, and also in Chapter IX, which is devoted to continuous processes related to Brownian motion.

SELF AVOIDING RANDOM WALK AND POLYMER GEOMETRY

In Chapter III, we alluded to the self avoiding random walk. This well-known process, by definition, goes forward with no regard to its past positions, except that it is prohibited from passing through a point more than once and from entering a region from which it will find it impossible to exit. We return to it, now that "ordinary" Brownian motion has become thoroughly familiar. Then we describe several different combinations of such walks: self avoiding polygons, which are walks overlapping only at their ends, and walks strung together to form a river network.

Given a lattice of points on a plane or in space (for instance, the rectangular lattice of points such that their coordinates are integers), let us consider a random walk that can go from one of these points to any neighboring point that was not yet visited and does not lead into a dead end. All the permissible directions are given equal probabilities.

On the straight line, such a motion necessarily continues in either one direction or the other but can never reverse itself, so this case is degenerate and easy.

In the non-degenerate cases of the plane and of space, on the contrary, the problem has turned out to be very difficult, in fact so difficult that to date no analytical study has been successful. Yet its practical importance in the study of macromolecules (polymers) is such that it has become the object of very detailed computer simulations. The result that interests us most is due to Domb, a good reference being found in Barber & Ninham 1970.

After $n \gg 1$ steps, the quadratic average of displacement R_n is of the order of magnitude of n raised to a power which Domb denotes by γ, but which we shall denote by $2/D$.

This result strongly suggests that

within a circle or sphere with radius R surrounding a site, one may expect the number of other sites to be approximately R^D. This is the first of several reasons for considering D a fractal dimension. Its value on a straight line is (trivially) $D=1$, which happens to be identical to $E=1$. In the plane, computer simulations yield the value $D=1.33$, which is far from the fractal dimension $D_B=2$ of Brownian motion. In 3-space, simulations yield $D=1.67$, again short of $D_B=2$. Only in the limit $E \to \infty$ does a limit argument due to Kesten establish that $D \to 2$.

◁ The values of D_B are close to $(E+2)/3$, which leads one to ask whether $D=2$ is an asymptotic limit or is already reached for $E=4$. ▶

The above values of D happen, moreover, to be quite sensitive to the details of the underlying assumption. If a polymer in 3-space is made of two different types of atoms, so that the walk is not constrained to a lattice, Windwer found that $D=2/1.29$, which he claims is significantly below Domb's value of $D=2/1.15$. In a polymer dissolved in a reacting solvent, the imbedding space is even less inert, in particular, D becomes interaction-dependent. The Θ-point is characterized as yielding $D=2$. In good solvents $D<2$ and D decreases with the solvent's quality; in particular, a perfect solvent yields $D=2/1.57$ if $E=2$ and $D=2/1.37$ if $E=3$. Even the worst solvent in 2-space cannot conceivably lead beyond $D=2$, but a bad solvent in 3-space yields $D>2$. Coagulation and phase separation set in, and a single chain is no longer a satisfactory model.

◁ The preceding paragraphs do nothing but transcribe known results into fractal terminology, but somehow I feel that this transcription has already helped to clarify their statement. It suggests that the elastic properties of polymers are related to fractal effects at the molecular level. ▶

It is interesting to sidetrack here to examine whether a self avoiding random walk satisfies the cosmological principle. Its first few steps do not. Eventually, however, a conditionally cosmographic steady state seems certain to be achieved (but I do not know of a proof).

THE JOSEPH AND NOAH EFFECTS

As an interlude, it is important to compare the behavior of M(R) in a Lévy flight and in a self avoiding random walk. The analytical form is the same, but the underlying reasons are radically different. Indeed, Lévy flight assumes the jumps to be mutually independent, and the fact that $D<2$ is due to the chance presence of very large values, those that separate distinct masses or clusters. Here the jumps are of fixed length, and $D < 2$ is due to the avoidance of previously occupied positions making the movement persistent.

In unpublished lectures on the topic of *New Forms of Chance in Science* (sketched in part in Mandelbrot & Wallis 1968 and in Mandelbrot 1973f), I suggested that these two dimension-decreasing symptoms be christened Noah and Joseph Effects, respectively, honoring two Biblical figures, the hero of the flood and the hero of the seven fat and seven lean years. Thus, Chapters IV and V are devoted to Noah Effects, while the study of the Joseph Effect led to the fractional Brownian models examined in Chapter IX.

◁ Discrete and continuous models often develop independently of each other. When I recognized (Mandelbrot 1965h) that Brownian motions, when made persistent, can account for the Joseph Effect, I was not aware that self avoiding walks were being studied by specialists in polymers. The correlation postulated there is a straightforward formal generalization of the correlation that Domb & Hioe 1969 were to observe between distant steps of a self avoiding walk. ▶

SELF AVOIDING POLYGONS

Let a polygon be chosen at random among all the n-sided self avoiding polygons. Sometimes it will be squarish, with an area about $n^{1/2}/4$. Sometimes it will be spindly and skinny, with an area about $n/2$. If one averages by giving to each polygon the same weight, numerical simulations indicate that the average area is about $n^{2/D}$ with $D \sim 1.33$ (Hiley & Sykes 1961). Hence a polygon behaves like two self avoiding random walks strung together.

The value of $D \sim 1.33$ is of direct interest from the viewpoint of the topic of Chapter II. It seems to qualify self avoiding polygons as models of coastlines that are somewhat more irregular than the average. We may rejoice at this finding, but only in moderation.

First of all, there is the problem of islands. The concept of dimension should at the same time account for the coastlines' irregularity, their fragmentation, and the relationship between irregularity and fragmentation. On the other hand, self avoiding polygons do not involve offshore islands.

Secondly, we know that no single value of D could suffice for all of Earth's coastlines.

Last but not least, when one goes to

the limit of n→∞ in order to increase the range within which our model is self similar, the intuitive meaning of self avoidance changes qualitatively. Suppose indeed that a very large self avoiding random walk is scaled down so that the lattice step decreases from 1 to a small value η. It is clear that two points of the walk that used to be distant by 1 will converge to the same limit point. The limit of the walk is no longer self avoiding, merely nonself intersecting, and it may involve self contacts. I do not believe we wish to see such points in a model of a coastline.

RIVERS' DEPARTURE
FROM A STRAIGHT COURSE

Toward the end of Chapter II, we mentioned Hack's empirical data to the effect that it is typical for a river's length to increase like the power of $D/2$ of its drainage area. If rivers had tended to flow straight through round drainage areas, stream lengths would be proportional to the square root of the drainage area, so the value of D would be $D = 1$. In fact, one finds D to be of the order of 1.2 to 1.3. In response, Chapter II described a model based on a plane-filling curve that yielded $D = 1.1291$.

In a very different stochastic attempt to explain the Hack effect, Leopold & Langbein 1962 make computer simulations of the development of drainage patterns in regions of uniform lithology. They propose a a two-dimensional random walk game in which both source locations and directions of propagation are chosen by chance. The game is played on a rectangular area divided into squares. The source of the first stream is a square chosen at random, and a channel is generated by successive moves into adjacent squares. On the first move, each of the four possible directions is equally probable. On subsequent moves the stream is not allowed to reverse itself, but each of the other possible moves has an equal chance. The first stream continues until it goes off the boundary area, so it is nearly a self avoiding random walk. Then a second source is chosen at random and another stream generated as before, except that it terminates either by going off the boundary or by joining the first stream. One often observes what might be called the Mississippi-Missouri syndrome, wherein the second stream, the Missouri, is longer than the portion of the Mississippi situated above their junction. It is not even excluded that the junction should occur at the first stream's source. The same procedure continues until all squares are filled in. In addition to these general rules, various rather arbitrary decisions avoid loops, snags, and inconsistencies. Under these conditions, it is possible for one or more streams to flow into a unit area, but only one can flow out.

Such random selection of directions turns out to generate a stream network having striking similarities to natural drainage nets. Divides develop, and the streams join so as to create rivers of increasing size. Leopold & Langbein argue that the random pattern represents a most probable network in a structurally and lithologically homogeneous region.

Computer simulation indicates that in this random walk model, river length increases as the 0.64 power of the drainage area. Hence $D \sim 1.28$. The closeness of this value to Domb's $D \sim 1.33$ may or may not be coincidental. It may turn out that the difference is nothing but a statistical sample variation due to the fact that the Leopold & Langbein simulations were less extensive than Domb's. But I am tempted to view the discrepancy as genuine, because the cumulative interference from other streams seems more accentuated than the interference from a single motion's past values, resulting in a smaller D.

Compared to actual maps, Leopold & Langbein rivers wander excessively. To avoid this defect, numerous improved models have been proposed. The model due to Howard 1971 postulates headward growth, from mouths placed on the boundary of a square toward sources placed inside. A large number of rivers succeed or fail in growing up together (actually, each grows up in turn according to various perfectly artificial schemes) and therefore each interacts very strongly with its own and other rivers' downstream portions. This procedure results in rivers that are markedly straighter than in the Leopold & Langbein scheme and hence presumably involve a smaller D.

Thus far, the study of random networks such as those of Leopold & Langbein and of Howard has not yet gone beyond limited computer simulations. It is a shame, and I wish to broadcast the great interest of these problems to mathematicians. The fact that the ordinary self avoiding random walks have proven extremely resistant to analysis should serve to warn off seekers of easy problems with a large payoff, but the present variant might be easier.

To repeat: the mathematical difficulties encountered in the past studies of self avoiding walks have basic roots. Similarly, a local change in a Leopold & Langbein network may result in a big river breaking through a dividing line into the neighboring basin. Such problems are global to an extreme degree. One would be happy to measure the intensity of the long-term interaction, so to speak, macroscopically. Naturally, I expect this parameter to be a fractal dimension.

FRACTALS IN PHYSICS

Originally this Essay involved only one phenomenon (a grand one) from the physics practiced by physicists: Brownian motion itself. Then another example or two were identified, until it seems today that fresh ones are revealed by every stone one chooses to turn. Mentions of a few are scattered around. Others are thrown together in the heterogeneous remainder of this chapter.

BERNOULLI PERCOLATION

When we discussed percolation in Chapter VI, we were working in a highly nonclassical environment. We shall now turn to it in its classical context, which — to avoid confusion — will be called Bernoulli percolation. (For surveys of the extensive literature on this topic, see Domb & Green 1972, especially a chapter by Essam; Shante & Kirkpatrick

1971; and Kirkpatrick 1973. For a semi-popular discussion, see deGennes 1976.)

Consider a slab bounded by two planes and crisscrossed by "capillaries" that form the lines of a regular lattice and intersect at points called sites. The flow of liquid from the top of the slab to its bottom is perturbed by plugging some sites that are selected by a Bernoulli process; this term expresses that sites are plugged (or unplugged) independently of one another, with a given probability $1-p$ (or p). Obviously, the liquid will be able to pass through if and only if at least one path through the slab is unobstructed.

The basic problem concerns the case in which the slab is many times thicker than the lattice separation. It was posed and investigated by Hammersley, who showed that the probability of percolation depends drastically upon the sign of $p-p_0$, where p_0 is called the critical probability. When $p < p_0$, percolation is almost surely impossible. When $p > p_0$, percolation has a positive probability that increases with $p-p_0$, and it is obvious that when $p \to 1$, percolation tends to become certain.

In this context, *percolation* is a self-explanatory term, and the usage adopted in Chapter VI is finally justified. Clearly, the property that "percolation is possible" is topological, since it expresses that the top and bottom layers of the lattice are connected. In fact, Bernoulli percolation may well be the earliest and simplest example of a successful investigation in statistical topology. The key of its success resides in the possibility of reducing a topological problem to a metric one. Indeed, around each unplugged vertex of the lattice, one can construct a "cluster" made of the nearest neighbors that are unplugged, then *their* unplugged nearest neighbors, and so on. As p increases, so does the average cluster size. For $p < p_0$, it turns out that a cluster is almost surely finite in size and therefore a sufficiently wide slab cannot be spanned. For $p > p_0$, however, there is a positive probability of the cluster being infinite and hence of being able to span any finitely thick slab.

Furthermore, in the neighborhood of the critical probability, the clusters of unplugged sites turn out to be all skin and no flesh. These complicated and bizarre shapes stimulate the physicists' linguistic imagination but have not yet received a proper name (one reads such terms as *ramified* and *stringy*). Specifically (Leath 1976) the number of sites situated outside of the cluster but next to a site within the cluster is roughly proportional to the number of sites within the cluster.

In addition, physicists have established that critical percolation possesses the analytic property of scaling, which (in this instance at least) is akin to analytic self similarity as practiced in the study of turbulence. Unavoidably, one is led to ask whether self similarity also extends to geometrical properties. The empirical investigation we carried out suggests that the answer is affirmative. The procedure is already familiar in other guises. Take a big eddy, which is simply a square or cubic lattice of side set to 1. Pave it with subeddies, squares (if $E = 2$) or cubes (if $E = 3$) of side r, count the number N of the squares or cubes that intersect the cluster, and evaluate $\log N / \log(1/r)$. Then repeat the process with each nonempty subeddy of side r by forming subsubeddies of side r^2. And so on as far as is feasible.

Immediately, we encounter the fact that the concept of cluster dimension is not determined uniquely. We must indeed recall that coastlines (Chapter II) involve two distinct fractals, the coast of one island by itself (a closed curve) and the coast of all of an island's neighbors taken together. In general, the two have different dimensions. Similarly, the study of percolation (both fractal and Bernoulli) must cover both single clusters (for example, the biggest among them) and all clusters taken together.

For the biggest cluster, computer experiments suggest that in the plane one has $D \sim 1.77$. This result is an independent corroboration and interpretation of a conclusion reached by an entirely different method in Stanley et al. 1976.

As to all clusters taken together, they eventually fill the plane and hence are of dimension $D = 2$.

◁ The data in Leath 1976 on the average cluster diameter as function of number of sites in a cluster agree with $D \sim 1.77$. However, in a section inspired by the French version of this Essay and that in turn triggered the investigation now being reported, Leath claims that $D = 2$ for individual critical clusters.

◁ Even before the measurements reported above, $D = 2$ seemed contradicted by the perceptual evidence, and indeed it turns out to be based on unfounded arguments. The first argument seems to confound the two dimensions we sought to distinguish. The second argument applies out of context the formula in Chapter II that links the human length of a fractal curve to the area it surrounds. This formula is inapplicable to clusters because we saw that they have no interior points and surround no area. ▶

THE "ISLANDS AND CONTINENTS" VOCABULARY. The usual procedure to decide which sites to plug is to attach to each site a random scalar U, uniform over $[0,1]$; if $U > 1-p$ ($U < 1-p$), the site is plugged (unplugged). The isolated clusters relative to $p < p_0$ are sometimes described as "islands," and the big cluster that arises for $p = p_0$ as a "continent." An increase in p is described as a lowering of overall water level until the critical moment when a continent emerges. These images are suggestive but potentially misleading, especially in this Essay's context. The critical cluster is indeed "stringy," while a continent is solid. The Bernoulli process is stationary but Chapter IX suggests that a continent's coastline is the zeroset of a nonstationary process. See STATIONARITY, 4 in Chapter XII.

PSEUDO FRACTAL SEQUENCES

◁ The method used in Chapter II to deal with fractal curves may not yet have exhausted all its potentialities nor revealed all its pitfalls.

◁ Basically, when a curve is so complicated as to be nonrectifiable, we deal with it through a sequence of approximating simple curves. (a) These approximations are known or trusted to converge in the limit to a curve – and serve to define this curve as a fractal. (b) The approximate length measured to the scale η is $\lambda \eta^{1-D}$.

◁ However, it *might* perhaps happen

that property (b) should hold even when (a) does not. Since practical-minded scientists seldom check convergence, they may be faced, so to speak, with pseudo fractal dimensions without fractals. More precisely, they should watch for exponents that "look like" fractal dimensions but are associated with sequences that have no limit and define no fractal. ▶

SINGLE-CRYSTAL SMECTIC FILAMENTS

The Apollonian model of smectics studied in Chapter VII is far from having exhausted the beautiful and mysterious geometry of liquid crystals. Another application of fractals seems possible in the case of the growth of single-crystal filaments, as described in unpublished work by F. Jones & R. B. Meyer.

As one cools a solution of an appropriate organic molecule, it begins by precipitating into small globules of the smectic A phase. While cooling continues, it precipitates further molecules that find no unplugged room on the globule, try to squeeze in, and thereby cause the globule to first grow a pimple, then shoot out in the form of a fairly straight filament composed of a fixed number of layers concentric to a core of ill-known structure. Cooling goes on and precipi-

tates still other molecules; they too find no unplugged room and try to squeeze into vacancies that may exist in the sheet due to spontaneous fluctuation. When the next fluctuation goes in the opposite direction, it provokes strains, with the effect that the straight filament buckles. When cooling is rapid, buckling leads to a tight and regular sinusoid. When cooling is slow, buckling leads to loops that are wider and more irregular. At any moment, the filament constitutes a single constant-radius crystal, which cannot intersect itself, and fills the solution more and more tightly. Eventually its structure becomes unstable and it collapses into a different shape that may be a globule, a disc, a loop, or a hollow sphere.

The lengthening and buckling of filaments is oddly reminiscent of the Koch construction and of Peano curves. Is the resemblance real?

SUPERCONDUCTORS

Our interest will now jump again to another specific subject matter. One type of superconductor (Type II) can be found not only in a normal and an active state but also in an "intermediate" state in which the basic forms are mixed.

The study of intermediate superconductors is one of those rare chapters of physics in which a geometric description was necessary. It was provided by Landau 1943. (My other source is Livingston & DeSorbo 1969.) The first stage model assumed an alternation of two kinds of essentially flat laminae, normal and superconducting. The second-stage model assumed that active laminae split as they near the surface, into twos, fours, and so on. Close to the surface they are taken to be so finely divided that no distinction is possible between the phases, and a new "mixed" state occurs.

Landau's second model warrants being described further to probe for possible fractal aspects. Let the z axis be perpendicular to the surface, so the latter is defined by $z=0$, and the y axis be parallel to the laminae. If one takes the flatness of the laminae seriously, it suffices to describe the geometrical consequences of the model on the section of the intermediate phase by the plane (x,y). In other words, it suffices to consider a succession of straight z-constant lines corresponding to decreasing values of z. Far enough from the surface, the superconducting phase subdivides into laminae of thickness σ separated by normal phase laminae of thickness ν. For smaller z, we have twice as many laminae of thickness $\sigma/2$, then four times as many laminae of thickness $\sigma/4$, and so on.. Very close to the surface, superconductive laminae of thickness $\sigma 2^{-k}$ alternate with normal laminae of thickness $\gamma 2^{-k}$.

This pattern seems to achieve asymptotic mixing, allowing Landau's image of the rough mixture to be based on normal and superconducting phases separated by a collection of infinitely tightly packed parallel lines. The geometry is vaguely reminiscent of a Peano curve, but in fact it misses the latter's main ingredient. As z decreases, the boundary between the two phases jumps around so widely that it fails to converge.

Since actual evidence became available, Landau's model has become increasingly controversial among the specialists. Distinct normal and superconducting regions turn out to come right out to the surface. Among the shapes they exhibit, one finds mazes with corrugated boundaries, honeycombs in which the superconducting phase is multiply connected, superconducting islands inside normal material, and so on. The corrugations seem to have a discrete scale repeated periodically, but other features appear to involve a mixture of scales. Perhaps they would respond to the fractal view of the way the world is put together. Perhaps, more specifically, the models of turbulence discussed in this Essay – including the thoughts concerning turbulent dispersion that were expressed in the caption of Plate 53 – could help analyze the geometry of superconductivity.

CHAPTER IX

Fractional Brownian Facets of
Rivers, Relief, and Turbulence

This chapter brings together selected additional facets of several diverse phenomena that were already encountered in earlier chapters. We first examine the long-term persistence in river discharges. Next we turn to Earth's relief and incidentally tackle the shape of coastlines. We finally turn to turbulence and deal with the shape of isosurfaces of temperatures. All these heterogeneous questions deserve a unified treatment because of one characteristic they have in common. Like Brownian motion and unlike the other phenomena discussed in this Essay, each can be described to a useful degree of approximation by a Gaussian curve or surface. Each involves an intrinsically interesting generalization of the familiar or "ordinary" Brownian sets – surfaces, zeroset, function, and trail.

One generalization considered below involves Brownian surfaces. The simplest among them, using the terminology of Chapter III, is the Brownian plane-to-line function. It may represent the relief on a flat Earth, since to every point of a two-dimensional Euclidean plane it associates a point on an altitude axis.

Another generalization is the Brownian 3-space-to-line zeroset: the surface on which a scalar function, such as the temperature defined for all points of 3-space, takes some prescribed value, ordinarily thought of as being zero.

Still another generalization is the Brownian sphere-to-line function: to each point on a sphere (actual latitude and longitude), it associates an altitude.

The preceding list may seem to promise smooth sailing without a need to justify each model. After all, that is how many fields of science proceed when they invoke ordinary Brownian motion models. Furthermore, ordinary Brownian motion being the very prototype of nonconstrained chance, one may wonder why this chapter is not the fourth in this Essay. The first reason for placing it here is that in the above listed generalizations, the classic simplicity of the ordinary Brownian motion is lost.

The second and more important reason is that the most important generalizations used here have not been mentioned yet. They are Brownian functions and trails. Their common characteristic concerns the presence of very marked, very long-range dependence. For example, fractional Brownian trails, contrary to the ordinary Brownian trails, take a highly nonnegligible account of their values in the distant past. Unlike the self avoiding trails, however, they are not ordered to avoid past sites altogether, only instructed to have a tendency to be persistent, that is, to avoid turning back. Alternatively, they may be antipersistent, that is, instructed to have a tendency to turn back more than ordinarily.

JOSEPH EFFECT AND HURST PHENOMENON

The Biblical story of Joseph to which we referred in Chapter VIII deserves to be taken very seriously. Although its exact wording doubtless involves much po-

etic oversimplification, it is likely to refer to some actual "run" of systematically high flood levels in the Nile followed by systematically low levels. Naturally, the magical "seven and seven" is not to be taken textually and (not so obvious) there is no periodicity in actual Nile records. On the other hand, it is a well-established fact that successive yearly discharges and flood levels of the Nile are extraordinarily persistent. The same applies to a good many other rivers. This persistence is as fascinating to scholars of all kinds as it is vital to those involved in the design of dams. For a long time, however, it remained beyond the scope of mathematical analysis.

The breakthrough occurred with the work of Harold Edwin Hurst (in Hurst 1951, 1955; see also Chapter XI) and with the explanation advanced in Mandelbrot 1965h and described with more detail and illustrations in Mandelbrot and Wallis 1968, 1969a, 1969b, 1969c.

The subtler points of Hurst's work would lead us too far astray, but it is interesting to present the crude summary that guided me in this study. Denote by $X*(t)$ the cumulated discharge through a river such as the Nile, between the beginning of year 0 and the end of year t. We assume that the average discharge will have been subtracted. Then define

R∗(d) as the difference between the maximum and the minimum of X∗(t) as t ranges from 0 to d. Hurst's data suggest that the X∗(t) are near Gaussian *and* that R∗(d) is proportional to d^H with H∼0.7. This value of H was entirely unexpected. When the increments of X∗(t) are statistically independent, X∗(t) is approximately a classical Brownian function (Feller 1951). Hence R∗(d) is proportional to the root mean square of X∗(d), which is ∼$d^{1/2}$. And the proportionality to $d^{1/2}$ also holds when the yearly discharges, for example, are Markovian.

PERSISTENT FRACTIONAL BROWNIAN MOTIONS

Nevertheless, a weaker inference can be made from the data. In order for R∗(d) to behave like d^H – with any desired H – it is *sufficient* – though of course not necessary – that X∗(t) be a non-Brownian self affine function. First used in Chapter III, this last term designates a property of scaling invariance that is closely akin to self similarity and also involves a transformation from a whole to its parts. It is not a similarity, which would reduce both coordinates in the same ratio, but an affinity, which reduces time in the ratio r and the other

coordinate in a different but related ratio h(r). The Hurst data makes us desire a self affinity with $h(r) = r^H$, while ordinary Brownian motion would force upon us a self affinity with $h(r) = r^{1/2}$.

Since I shared the general bafflement about Hurst's findings, I attempted to subdivide the problem it raises into two parts. The most urgent in my view was to identify a self affine function X∗(t) with the proper expression h(r). The task of explaining the X∗(t) could be tackled later.

The purely formal search for X∗(t) proceeds through its delta root mean square (delta r.m.s.). Its square is the delta variance of the function, an expression that we have defined as the variance of the function's increment ΔB during the time increment Δt. In the case of ordinary Brownian function, the delta r.m.s. is $|\Delta t|^{1/2}$. To account for the Hurst effect, it would be sufficient to set the delta r.m.s. to $|\Delta t|^H$. The resulting Gaussian random process I termed *fractional Brownian line-to-line function* and represented by $B_H(t)$.

The exponent value $H = 1/2$ leads back to the classical case where the function's increments over nonoverlapping line increments are independent. On the contrary, each exponent such that $H \neq 1/2$ expresses a very specific form of very long-term dependence. The resulting persistence increases progressively when H increases from $1/2$ to 1. The annual discharges of the Nile, being far from independent, turn out to be represented by annual increments of a fractional Brownian line-to-line function with the parameter $H=0.9$. For the rivers Saint Lawrence, Colorado, and Loire, H is somewhere between 0.9 and $1/2$; for the Rhine, we find that $H=1/2$, within experimental error. The limit case corresponding to the maximum exponent $H=1$ reduces to a straight line with a random Gaussian slope.

The vital property of the function $B_H(t)$ is that as a result of its self affinity the persistence in its increments extends to all scales. Therefore, insofar as the evidence embodied in the Hurst phenomenon responds to such a model, the persistence encountered in water records is not limited to short time spans (like the term in office of Pharaoh's ministers). It extends over centuries (some are wet, others are dry) and even millennia. The persistence in discharge depends on the single parameter H. Persistence in the increments of B_H is synonymous with a decreased irregularity in the graph of the function $B_H(t)$. We shall indeed see later in this chapter that $B_H(t)$ can be obtained by modifying the ordinary Brownian motion in Plate 11, so as to soften it and make it less irregular *at all scales*.

ON EXPLAINING

Why should this model work so well? Several possible reasons will be listed later on, but somehow the explanations are much more complicated hence less convincing than the self similarity they claim to explain.

Postscript. These imperfect explanations turn out to be of unexpected antiquity, as described in Chapter XI, in an entry concerning the Joseph Effect.

FRACTIONAL BROWNIAN RIVER OUTLINES

As an aside, let us backtrack a bit to a problem first tackled in Chapter III. We described the first Leopold & Langbein model of a river's course, in which it is assumed the river can only flow in directions other than, say, west, the river's course being a Brownian line-to-line function. An easy argument gave a rough justification for Hack's empirical law with the exponent $4/3$.

The availability of fractional Brownian function now makes it feasible to generalize this model by requiring that whichever course (other than straight west) the river has engaged, it should persist in it.

Then the very same affinity argument used in Chapter III yields Hack's law with the exponent $2/(1+H)$, which can lie anywhere between $4/3$ and 1.

FRACTIONAL BROWNIAN PLANAR TRAILS

Unfortunately, like the Brownian function of Chapter III, the fractional Brownian function in the preceding section is nonisotropic. We continue our study by searching for an isotropic continuous curve such that its direction tends to persist at all scales. Persistence includes an appropriately intense tendency, but not an obligation, to avoid intersection. If we also want to preserve self similarity, as we do in the present Essay, it is simplest to assume that the two coordinate functions of P(t), call them X(t) and Y(t), are two fractional Brownian line-to-line functions of time and that they are statistically independent and have the same parameter H. In this way, one obtains a fractional Brownian line-to-plane trail.

Examples of curves thus obtained are represented on Plate 209. Had we drawn each of the coordinates as a function of time, the curves' appearance would have differed only slightly from that of Plate 87 from Feller 1950. In two dimensions, the effect of the choice of H carries much greater emphasis.

In the first drawing of Plate 209, H takes on the value 0.9000, which has been said to account for the Joseph Effect for the Nile. Having a very strong tendency to continue in any direction upon which it has embarked, a fractional Brownian trail with a high H diffuses much faster than the ordinary Brownian trail and its self intersections are practically invisible. As a result, to return to the question discussed in Chapter II, our curve would be a priori a sensible image of the shape of coastlines in all cases when they are not too irregular.

Moreover, this hunch is confirmed by the value of its dimension. In general (certain mathematical details remain to be settled, but the result cannot be doubted), it turns out that the fractal dimension of the fractional Brownian trail is equal to $1/H$. In particular, it is at least $1/1=1$, as must be the case for a continuous curve. Also, this dimension is at most $1/(1/2)=2$, again a result in intuitive agreement with the fact that such a curve fills the plane less "densely" than for ordinary Brownian motion. The Figure to the left of Plate 209 is very close to the lower bound $D=1$, since we have $D=1/0.9000=1.1111$.

In the Figure to the right of the same Plate, we retain the same pseudo random generative seed while decreasing H. It is obvious that as H decreases and D increases, irregularity also increases, and the tendency to avoid self intersection weakens very quickly. Therefore, we do not know as yet how to represent other than the most regular coastlines. Our search for a model of coastlines is making progress but has not reached its goal.

◁ As was announced in a digression in Chapter VIII, the correlation structure of a fractional Brownian line-to-plane trail generalizes that of the self avoiding random walk. ►

BROWNIAN RELIEF ON A FLAT EARTH, AND ITS COASTLINES

Until now, we have kept looking for a shortcut that might have allowed us to represent a coastline without regard for the other aspects of the relief to which it is attached. However, it is time to acknowledge that this quest was unreasonable and to attack the problem by way of the entire relief.

It is prudent, given what we have learned in the preceding section, to try and approach the relief by way of characteristic curves that *cannot* self intersect. We need not look further than the vertical sections. In fact, as already indicated in Chapter I, one of the sources of this Essay was a feeling I acquired long ago, that a scalar random walk is a rough first approximation of the profile of a mountain, or perhaps of its cross-section. Indirect evidence in the latter direction comes from papers like Balmino et al. 1973.

Therefore let our recently acquired knowledge about fractional Brownian persistence be filed away for a moment, and let us in a first approximation relapse into ordinary Brownian terms. Question: Doesn't the tool box of the Guild builder of statistical models include a random surface such that its vertical sections are Brownian line-to-line functions? Until now, no such tool was to be found in the box, but one has been defined; it is classical if somewhat obscure and is available for adoption.

It is the Brownian plane-to-line function of a point, $B(P)$, as defined in Lévy 1948. Without a doubt, its discoverer had perceived its form with his mind's eye. However, in order to become familiar with it on shorter notice and to become better prepared to apply it concretely, there is no substitute for a careful examination of actual simulations – as exemplified by the top Figures on Plates 211, 213, and 215. Our imaginary landscape is of dimension $D = 5/2$. Its general resemblance to the surface of Earth is no more than approximate; it is definitely too rough, but isn't it a beautiful long jump forward!

The reader will notice that the remaining Plates in the chapter form a "portfolio" within which we do not observe the order in which they are mentioned in the text. Any resulting confusion should hopefully be counterbalanced by an enhanced "feel" for the dependence of each separate facet of form upon fractal dimension.

The Brownian model of the landscape is sufficiently realistic to encourage us to stop and examine more closely to what extent we are making progress in the study of ocean coastlines. Thus far, it had not been necessary to seek a formal definition, but now we must do so. We want the coastline to be determined entirely by the relief, independently of wind and tides. It will simply be the curve formed by the points located at ocean level, inclusive of points situated on offshore islands. In other words, a coastline is a zeroset. A sample is seen on the top Figures of Plates 211, 213, and 215. It finally provides us with our long-sought example of a curve that (a) is devoid of self intersections, (b) is practically devoid of self contacts, (c) has a fractal dimension clearly greater than 1, and (d) is isotropic.

More precisely, the dimension of these coastlines is $3/2$, which is of course much higher than most of Richardson's values of Chapter II. The Brownian artificial coastline is therefore limited in its applicability. It recalls northern Canada, Indonesian Islands, perhaps western Scotland, and the Aegean. It is applicable to other examples as well, but certainly not to all. Because of the Richardson data, it would have been

foolish, anyway, to expect any single D to be of universal applicability.

It is a pity, because a relief of dimension $D = 5/2$ and coastlines of dimension $D = 3/2$ would have been easy to explain. Indeed, the Brown-Lévy function is an excellent approximation of the relief that would have been created by superimposing independent rectilinear faults. The generative model proceeds simply as follows. Starting with a horizontal plateau, break it along a straight line chosen at random to introduce a kind of cliff, a random difference between the levels of the two sides of the break. Then start all over again, ad infinitum. The process merely generalizes the Poisson process. With no need for mathematical or physical details, we can see that the argument seizes at least one aspect of tectonic evolution and leads us to conclude that the list of primary chances that we discussed in Chapter III should be enriched by the Brown-Lévy model.

Because of this model's simplicity, it would be comforting to believe that in some "initial" and hence especially "normal" state of affairs, Earth's relief was Brownian with $D = 5/2$ throughout. One would add that the persistence leading to D between 2 and $5/2$ came later, through processes that vary from region to region. But this topic must be withheld to a later section.

◁ We must digress for a warning.

The proliferation of variants of Brownian motion being endless and terminology being casual, the Brownian plane-to-line function could be confused with the Brownian sheet, which is an entirely different process. For our purposes, the best reference on its account is one that includes an illustration, Wellner 1975. ►

GLOBAL EFFECTS IN BROWNIAN SPACE-TO-LINE FUNCTIONS

Exceptionally attentive readers may have noted that a complication that had previously been billed as nearly unsurmountable has somehow vanished. Chapters III and VIII stressed that since a coastline must avoid intersections, its parts must be globally interdependent. Now we manage to pull a nearly acceptable coastline out of some kind of Brownian function. Since it is Brownian, is it not obvious that this function's different parts are independent?

The answer to this paradox is that the seemingly obvious happens to be untrue. In the present case, the parts are in fact very far from independent. In other words, in order to imbed the Brownian line-to-line function in a Brownian plane-to-line function, it is necessary to give up one aspect that until now had been the characteristic virtue of Brownian chance: independence of its parts.

◁ Let us consider a north-south section of the relief, as well as two points located, respectively, east and west of that line. The relief along this section is a Brownian line-to-line function, hence slopes at different points are independent. It is true, as seems obvious, that any degree of knowledge of the relief along the section provides information as to the relief at the western point. But the next obvious conjecture is untrue. We would expect our north-south line to act as a kind of screen, in such a way that knowledge (or a lack thereof) of the relief at the eastern point is not expected to affect the relief's distribution at the western point. (If such were the case, the relief would have been Markovian.) In fact, the direct influence of the western relief on the eastern relief is highly nonnegligible. Its presence expresses that the generative process inevitably manifests a strong overall dependence.

◁ This overall dependence implies that in contrast to the Brownian line-to-line function, it is impossible to construct a Brownian plane-to-line function by beginning with a rough grid and then filling in each square independently of all others. It is also impossible to construct it layer by layer. More generally, all the algorithms that seem to promise an easy multi-dimensional generalization of the Brownian line-to-line construction inevitably turn out to be invalid. In every

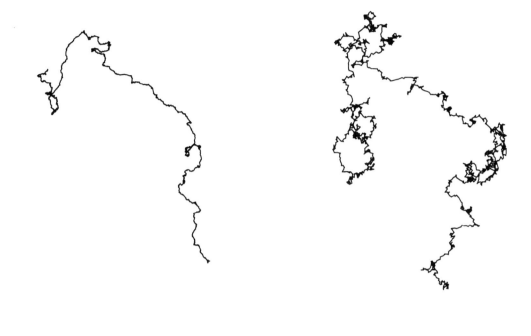

Plate 209

FRACTIONAL BROWNIAN TRAILS
(DIMENSIONS D=1.1111, D=1.4285)

The Figure to the left constitutes an example of a statistically self similar fractal curve with the dimension $D = 1/0.9000 = 1.1111$. Its coordinate functions are independent fractional Brownian functions of exponent $H = 0.9000$, which implies that self intersections, though not strictly prohibited, are greatly discouraged by forcing the curve to be very persistent. Thinking of complicated curves as the superimpositions of large, medium, and small convolutions, it may be said that in the case of high persistence and dimension close to 1, small convolutions are barely visible.

The Figure to the right uses the same generating program and the same pseudo random seed but a dimension increased to $D = 1/0.7000 = 1.4285$. Without having to change the shape of the convolutions which add up to form this trail, we have increased the relative importance of the small ones, and to a lesser extent, of the medium ones. Previously invisible details become very apparent. The general underlying shape, however, remains easily recognizable.

Plates 210 and 211

BROWNIAN "LANDSCAPES," ORDINARY AND FRACTIONAL (DIMENSIONS 9/4, 5/2, 8/3, 17/6)

The top of Plate 210 is a model landscape, and the other Figures are extrapolated Earth landscapes. As one proceeds clockwise, the relative importance of small-scale detail is increased gradually.

The first figure on top of Plate 211 is the simplest to describe. It is a view of an ordinary Brownian surface; the altitude as function of longitude and latitude is such that the straight section along every line in the plane is an ordinary Brownian function. To help enhance the relief, all the areas where the altitude falls below a certain threshold have been filled with water (or is it really snow? Perhaps sand or alluvial fields?).

Compared to reality, an ordinary Brownian landscape is too erratic. In numbers, the poor fit is due to the fact

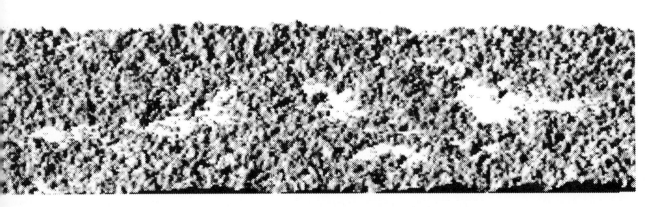

that its dimension D = 2.5000 is too large.

The fit is much improved in the top Figure on Plate 210, where the ordinary Brownian function is replaced by a *persistent* fractional Brownian function with H = 3/4. Hence the relief is of dimension D = 2.2500 and the coastline takes the sensible dimension D = 1.2500. The clear-cut ridges in the Figure are entirely compatible with the fact that it was generated by an isotropic mechanism.

Continuing clockwise from H = 3/4 through the H = 1/2, the bottom Figures on Plates 210 and 211 involve two *antipersistent* fractional Brownian surfaces with H = 1/3 and H = 1/6. Both are of limited use in landscape modeling (except for flooded alluvial plains). However, H = 1/3 will be seen on Plates 222-223 to be of direct interest in the study of turbulence.

Further details on these two Plates are given in the caption for Plates 212 and 213.

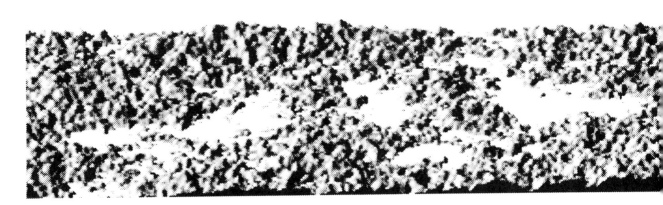

Plates 212 and 213

BROWNIAN "LAKE LANDSCAPES" AS SEEN FROM THE ZENITH

On these figures some features of the landscapes of Plates 210 and 211 are clarified by inspection from the zenith.

Additional comments on Plates 210 through 215. All these Figures use the same program and the same pseudo random seed, so the change of appearance as one proceeds clockwise along any of the three sets of Plates is solely due to progressive variation of D.

The areas covered with water are not lakes but drowned *deadvalleys* in the self-explanatory terminology introduced in the middle of Chapter IX.

In order to enhance the relief, it is lit through computer simulation from a source placed at 60° above the left horizon, and the result is viewed from an an-

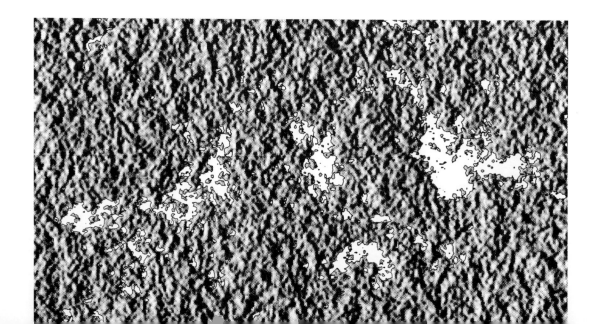

gle of 25° above the front horizon.

Comment specific to Plates 210 and 211. The computation of the altitude was limited to longitudes and latitudes within a finite rectangular domain. Hence the black stripes at the bottom of the four Figures on these Plates are topped by jagged lines that are vertical sections of the relief; hence they are samples of either ordinary or fractional Brownian functions.

The upper contours are a more complicated matter, because some of their points belong to back vertical sections and other points belong to higher nearby mountains. In any event, and contrary to appearances, none of the upper contours is a real "horizon." An actual horizon can be seen on an alternative view of the top of Plate 210 given in Mandelbrot 1977m.

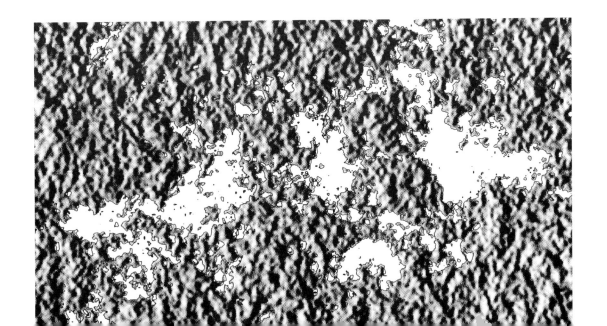

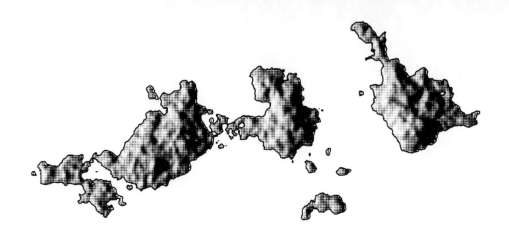

Plates 214 and 215

BROWNIAN "ISLAND" LANDSCAPES
SEEN FROM THE ZENITH

These figures are in a way mirror images of those in Plates 212 and 213. The signs of the altitudes above water level were inverted so that all the points that used to be hidden become visible and conversely. This procedure uncovers the behavior of the various Brownian func-

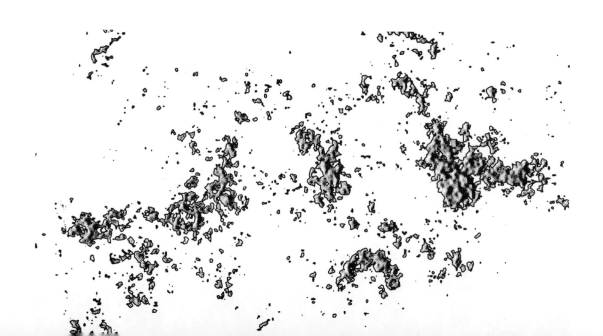

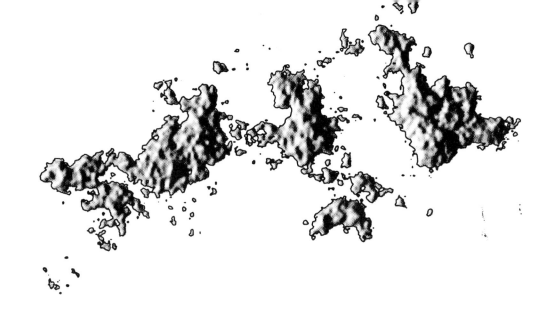

tions near an extremum, and it helps one see the details of the coastlines, especially for the higher values of D.

The relationship between dimension and degree of space filling. It is clear that even after the dimension has been fixed, the degree of space filling by the coastlines continues to be adjustable at will; it suffices to move the sea level up or down. Hence space filling and D are only related when sea levels are comparable – which is not the case in these illustrations.

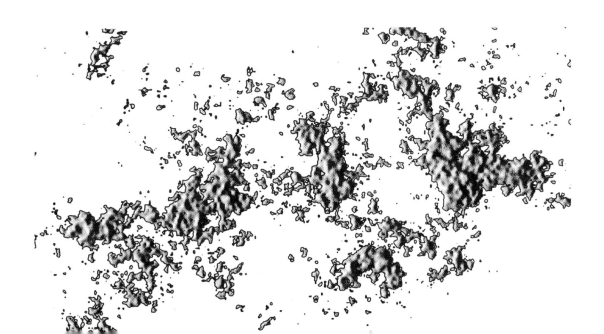

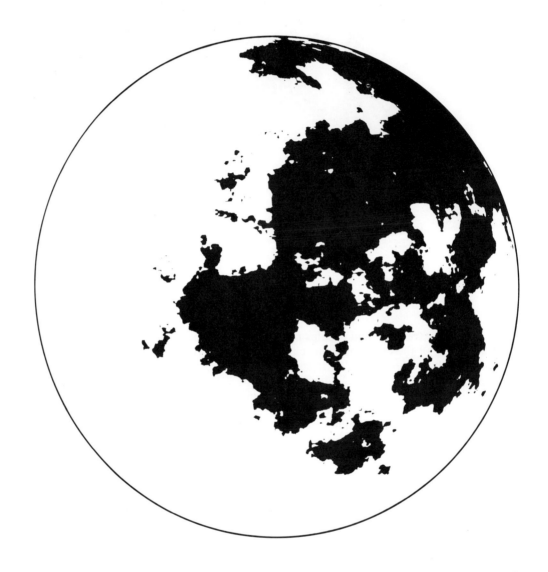

Plates 216 and 217

BROWNIAN PANGAEA
(COASTLINE DIMENSION D=1.5000)

The present Plates are concerned with global properties of the Brownian model, so they postulate a generally spherical Earth.

Plate 216 represents a fictitious fractal Pangaea seen from far away. Its relief was generated by implementing on the computer (to the best of my knowledge, for the first time) a random surface

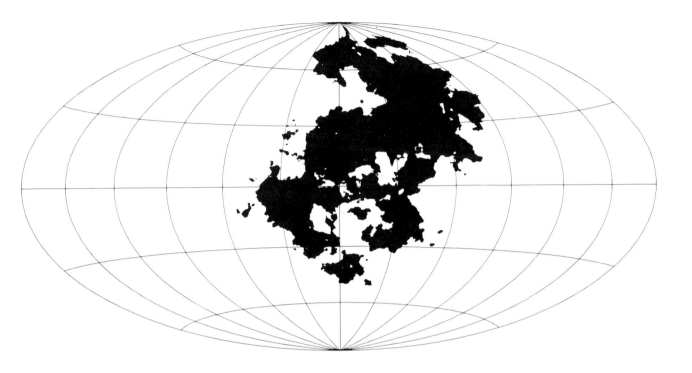

due to Paul Lévy: a fractional Brownian function from a point on the sphere (the latitude and the longitude) to a scalar (the altitude). Sea level was adjusted so that three quarters of the total area is under water. The coastline was obtained by approximative interpolation and projected on a plane tangent to the sphere.

Plate 217 shows the same Pangaea on a Hammer map – a projection favored by students of Wegener's theory of the continental drift.

How closely does this model Pangaea resemble the "real" one? Of course, the specific local detail is not expected to be right, only the degrees of wiggliness, both local and global. In other words, the question of resemblance concerns the value of the dimension. The resemblance is imperfect, as one may have expected. Indeed this model Pangaea satisfies $D = 1.5000$, while the drawings in books of geology attribute to the real Pangaea the same D as observed for today's continents, $D \sim 2.3$. If the evidence turns out to be compatible with $D = 1.5$, the geometry of Pangaea would be incomparably easier to explain. One could account for it with the help of rather elementary tectonic assumptions.

Plates 218 and 219

CONTOUR LINES IN FRACTIONAL BROWNIAN LANDSCAPES

Each of these Plates combines two or three contour lines (the bold lines being easily visualized as being coastlines) relative to distinct samples of fractional Brownian functions.

Plate 218 combines two figures relative to different dimensions but the same program and seed: on the top figure, D=1.3333, and on the bottom figure, D=1.1667. By inspection, both dimensions are credible from the viewpoint of geography, but the former is on the high side while the latter is on the low side. (Recall that the dimension D=1.129 of the near hexagons of Plates 46 and 47 was also perceived to be on the low side; this impression is reinforced when we introduce randomness.)

Plate 219 represents a different sample for D=1.3333. The fact that it seems much more "rugged" than the corre-sponding Figure of Plate 218 is fully compatible with its having the same D. Indeed, Plate 219 happens to correspond to a sample in which every section far enough from the direction NW-SE exhibits a very strong maximum and the overall variation is exceptionally slight. The landscapes on top of Plate 218, on the contrary, happen to correspond to the side of a huge mountain in the sense that the corresponding relief exhibits strong overall variation. Plate 218 is close in its "generic" appearance to a blown-up version of some particularly rugged small piece of Plate 219.

By comparing these different contour lines, we become better aware of the wide margin that is left for the interplay between irregularity and fragmentation even after D has been fixed.

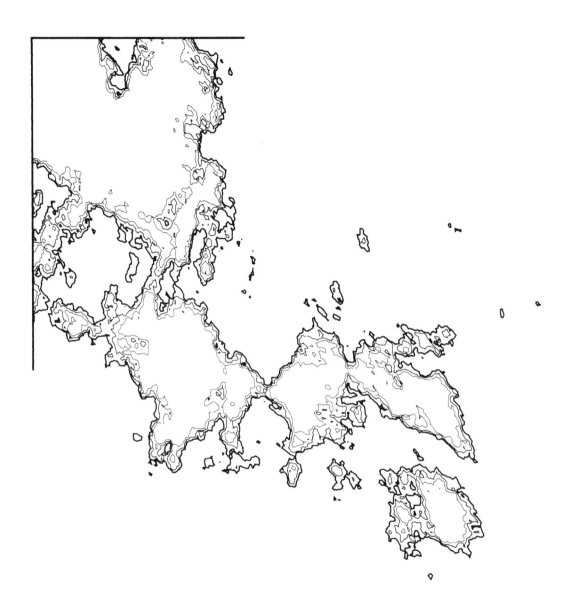

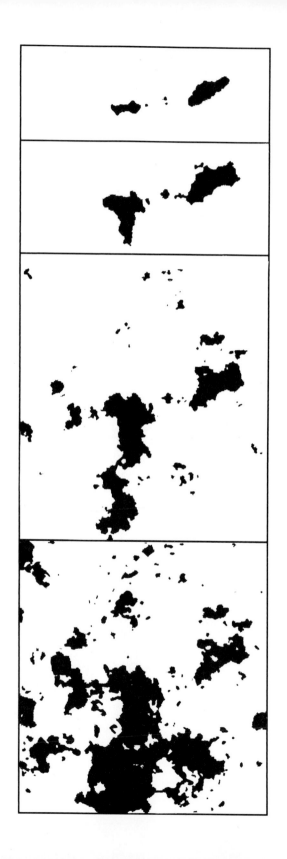

Plate 220

THE FIRST BROWNIAN COASTLINES
(ORDINARY AND FRACTIONAL)

The claim that appropriately selected fractional Brownian functions are reasonable models of Earth's relief is confirmed when we look at these four model coastlines (which are, like Plate 221, a carry-over from the French version).

When D is close to 2, as on the bottom Figure, the coastline tends to fill the entire plane. When D is near 1, as on the top Figure, the coastline is too "straight" to be of conceivable use in geography.

On the other hand, it is difficult indeed, while examining curves of intermediate dimension, not to be reminded of some almost real Atlas. Such is particularly the case with the coastline corresponding to D = 1.3000, as on the second Figure from the top. We see here an unmistakable echo of Africa (big island to the left), of South America (big island to the left, as seen in mirror image), Greenland (big island to the right, after the top of the page is turned from twelve o'clock to nine o'clock). Finally, if the top of the page is turned to three o'clock, both islands together simulate a barely undernourished New Zealand. When D rises to 1.5000, as on the third Figure from the top, the guessing game becomes less easy to play. When D increases again, the game becomes even more difficult and eventually it becomes impossible.

Other seeds yield the same result: the Atlas is closely simulated by curves of dimension D~1.3000.

Plate 221

THE FIRST FRACTIONAL
BROWNIAN ISLAND
(DIMENSION D=2.3000)

Inclusion of this Plate may be an instance of sentimental overkill, because it does not say anything that is not also expressed by other Plates in this portfolio. On the other hand, it is the first illustration of a fractional Brownian surface I arranged to be drawn; it was featured in this book's first (French) version, and I have become fond of it. Four views correspond to various sea levels.

In this instance (as opposed to Plates 210 through 215), no deliberate simulation of side-lighting was required. As luck will have it, the graphic process spontaneously created the impression that the sea is shimmering toward the horizon.

Constantly, I lapse into wondering during which trip (or in which travel film) I actually saw the last view in the sequence, with its vista of small islands scattered like seeds at the tip of a narrow peninsula.

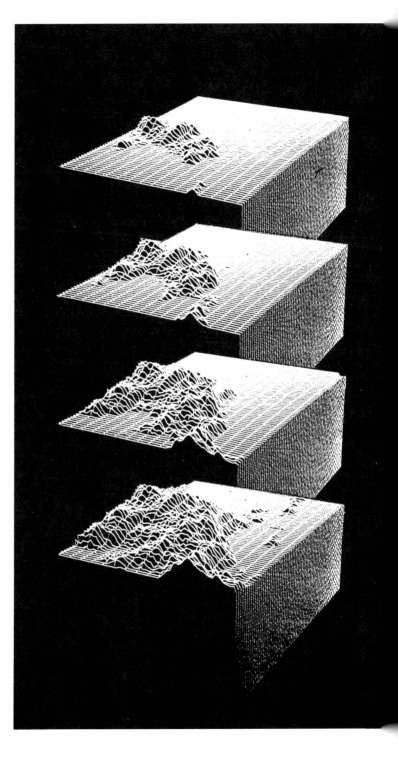

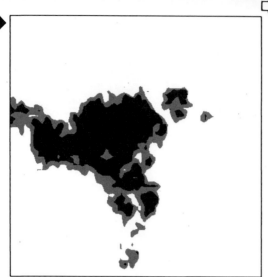

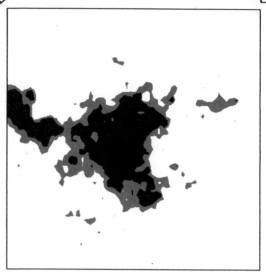

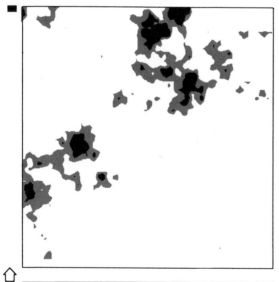

Plates 222 and 223

SLICES OF FRACTALS MODELING TURBULENT ISOSURFACES

Let us first add a word concerning the fractal flake on Plate 9. It is the volume wherein a fractional Brownian 3- space-to-line function exceeds a certain threshold. We chose H=0.7500, so that our flake's surface is of dimension 2.2500 and its planar cross-sections are the curves recommended as coastline models. Moreover, this flake is moderately irregular and moderately fragmented; in other words, well defined and therefore susceptible of being seen from afar as an individual object.

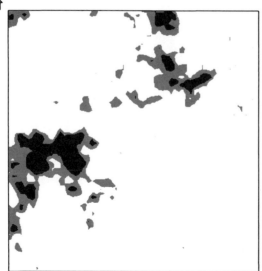

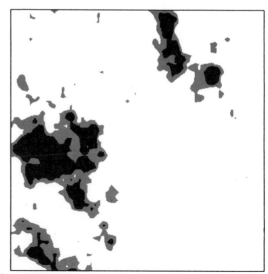

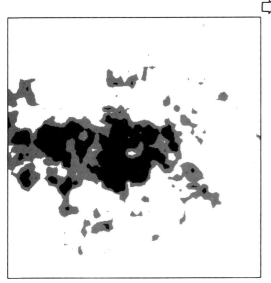

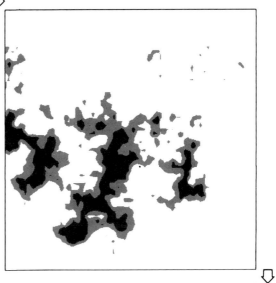

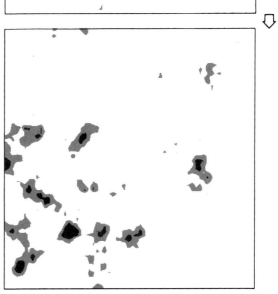

By contrast, homogeneous scalar turbulence is characterized by $H = 1/3$, and
its isosurfaces by $D = 8/3$. The resulting
irregularity and fragmentation are so
strong one must resort to plotting sections. We do so for a value exceeded
only in small scattered areas, represented
in black, and for a value exceeded over a
fairly large area. In order to scan the
two objects at increasing depths, follow
the Figures clockwise starting with the
top left Figure of Plate 222.

For a different sample with $H = 1/3$
and a similar illustration with $H = 1/2$,
see Mandelbrot 1975f.

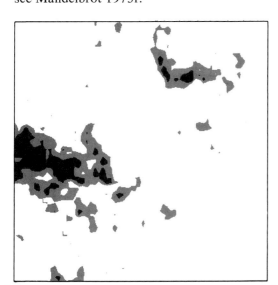

correct step-by-step construction, each step affects the function globally. ►

BROWNIAN RELIEF
ON A SPHERICAL EARTH

We have assumed thus far that the base surface of Earth's relief is a plane, but it is known to be nearly a sphere. Fortunately, the corresponding Brownian sphere-to-line function B(P) has also been provided in Lévy 1970. It is easy to describe, it is fun, and it may even be significant, so we shall dwell on it for a moment. But we shall see it is not really realistic either, because it, too, predicts coastlines with the dimension D = 1.5000. An incorrect dimension is a very serious drawback – not a detail that minor changes in the model could correct.

◄ The simplest definition of B(P) uses noise terms that are familiar to many readers. One lays on the sphere a blanket of white Gaussian noise, and B(P) is simply the integral of this white noise over a half sphere made of points such that their nearest angular distance from P is below $\pi/2$. ►

Locally, that is, within angular distances less than 60°, the present Brownian B(P) looks very much like a Brownian plane-to-line function. Globally, however, it does not. B(P) has the striking property that when P and P_0 are antipo-

dal points on the sphere, the sum $B(P) + B(P_0)$ is independent of P and P_0. ◄ Indeed, this sum is simply the integral taken over the whole sphere of the white noise used to build B(P). ► In terms of relief, this result means that every big hill at the point P must be counterbalanced by a big hole located precisely at the antipodal point P_0, and conversely. Such a distribution would seem to have a center of gravity distinct from the center of the base surface, and it should hardly be in a stable equilibrium, but in fact it is saved from static instability – and hence from early dismissal as a model – thanks to the theory of isostasy. According to this theory, Earth's near-solid crust is very thin at the ocean's deepest points and very thick below the highest mountains, in such a way that a sphere concentric to Earth and drawn a bit below the Ocean's deepest point nearly bisects the crust. After it is agreed that each mountain's visible crest must always be considered in conjunction with its invisible root under the reference sphere, the identity

$$B(P) + B(P') = \text{constant}$$

no longer seems to imply gross static unbalance.

Having been reassured about the basic statistics, we may investigate to what extent the above variant of the Brownian relief fits the evidence. On the basis of today's continents and oceans, the fit is very poor indeed. (Also, our extreme familiarity with the continents' detailed shapes leads us to demand high precision in the fit.)

On the other hand, the theory of continental split and drift suggests that the test of adequacy should be carried out with the primeval Earth as it appeared 200 million years ago. Also, the evidence being flimsier, the test is less certain to fail in this case. We are told, for example in Wilson 1972, that once upon a time all the continents were linked within an essentially undivided supercontinent, Alfred Wegener's Pangaea, while all the oceans formed an essentially undivided superoceanic Panthalassia. As to the total respective extents of land and sea, they were approximately the same as today, in the rough ratio of 1 to 2.

Since the altitude of Earth's mountains is small compared to its radius, a visual test of the present variant must rely upon the level line that corresponds to the actual relative proportions of land and ocean. Of course, the test should not focus on any specific detail but upon a kind of overall kinship with Pangaea.

Such a randomly generated model is shown on Plates 216 and 217. It is a big blob of land, dented here and there by broad sinuses. The first-glance resemblance is too good to be true (in other words, quite misleading) because the first glance pays excessive attention to very large-scale detail. The latter is almost completely due to the combination of the geometry of the sphere and the fact that on the sphere the Brownian rules of dependence involve a strongly positive correlation for angles below $60°$, with strong negative correlation between antipodal points. Highly non-Brownian rules of dependence would also generate a Pangaea-like blob if the basic correlation had the same gross behavior. Hence, testing requires a more attentive second glance focused on less global features. At those scales, the fit deteriorates; for angles below $30°$ (say), a coastline on the sphere becomes indistinguishable from one on the plane, of the sort we have already investigated. All the defects of that model float back to the surface.

◁ A fractal flake in which the altitude would be analytically the same as on the above Pangaea but with a scale of the order of magnitude of half the radius would have a surface of dimension $D = 5/2$ and would look overall like one of the irregular moons of the outer planets. In contrast to the fractal flake illustrated on Plate 9, the dimension of such a Moonlike flake would be a measure of irregularity alone and not of fragmentation. Indeed, this flake would be a single volume unaccompanied by any flotsam or jetsam. Unfortunately, drawing of shadows in this model raises rough questions, and an acceptable illustration is not ready yet. ▶

THE HORIZON

The notion of *horizon* is easy enough to define, and it is interesting (on R. W. Gosper's suggestion) to dwell on it for a moment. An observer "standing" at a finite distance from Earth's surface investigates in turn every direction on the compass, and among all nonhidden points on the surface he identifies the point of greatest apparent height. These points are said horizon.

When the relief is a Brownian perturbation upon a flat horizontal plane, the shape of the horizon is non-obvious, but it turns out that the horizon is almost surely at a finite distance from the observer. (A priori, one may have feared that each mountain would be backed in a distance by a higher mountain further away ad infinitum, but in fact very distant mountains do not affect the horizon. Indeed, as the distance R between an observer and a mountain increases, the mountain's actual height is of the order of $\sqrt{R}$ so its apparent height in degrees above the horizontal plane is about $1/\sqrt{R}$ and tends to 0.)

When the relief is an arbitrary perturbation upon a spherical Earth, the horizon is also at a finite distance, obviously.

The differences between the two models' predictions are therefore rather subtle. First of all, let us examine the horizon through its relative shape, as defined by the ratio of distance to the average distance. On a flat Earth, the relative shape is highly variable and is furthermore statistically independent of the observer's height. On a round Earth, to the contrary, the horizon tends to a circle as the observer grows taller. A second difference concerns the horizon's average height. On a flat Earth, the whole horizon lies *above* a plane passing through the observer, independent of the latter's height. On a round Earth, the horizon falls *below* such a plane as soon as the observer is made tall enough.

In summary, there is truth to the assertion that the observed properties of the horizon imply that Earth must be rounded, but the matter is unexpectedly involved and interesting.

FRACTIONAL BROWNIAN RELIEF ON A FLAT EARTH

The trouble with either Brownian model of the relief is that any theory that yields $D = 1.5000$ for the coastlines can be no more than a first approximation, because real coastlines generally have a different dimension, ordinarily a lower one. Thus our search for a more widely applicable model acquires an unexpected flavor. Long ago, we determined that $D > 1$ and we started looking for ways to force D to rise above 1. From now on, we must try to squeeze it below 1.5000. To obtain less irregular coasts, we must have a less irregular relief and less irregular vertical sections.

Fortunately, earlier parts of this chapter have prepared us well. To achieve a model of vertical sections, we should endeavor to replace the usual Brownian line-to-line function by its fractional variant. Random plane-to-line functions $B_H(P)$ possessing such sections do indeed exist. In addition, I have perfected an algorithm that makes it possible to simulate them on a computer. Though mathematical details require further attention, there seems to me no doubt that the dimension of these surfaces is $3-H$ and that the dimension of their constant level lines and vertical sections is $2-H$.

Therefore, there is no longer any difficulty in modeling and simulating any dimension that the empirical data fancy to require. We expect coastline dimension to be around $D = 1.3$, and we can therefore go a long way with $H = 0.7$ – a value that finally justifies the top Figures on Plates 210, 212, and 214.

Every approach suggests that, while $D \sim 2.3$ is a widespread value of the relief's D, other values may be needed to account for specific areas of Earth. Values of $D \sim 2.1$ or so would account for areas where relief is dominated in relative terms by its very slowly varying components. When this component is a big slope, the relief is an inclined uneven table and the coastline differs from a straight line by no more than mild irregularities. Near a major maximum, the relief is an uneven cone and the coastline a mildly irregular oval (often a double-headed cone and two ovals). All these configurations need not be illustrated because they are quite dull to look at. (Moreover, they happen to be comparatively expensive to produce.)

Values of D near to 3 are also potentially useful. (See the bottom Figures of Plates 210 through 215; one of them resembles a flooded alluvial plain.) Therefore, all values of H will have to find a place in the tool box of the Guild of professional builders of statistical models.

An alternative and less artificial construction of $B_H(P)$ proceeds as for $B(P)$, by superimposition of rectilinear faults. However, the fault's profile must no longer be a cliff; its slope must increase as one approaches the fault, then diminish. It is possible to obtain $B_H(P)$, as long as the profile is appropriately chosen.

However, the shape that must be postulated is not, a priori, very natural.

FRACTIONAL BROWNIAN MODEL OF RIVER DISCHARGE, MOTIVATED

An unattractive feature of the two ways we used to introduce $B_H(P)$ resides in their being guided not by a theory but solely by one scientist's personal experience of which mathematical and graphical tricks are most likely to work. One could argue that a lack of motivation in a model that works well is much preferable to a lack of fit in a model that is well motivated, but scientists are greedy and prefer to have both. Therefore, we must try and outline various forces capable of effecting the desired increase of H.

We begin, by way of digression, with the promised discussion of methods that achieve this result in the case of water discharges of rivers (Joseph Effect). In this instance, engineers began in terms of a crude model in which successive yearly discharges are independent and the cumulative discharge is in effect represented by a graph of dimension 1.5. They were of course not aware of this matter of dimension. In the next step they tried to inject some persistence by taking into account the fact that a certain quantity of water is stored in natural reservoirs and thus carried over from one season to the next. As expected, storage does prevent a river's annual discharge from jumping from year to year as brutally as it would under the assumption of independence. However, such smoothing and "planing" of the records, being exclusively local, could not account for Hurst's data. It can only introduce a local form of persistence, and from the long-term viewpoint the graph of the cumulative discharge remains "physically" of dimension 1.5.

The belief that the global persistence present in the fractional Brownian model of the Nile's discharge merely results from smoothing can be implemented, but it requires a whole hierarchy of smoothing processes, each with its own different scale.

We might, for example, picture the Nile as being the additive superposition of the following sequence of independent processes. The first takes into account the natural reservoirs of the kind we have already mentioned, which only imply interaction from one year to the next. The second takes account of microclimatic changes, the third of climatic changes, and so forth, introducing a large number of parameters. From an entirely theoretical point of view, one must continue this process to infinity, using infinitely many parameters. As a result, this approach does not resolve the basic issue, but only moves it somewhere else. It remains necessary to explain why the various contributing processes add up to a self affine outcome. At one point of the discussion, one has the choice of explaining why a certain curve *is* a hyperbola or, having developed said curve into a sum of exponentials, explaining why they *add up* to a hyperbola. The two tasks are of precisely equal difficulty.

Of course, the practicing hydrological engineer will go no further than a finite outer time scale of the order of magnitude of the horizon of the longest engineering project. Hence the contributing factors with which he will deal are in finite numers, nevertheless too numerous to be acceptable.

◁ *Postscript.* The preceding argument about the hyperbola and its component exponentials turns out to be of unexpected antiquity. See JOSEPH EFFECT in Chapter XI. ▶

FRACTIONAL BROWNIAN MODEL OF THE RELIEF, MOTIVATED

Returning from rivers to the relief, one should begin by observing that to search for explanations for the observed D is effectively a search for reasons for changing the dimension from $D = 2.5000$ down to about $D = 2.3000$. One would think of invoking exclusively local smoothing, but this method is inadequate even in cases where it is extremely brutal. On the one hand, it causes all detail to be completely smothered, transforming a surface that was (theoretically) of infinite area into a surface for which area is well defined and finite. On the other hand, it leaves large features completely unaffected. Therefore, local smoothing can at best replace an object having a well-defined dimension throughout by an object that must be handled like the skein of thread discussed in Chapter I: it exhibits two successive zones characterized by a global dimension of 2.5000 and a local dimension of 2.

More generally, if one performs K distinct smoothings having different fundamental scales, one ends up with $K + 1$ zones of distinct dimensions connected by transition zones. However, by handling everything carefully, one may end up with zones that alternate between different dimensions, and the whole may become indistinguishable from a single zone with an intermediate dimension, say, 2.2.

In other words, a superposition of phenomena with well-defined scales may well mimic scalelessness. We know, on the other hand, that a scaleless phenomenon is often spontaneously analyzed by the mind into a seeming hierarchy in which each level has a scale. (The stellar clusters of Chapter V are a conspicuous example.) Given these pitfalls, is it safe to try to follow Descartes's recommendation and begin by subdividing a difficulty in as many parts as possible? Can we trust our mind when it spontaneously analyzes geomorphological configurations into superpositions of absolutely distinct features? Perhaps we ought to exert caution in accepting these as real. This issue may be characteristic of the saying that reality does not become real until sanctified by a believable theory.

In one respect, fortunately, the parallelism between the surfaces representing the relief and the lines representing cumulative discharge as a function of time is incomplete. Indeed, relief has the seldom encountered characteristic of involving a natural outer scale: the support of Earth's relief is round. Therefore, it is safe to assume that the various planings the relief has undergone throughout geological history involve spatial scales that only go as far as the order of magnitude of the continents. Were we to think that all of Earth corre-

sponds to a single value of H, we would have to add that the relative intensity of each planing is universal in character. But, if we make the more realistic assumption that H varies from place to place, then we must assume that this planing's relative intensity is also local in character.

PROJECTIVE ISLAND SURFACES AND THE KORČAK LAW

Still another test of the adequacy of the fractional Brownian model may be obtained by comparing the theoretical and empirical distribution of the projective surfaces of ocean islands. This issue was already touched upon in Chapter II, where we stressed that the variability of the island surfaces S is obvious when we look at a map and may even be more striking than the shape of the coastlines. We reported that Korčak 1938 discovered that the distribution of S is hyperbolic: $\Pr(S>s)=\sigma s^{-B}$. Finally, we showed that in order to account for this empirical result, it is sufficient to assume that the coastline is self similar. We are now in a position to add that it is a fortiori sufficient to assume that the relief is self similar. (This idea was among several that motivated the fractional Brownian model.)

The first question raised by this model is whether the relationship 2B = D extends from regular self similar coastlines to fractional Brownian zerosets. The answer is in the affirmative: a simple argument (about the validity of which I have no worry though it is still partly heuristic) shows that 2B = 2−H. Hence 2B = D, just like in the case of nonrandom Koch coastlines examined in Chapter II. The distribution predicted as corresponding to the fractal relief with H = 0.7 comes really very close to the empirical data regarding all of Earth.

A second question relates to the dimension D_0 of each island taken by itself. I do not know its value.

DEADVALLEYS AND LAKES

Those authors who have analyzed island areas naturally did the same for lakes, and they concluded that the hyperbolic law again provides a very good representation. Therefore, a superficial analysis might lead us to dismiss the issue as involving nothing new. At second thought, however, this added scope begins to look too good to be true, because the definitions of lakes and ocean islands are by no means symmetric.

The conceptual symmetric of an ocean island is an area, enclosed by a continent, in which the altitude is below the level of the ocean. We shall denote such areas by the self-explanatory mixed term *deadvalleys*. Some contain water – ordinarily to a level below that of the ocean. Examples that come immediately to mind are the Dead Sea (filled up to −1280 ft.), the Caspian Sea (−92 ft.), and the Salton Sea (−235 ft.). Other deadvalleys are dry, like Death Valley (bottoming at −282 ft.) or the Qattara Depression (−436 ft.).

It would be nice to have information concerning the projective areas within deadvalleys' contour lines at the ocean level, but no data could be located, and nothing indicates that it has ever been collected to any degree of precision. We must be content, therefore, with the impression obtained by looking at a map, that deadvalleys are fewer in number than islands. In the context of the model that assumes Earth to be flat except for an added Brownian plane-to-line relief, this asymmetry between island and deadvalley would be incomprehensible. Within the Brownian sphere-to-line model, on the contrary, said asymmetry is simply a corollary of the asymmetry between Pangaea and Panthalassia. The lesser area (land) is more cut up in pieces than the greater one (water). On the other hand, areas much smaller than continents are not affected by Earth's roundness, and their statistical distribution should be the same as that of the ocean islands.

A different lakelike structure, which may be called a cup, is a basin surrounding a local minimum. One can think of it as being filled to the brim, that is, up to the level of the nearest saddle-point.

As to the real lakes, they raise entirely new issues. While an island, a cup or a deadvalley appear wherever the relief calls for one, the presence of a lake also depends upon a thousand other factors. A model that disregards actual geological and geomorphic evolution is no longer sensible, even as a first approximation. For example, water flowing over the brim of a cup wears it off, thus lowering the cup's water level and decreasing its area. Furthermore, a lake stays in its basin only if its basin is watertight. Its area (remember the Dead Sea and Lake Chad) varies with rain and evaporation, and thus with wind and surrounding temperature. In addition, sediments affect the terrain and soften a lake's shape.

Therefore, the empirical observation that lake areas do seem to be hyperbolically distributed requires special thought. One may even go so far as to question the result to be explained. Optimists such as I, on the other hand, will simply conclude that it might well be that all influences on lake areas, other than the relief, are totally independent of the area. ◁ This assumption would be sufficient because the product of a hyperbolic random multiplicand and an almost completely arbitrary multiplier happens to be hyperbolic. ▶

Anyway, pure mathematicians ought to take an interest in the structure of cups, if only in the ordinary Brownian case, $H = 1/2$.

ANTIPERSISTENT FRACTIONAL BROWNIAN MOTIONS

Thus far in this chapter (except for asides), the fractional Brownian motions have only been studied for $H > 1/2$, but their definition through a delta variance equal to $|\Delta P|^{2H}$ continues to be valid for $0 < H < 1/2$.

An antipersistent fractional Brownian line-to-plane trail is a curve whose coordinates are fractional Brownian line-to-line functions with $H < 1/2$. It has been so far of limited practical interest, but it raises an interesting theoretical point concerning the nature of fractal dimension. Such a trail diffuses *more slowly* than an "ordinary" Brownian trail does. It is called antipersistent because it turns back constantly. When constrained to the plane, it fills it more thoroughly even than the ordinary Brownian trail; not only does it go nearly everywhere, but it tends to cover the same territory again and again.

As to the dimension, the brute force application of the formula $D = 1/H$ would yield a dimension in excess of 2. In the case of the plane $E = 2$, such a dimension is of course inconceivable, and indeed the formula $D = 1/H$ is only applicable for $H > 1/2$. For $E = 2$ and $H < 1/2$, the fractal dimension is fixed at its greatest conceivable value, namely, $D = 2$. This fact is a new example of an already observed phenomenon, that the ordinary Brownian trail is, so to speak, latently, capable of at most the dimension $D = 2$ and indeed realizes it when $E \geq 2$. However, when squeezed into a real line with $E = 1$, it must accommodate itself to $D = 1$. By the same token, the following conjecture seems unavoidable: a fractional H Brownian line-to-E trail is capable of the dimension $1/H$, but in order to realize this dimension, it is necessary to imbed the trail in a space such that $E \geq 1/H$.

When $H = 1/3$, the trail barely fills the ordinary 3-space. We shall see that this H is linked to turbulence.

◁ Incidentally, returning to the topic of stellar matter treated in Chapter IV, we see that the generalization of the Brownian motion sites could have proceeded differently. We had used Lévy motions, which generalize D to lie between 0 and 2 and involve "dusts" of totally disconnected points. An alternative would be to use fractional Brownian motions with $1/3 < H < 1$, which generalize D to lie between 1 and 3 and involve continuous curves. The obvious feature that makes this alternative undesirable is that we did want to obtain dusts. This feature is not fractal but topological. ▶

ISOSURFACES OF TURBULENT SCALARS: THE PROBLEM STATED

We now proceed to fractional Brownian functions rather than trails, which we shall allow to satisfy $H < 1/2$. The case $H = 1/3$ has a very important application. Indeed, one of the manifold aspects of turbulence involves fractional Brownian plane-to-line functions and the surfaces which constitute their zerosets and other constant level sets.

When a fluid is turbulent, the region in which its temperature exceeds a threshold such as 45°F could be deformed continuously (without break) into a collection of balls, and similarly the isotemperature region in which the temperature is exactly 45°F is topologically a collection of spheres. However, it is intuitively obvious that this surface is by far more irregular than a sphere or the boundary of any solid described in Euclid. We are reminded of a passage in Perrin's text, quoted in Chapter I, that describes the form of a colloid flake obtained by salting a soap solution.

◁ As a matter of fact, the resemblance may perhaps extend beyond mere geometric analogy. It may be that a flake fills the zone in which the soap concentration exceeds some threshold, and that in addition this concentration acts as an inert contaminant of a very mature turbulence. ▶

However superficial it may turn out to be, the analogy with colloid flakes suffices to suggest that the isosurfaces of temperature are approximate fractals. We would wish to know whether or not this is indeed the case, and if it is, to evaluate the fractal dimension. To do so, we need to know the distribution of temperature changes in a fluid. Corrsin 1959d, among others, has reduced this question to a classical one, which Kolmogorov and others had faced in the 1940s. In part, these authors had triumphed to an extraordinary extent; in part, they had failed. A review of these classical results is inserted here for the sake of the nonspecialist.

THE KOLMOGOROV AND BURGERS DELTA VARIANCES

Kolmogorov's triumph has been to show that the delta variance of velocity between two given points P and $P' = P + \Delta P$ must fall within one of only two possibilities. Either it is universal, that is, takes the same form regardless of the conditions of experiment, or it is an unholy mess. To be universal, the delta variance must be proportional to $|\Delta P|^{2/3}$.

The punch line is that it was eventually verified that the turbulence in the ocean, the atmosphere, and all other large vessels indeed fulfills Kolmogorov's predictions. See, for example, Grant, Stewart & Moillet 1959. This verification constituted one of the most striking triumphs of a priori thought over the messiness of raw data. It deserves (despite numerous qualifications; Chapter V added fresh ones) to be better known outside of the circle of specialists.

It has superseded, among others, a very crude earlier theory that argues that the delta variance of velocity is proportional to $|\Delta P|$. This last postulate, advanced in the late 1930s by J. M. Burgers, is, however, so simple that students of the topic refuse to let it die. At least since von Neumann 1949-1963, it has been customary to view it as defining *Burgers turbulence*.

A precise mathematical model of a Burgers function is the Poisson function which results from an infinite collection of steps with directions, locations, and intensities given by three infinite sequences of mutually independent random variables. This description should ring a bell. Except for the addition of the variable z to x and y, a Gaussian Burgers function is identical to the ordinary Brownian model of Earth's surface described earlier in this chapter.

J. von Neumann (1949-1963, p. 450) stated: "It would appear [the Burgers approach] describes a fixed number of shocks of fixed size correctly, but it seems questionable whether its conclusions still apply to an asymptotically (with time) increasing number of (individually) asymptotically weakening shocks. Yet, this is probably the pattern of hydrodynamical shocks when they combine with turbulent motion." It is not obvious which "conclusions" the author had in mind, but the mathematical construction of a field made of such shocks is not only possible but easy. As we already know, shortly before von Neumann wrote those lines, Lévy 1948 had defined the Brownian function B(P) of a point P, which is the scalar version of a Gauss field having the required properties.

Now the Gaussian function with the Kolmogorov delta variance should also ring a bell. It is nothing but a fractional H Brownian space-to-line function of three spatial variables, with a value of the exponent H smaller than $1/2$, namely, equal to $1/3$. Thus the Kolmogorov field involves antipersistence, while Earth's relief favors persistence.

◁ Another difference is that while the value of H required to represent Earth's data is purely phenomenological so far, the Kolmogorov value $H = 1/3$ is rooted in the geometry of space. ▶

IN HOMOGENEOUS TURBULENCE, THE ISOSURFACES ARE FRACTALS

Considering his triumph in predicting that $H = 1/3$, the major continuing failure of the Kolmogorov approach is that the distribution of the differences of velocity or of temperature in a fluid remains unknown, or, rather, the only information about it is purely negative. For turbulence the Gaussian assumption is untenable. Such negative results are most awkward, but they rarely suffice to force a convenient assumption to be abandoned. At most, the students of turbulence must be cautious when investigating a Gaussian model: if and when a calculation yields a logical impossibility, they must abandon the model. Otherwise, they forge ahead. In particular – and now we return to our study of temperature – the Gaussian assumption is an integral part of the investigation of Mandelbrot 1975f, which combines it with the Burgers and the Kolmogorov delta variances. As usual, it is not known whether or not the resulting conclusions would remain correct if one did not include the Gaussian assumption, but one can hope that they are sound, because they use little more than continuity and self similarity.

In the 4-dimensional space of coordinates x,y,z,T, the temperature T defines a function $T=T(x,y,z)$. When T is a fractional H Brownian function, its graph is a fractal of dimension $4-H$. (The fact has been proved completely for $H=1/2$, and as yet incompletely but to any physicist's satisfaction for $H=1/3$.) Unfortunately, 4-dimensional surfaces cannot be visualized, which makes it necessary to proceed through lower-dimensional sections. Most turn out to be fractals we know well, and the last one is new and interesting. Let us run through them.

Linear sections. The section for fixed y_0, z_0, and T_0 is made of those points along a spatial axis where a certain value of a temperature is observed. They form the zeroset of fractional Brownian motion, and their fractal dimension is $1-H$.

Planar sections. For fixed y_0 and z_0, the section is the curve representing the variation of temperature along the x axis. It is a fractional Brownian line-to-line function, and its dimension is $2-H$. For fixed z_0 and T_0, the implicit equation $T(z_0,x,y)=T_0$ defines a temperature isoline in a plane. These isolines are of dimension $2-H$, and examples are shown as lakes or island shores on the bottom Figures of Plates 211, 213, and 215. Except for the value of D, they are identical to the coastlines we studied earlier in this chapter.

Spatial sections. For fixed z_0, the section is the graph of $T(x,y,z_0)$, a fractal of dimension $3-H$. For $H=1/2$, it is identical in definition to the Brownian relief on the top Figures of Plates 211, 213, and 215. For $H=1/3$, it is the fractional Brownian relief in the bottom Figures of the same Plates.

For fixed T_0, the section is an isosurface defined by the implicit equation $T(x,y,z)=T_0$. They are three-dimensional generalizations of the coastlines and introduce us to a new kind of fractal $3-H$, that is of dimension $D=3-1/2$ in Gauss Burgers turbulence and $D=3-1/3$ in Gauss Kolmogorov turbulence.

Unfortunately, while the shape of the function $T(x,y,z_0)$ can be represented graphically because this function is single-valued, the implicit solution of the equation $T(x,y,z)=T_0$ is wildly and variably multiple-valued. The best is to draw the "coastlines" defined by the implicit equations $T(x,y,z_0)=T_0$ for a sequence of z_0's. Such sequences are materialized in Plates 222 and 223.

CHAPTER X

Miscellany

The concepts handled thus far oscillate between arbitrariness and necessity. Indeed, this Essay's broadest aim has been to inject some mathematical tools that had seemed to be pure creations of Man into the study of natural phenomena that are entirely beyond Man's control. All those phenomena involve "systems" made of many "parts" articulated in a self similar fashion and much of our study concerned the rules of articulation rather than the intrinsic properties of the parts.

In this perspective, it is only proper to wind up with miscellaneous examples of applications of fractal analysis to "in-between" systems that are neither completely arbitrary nor necessary, such as man-made machines or aspects of his language or society. Moving in this direction, the present chapter will examine a self similar facet of computer design, then one of linguistics. Finally, I shall

sketch several self similar aspects of economics.

Since machines are imbedded in the same space as natural phenomena, they have a genuine geometric aspect. In the other two broad fields, on the contrary, geometry enters figuratively through abstract graphs. Nevertheless, the problems present real analogies to those tackled in earlier chapters.

COMPUTER ORGANIZATION

In order to achieve a complex computer circuit, it is necessary to subdivide it into numerous modules. Each contains a large number of elements, say, C, and is connected with its environment by a large number of terminals, say, T. The empirical link between these two quantities is $T = AC^{1-1/D}$. The forecast error between the computed and the actual values of T rarely exceeds a few per cent.

The form in which the exponent is written will be justified in a moment. (Within IBM, the above rule was first advanced by E. Rent, and my information comes mostly from unpublished memoranda by W. E. Donath.)

The earliest raw data suggested that all circuits without exception had $D \sim 3$, but it has since been determined that the value of the exponent D increases with performance. Performance, in turn, reflects the degree of parallelism that is present in the computer logic. Assuming a computer logic with extreme characteristics, one obtains extreme values of D. In a shift register, the modules form a chain and it turns out that $T = 2$, independently of C, so that $D = 1$. When there is integral parallelism, with each element requiring its own terminal, it is clear that $T = C$. Thus one can say in this case that $D = \infty$.

The cases where $D = 3$ were soon explained (by R.W. Keyes) by noting that the circuits in question are arranged within the volume of the modules, while the connections go through their surfaces. To show that this observation demands Rent's rule, it suffices to express the latter in the form $T^{1/(D-1)} \sim C^{1/D}$. On the one hand, all the various elements have roughly the same volume v. Consequently, C is the ratio total volume of the module divided by v, and $C^{1/D} = C^{1/3}$ is roughly proportional to the radius of the module. On the other hand, the various terminals require roughly the same surface σ. As a result, T is the ratio total surface of the module divided by σ, and $T^{1/(D-1)} = T^{1/2}$ is also roughly proportional to the radius of the module. Conclusion: When $D = 3$, the proportionality between $C^{1/D}$ and $T^{1/(D-1)}$ simply expresses the equivalence of two different measures of the radius.

Note that the concept of the module is ambiguous and almost indefinite, but Rent's rule is quite compatible with this characteristic, insofar as, in any module on any given level, the submodules are interconnected by their surfaces.

It is just as easy, in this context, to interpret the cases leading to the values $D = 1$ and $D = 2$. As already indicated, $D = 1$ corresponds to a linear structure and similarly $D = 2$ corresponds to a planar structure.

On the other hand, if the value of D is not an integer, it does not seem possible to interpret C as an expression of volume and T as an expression of surface. Yet these interpretations are very useful, and it is good to hold on to them. The reader has probably long since guessed that when $D < 3$, it may be supposed that the structure of the circuits is materialized by a fractal set of fractional dimension. If $D > 3$, it must be more complex, and we don't know how to handle it.

To visualize the passage from $D = 2$ to

D = 3, consider a subcomplex of dimension D = 2 in terms of metallic circuits on an insulated board. To increase performance, new interconnections must be established. But very soon one runs out of room, in the sense that any new interconnection would intersect connections already printed. Each time this happens, one needs wires that go outside the board. It has become the practice to use yellow wires. In certain cases, yellow wires simply indicate that the circuit was badly designed. The fact remains, however, that even a good design requires yellow wires, and that their minimum number increases with the desired performance. One would expect Rent's rule with $2 < D < 3$ in all cases where performance does not force the designer to work in 3-space, although it does force him to go outside the plane. Moreover, if the total system incorporates a self similar hierarchy, everything happens "as if" the architect were working in a space of fractional dimension.

LINGUISTICS AND ECONOMICS

The remainder of this chapter will be devoted to certain *specific* phenomena of linguistics and economics which are somewhat unusual in that empirical evidence is both abundant and sound. In two phenomena with which we begin, it has long been known that said evidence

is very well represented by hyperbolic laws, associated with such names as Vilfredo Pareto and George Kingsley Zipf. It should not (at this stage of the Essay) appear odd that among the models advanced to account for laws of this form, some at least involve self similarity.

The argument is simple and spotlights self similarity and dimension in their most "disincarnated" forms. But it must be stressed that we are not embarking on an attempt to squeeze two social sciences into a mold designed for physical phenomena. Quite to the contrary, the example from linguistics was the object of the first paper I ever wrote, well before I became aware that the topic has a fractal facet. The first example from economics also came before the present study of fractals. In addition it turns out to provide a smooth transition to a tidbit that had to find its place somewhere in this Essay, namely, an allusion to the role played in my work in economics by a notion of scaling close to that of physics.

LEXICOGRAPHIC TREES AND THE LAW OF WORD FREQUENCIES

In investigations of the distribution of word frequencies, it is customary to arrange the words in decreasing order of frequency. It will be agreed that a word is simply a sequence of proper letters terminating with an improper letter called space. Whenever there are several words of identical frequency, the order in which they are listed will be arbitrary. In this classification, the letter ρ will designate the rank assumed by a word of probability P. Thus *distribution of word frequencies* designates the relationship between ρ and P.

A priori, one might expect this relationship to vary wildly according to the language and the speaker, but in fact it does not. An empirical law made widely known by Zipf 1949-1965 (see Chapter XI) asserts that the functional relationship between ρ and P is nearly universal. A first approximation claims universal validity; it involves no parameter and takes the form

$$P \sim 1/\rho .$$

And in a second approximation the differences between languages and subjects boil down to two parameters, with

$$P = P_0(\rho + V)^{-1/D} .$$

Here, D and V are independent parameters, and P_0 is determined by them because all probabilities must add up to 1.

The main parameter is D. In order to measure the richness of a subject's use of vocabulary, it is sensible to examine the frequency of his use of rare words; for example, to examine the frequency of the word of rank $\rho = 1000$ compared to that of the word of rank $\rho = 10$. When

the above law applies, one can just as well measure richness of vocabulary by the parameters. Since rare words' relative frequency increases with D, and since V is about the same for all subjects, a subject's richness of vocabulary use is mostly measured by D.

Why is the above law of such universality? Since it is near perfectly hyperbolic, and granted all we have learned so far in the Essay, it is eminently sensible to try and relate Zipf's law to some underlying regularity akin to self similarity. (The procedure seemed less obvious in 1950 when I first tackled this topic.)

An "object" that one could assume to be self similar does indeed exist in the present case: it is a lexicographical tree. We first define it and describe what regularity means in its context. Then we prove that when the lexicographical tree is regular, word frequencies follow the two-parameter law written above. (As a matter of fact, said form of the law was first obtained theoretically during an unsuccessful attempt to derive the parameter free law $P \sim 1/\rho$.) In final comments, we discuss the validity of the explanation. Then we point out the interpretation of D as a dimension. ◁ And in parenthesis we mention the interpretation of D as a temperature. ►

Trees. A lexicographical tree's definition is obvious. It has N+1 trunks, numbered from 0 to N. The first trunk corresponds to the "empty word" constituted by the improper letter "space" taken by itself, and each of the other trunks corresponds to one of the proper letters. The "space" trunk is barren, but each of the other trunks carries N+1 leaders corresponding to the space and to N proper letters. Again, the space leader is barren and the others branch out into N+1 as before. Hence the barren tip of each space leader corresponds to a word made of one proper letter followed by a space. And the construction continues ad infinitum. Each barren tip corresponds to a word and conversely. It will be assumed that this tip is inscribed with this word's probability. And the tip of a non-barren branch will be inscribed with the total probability of the words that begin with the sequence of letters that determines said branch.

Regularity. Such a tree can be termed *self similar,* or (more modestly) *regular,* if each branch taken by itself is in some way a reduced-scale version of the whole tree. Hence our first conclusion is that a regular tree must branch out without bound. In particular, contrary to untrained intuition, the total number of words is not a sensible way of measuring richness of vocabulary. (In more concrete terms, everyone "knows" so many more words than he uses, that his vocabulary is practically infinite.) A further argument (which we shall skip) determines the form that must be observed for the probability of a barren branch growing on top of k live ones.

In the very simplest case, the branch must have a probability of the form

$$P = (1-Nr)r^k$$

where r is some real number much less than $1/N$ and a fortiori less than 1.

If we accept the regularity of the lexicographical tree, the promised derivation of the generalized Zipf law is straightforward. It suffices to note that, at level k, the lower bound of ρ varies between the value

$$1+N+N^2+...+N^{k-1} = (N^k-1)/(N-1)$$

(excluded) and the value

$$(N^{k+1}-1)/(N-1)$$

(included). Writing

$$D = \log N / \log(1/r)$$

and

$$V = 1/(N-1),$$

and inserting

$$k = \log(P/P_0)/\log r$$

in each of the bounds, we have

$$(P^{-D}P_0^{D})-1 < \rho/V \leq N(P^{-D}P_0^{D})-1 .$$

Finally, the desired result is obtained by approximating ρ through the average of its two bounds.

The above simplest regular tree corresponds to discourse that is a sequence of statistically independent letters, the probability of each proper letter being r and that of the improper letter "space" being the remainder (1−Nr). Less simple regular trees correspond to letter sequences generated by other stationary random processes and later cut into words by the recurrences of the space. The argument becomes more complex but the final result is essentially the same.

Comments. The preceding argument is from Mandelbrot 1951. A summary is given in Mandelbrot 1965z or 1968p, and details about the cases other than the simplest are in Mandelbrot 1955b.

Conversely, as with many such arguments, one might attempt to assert that Zipf's data show the lexicographical tree to be regular using ordinary letters. But such a conclusion would be absurd to the extreme, since many short sequences of letters never occur and many long sequences are fairly common. Actual lexicographical trees are quite irregular. Perhaps regularity refers to coding at a fundamental level, using phonemelike cerebral symbols. The details become less and less verifiable and hence hardly worth thinking about. But on the main it is widely felt that my argument suffices to establish that the generalized Zipf law might have been expected.

◁In effect, those interested in this topic have been led to accept the standards which physics adopts when it views the reduction of thermodynamics to microcanonical equiprobability as an explanation. In the case of molecules, equiprobability elicits no emotion, but words are known individually, so equiprobability is hard to swallow and does not taste good.

◁In the same vein, one might mention that it had originally been hoped that Zipf's law would contribute to the field of linguistics. Now this law is recognized as linguistically very shallow. Linguistics demands a detailed knowledge that physics never seeks. ►

◁The generalized Zipf law also holds within certain restricted vocabularies. Take the example of the esoteric discipline styling itself hagioanthroponymy, which investigates the uses of names of saints as surnames of humans (Maître 1964); it has established that the Zipf law applies to surnames. Also, Tesnière 1975 finds it applies to family names. Does this suggest that the corresponding trees are regular? ►

In the ordinary case where $D < 1$, D *is a fractal dimension.* It cannot have escaped the reader's attention that in the above model one always has $D < 1$, and that D is formally a similarity dimension. This formal analogy (which had not been noted in my papers on this subject) is not as shallow as one may fear. Indeed, if one precedes it with a decimal point, a *word* as we defined it is nothing but a number between 0 and 1 written in the counting basis $(N+1)$ and containing no zero except at the end. Let the set of such numbers be marked on the segment $[0,1]$ and add the limit points of this set. The construction amounts in effect to cutting out of $[0,1]$ all the numbers that include the digit 0 otherwise than at the end. One finds that the remainder is merely a Cantor set, the fractal dimension of which is precisely D.

As to the regular lexicographical trees other than the simplest ones, to which we have alluded ominously as providing a generalized proof of the Zipf law, they also correspond in the same way to generalized Cantor sets of dimension D. The equation for D in Mandelbrot 1955b is a matrix generalization of the equation $N \exp(D\log r)=1$, which is just a way of writing the familiar definition of the similarity dimension.

The case $D>1$. A curious aspect of actual data is that the condition $D<1$ is not universally fulfilled. The instances where the generalized Zipf law holds but the estimated D satisfies $D>1$ are rare but unquestionable. To describe the role of the special value $D=1$, let us assume that the law $P=P_0(\rho+V)^{-1/D}$ holds only up

to $\rho=\rho\star$. If $D\geq 1$, and consequently $1/D\leq 1$, the series $\Sigma(\rho+V)^{-1/D}$ diverges. Since one must have $\Sigma P=1$ and $P_0>0$, it is necessary that $\rho\star$ be finite, signifying that the dictionary must contain a *finite* number of words. If, on the contrary, $D<1$, then there is no difficulty with the infinite dictionaries suggested by the theoretical argument.

It turns out, indeed, that the cases where $D>1$ correspond to instances where the vocabulary is known to have been severely limited by artificial extraneous means. These cases are of interest, but we cannot dwell on them; they are discussed in my papers on this subject. Suffice it to say that a construction limited to a finite number of points can never lead to a fractal, hence $D>1$ is not interpretable as a fractal dimension.

◁ *In all cases,* D *is a temperature of discourse.* Parenthetically, another very different interpretation leads one to consider D as being "the temperature of discourse." The "hotter" the discourse, the higher the probability of use of rare words. The case when $D<1$ corresponds to the familiar case where there is no upper bound for the formal equivalent of energy.

◁ On the other hand, I recognized that the case when words are so "hot" as to lead to $D>1$ is equivalent of the highly unusual imposition of a finite upper bound on the energy. A few years after this sharp dichotomy was described in language statistics, it was independently recognized in physics. Among ordinary bodies, the inverse temperature $1/\theta$ is smallest – namely, zero – when the body is hottest, and N. Ramsey recognized that if the body is to become hotter still, $1/\theta$ must become negative. See Mandelbrot 1970p for a compact discussion of this parallelism.

◁ When thermodynamic concepts are taken concretely, the preceding analogy may sound forced. It becomes, however, entirely natural within certain more general approaches to thermodynamics. At the risk of overquoting myself on items that are peripheral to this Essay, one such formalism is given in Mandelbrot 1964. ▶

HIERARCHICAL TREES AND THE PARETO LAW FOR SALARIES

Another example of an abstract tree is found among hierarchic human groups. We deal with the simplest regular hierarchy if (a) its members are distributed among levels in such a way that except on the lowest level, each member has the same number N of subordinates and (b) all of the subordinates have the same "weight" U, which is equal to r times the weight of their immediate superior. The analogy between the form of regularity and self similarity is obvious.

It is most convenient to consider this weight to be an income. To be interpreted as a weight without artificiality, an income must be something like a salary.

When diverse hierarchies are to be compared from the point of view of the inequality in the distribution of incomes, it seems again reasonable (a) to classify their members in the order of decreasing income (the classification within each level remaining arbitrary); (b) to designate each individual by his rank ρ; and (c) to evaluate the rate of decrease of income as a function of rank, or vice versa. The more rapid the decrease in income when rank increases, the greater the degree of inequality.

The formalism already used for the frequency of words applies without change except that it is good to replace the P_0 by U_0. It shows that the rank ρ of an individual of income U is described approximately by the formula

$$\rho = -V + U^{-D} U_0{}^{D}.$$

The finding that this expression indeed represents salarial incomes very well is due to Vilfredo Pareto, and the present derivation is due to Lydall 1959.

This relation shows that the degree of

inequality is mostly determined by the formal similarity dimension

$$D = \log N / \log(1/r),$$

which does not seem to have any fractal interpretation worth writing down. The greater the dimension, the greater the value of r and thus the lower the degree of inequality.

It is possible (as in the case of word frequencies) to generalize the model somewhat by assuming that within a given level k the value of U varies from one individual to the other and that U is equal to the product of r^k by a random factor, the same for everyone. This generalization modifies the parameters V and P and hence D, but it leaves the basic relationship unchanged.

Note that the empirical D for all wage earners in a country is ordinarily near 2. In cases where it is exactly 2, let inverse income be plotted on an axis pointing downward. One obtains an exact pyramid (base equal to the square of the height). In this case, the income of a superior is the geometric mean of the income of all his subordinates taken together and of that of each subordinate taken separately. If $D = 2$ and $N = 2$, the model calls for $1/r = \sqrt{2}$. This value seems unrealistically high. It would seem that Lydall's model can only account for income distribution in hierarchies in which $D > 2$, and that the overall D is 2 because income differences *within* hierarchies pale in comparison with those *between* hierarchies. The study of the latter requires an entirely different approach.

ECONOMICS AND "SCALING"

Nonsalaried forms of income. Lydall's model, therefore, is not entirely convincing, but it does show that one should expect the law of Pareto to apply to salaries. However, the same formal law also holds for speculative incomes, which cannot possibly involve hierarchical trees and require separate treatment. Nevertheless, models of the law of Pareto are well advised to invoke some kind of self similarity, and such is indeed the approach taken in Mandelbrot 1960e and explored in Mandelbrot 1961e, 1962e, 1963e. The basic tools are the Lévy stable distributions that star in this Essay in Chapters IV and V.

Price variation. The biographical sketch LOUIS BACHELIER in Chapter XI tells the story of how that investigator was led in 1900 (before the physicists) to invent the mathematical Brownian motion process. The problem Bachelier tackles concerns the variation of prices on the Stock Exchange. One may (only half in jest) call it a geometric problem, since the so-called "chartists" claim that the future can be predicted from the details of the market charts' geometry. Bachelier's counterclaim was that these charts are perfectly random, so that, to take one example, the effectiveness of various trading rules depends on the *geometry* of the level crossings of a Brownian motion.

Of course, Bachelier's work did not take root in its time, and it took economists 60 years to reinvent his model and longer to acknowledge his contribution. ◁It took them even longer to make use of martingales, a sort of random process that Bachelier had introduced in passing and that has lately become the core of the newly relabeled notion of *stock market efficiency.* ▶

As suggested by the above-mentioned work on income distribution, my interests around 1960 were in statistical economics, and I was active in the Bachelier revival. More important, after evidence demonstrated his Brownian motion model to be a rough approximation, I had the good fortune to think up first one radical improvement, then another. The first (Mandelbrot 1963e, 1963b) was again related to Lévy stable laws. The second, starting with Mandelbrot 1965h, was re-lated to the fractional Brownian motions, which star in Chapter IX of this Essay. Both improvements involve self similarity, and assume that the daily, weekly, and monthly changes in a price (rather, in its logarithm) follow the same statistical distribution except for scale.

Scaling. Many readers will have recognized that said models of price variation are intimately related in their motivation to the scaling laws that physicists introduced soon afterward into the theory of critical phenomena (see Chapters VII and VIII). Thus Bachelier's feat of making economics move faster than physics (not much faster, but a positive difference is so rare as to be newsworthy) was apparently repeated.

More specifically, it is amusing that Lévy stable laws should appear ready to invade physics. See Jona-Lasinio 1975a, 1975b.

COUNTERINTUITIVE INSTANCES OF STATISTICAL STATIONARITY

The usual notion of statistical stationarity (restated in Chapter XII) cannot be quite as satisfactory as mathematicians believe it to be. Indeed, the stationary processes mentioned in the preceding section have been all too often described by serious practitioners of science as contrary to their intuitive notion of stationarity.

Let it be recalled that the first of the processes in question has an infinite population variance and the second has a finite variance, but its spectral density is f^{-B}, with a parameter satisfying $0 < B < 1$, so this density is infinite for $f = 0$.

Bachelier having been forgotten, the early models of price change were indeed mathematically nonstationary and the same was true implicitly of the models of river discharges. Therefore it must have indeed seemed surprising that the same phenomena could be modeled by stationary processes. On the other hand, the recurrent revivals of models using fully formulated nonstationary random processes are harder to understand (quite apart from any irritation provoked by seeing the old parade as the new). Some recent models of weather-related time series, for example, are based in an essential fashion upon the notion of a nested hierarchy of climatic or microclimatic changes and boast of this ingredient.

Perhaps the practitioners are simply trying to tell the mathematicians something about intuitive stationarity. In this light, it is useful to ponder the continuing use of heuristic arguments to the effect that a stationary process necessarily satisfies the following theorems: (a) the law of large numbers, (b) the central limit theorem with the classical denominator $\sqrt{N}$, and (c) a property that is seldom written down but is widely considered as valid: that events relative to far removed portions of a stationary random function are asymptotically independent.

The above arguments have no merit (Grenander & Rosenblatt 1957, p. 181; Mandelbrot 1969e), but they lead inevitably to the conclusion that when working scientists seem to accept the mathematical definition of stationarity, they may merely pay lip service to it. On more intimate acquaintance, they find they would wish a truly stationary process to satisfy the above three conditions. Increments of a Lévy flight fail condition (b), and increments of a fractional Brownian function fail conditions (b) and (c), but these "failings" are precisely what makes these models effective.

John Maynard Keynes made a telling comment in 1940 about the early econometric models and their use of cycles, by asking: "How far are these curves meant to be no more than a piece of historical curve fitting and description and how far do they make inductive claims with reference to the future as well as the past?" (quoted and discussed in Mandelbrot 1963e, p. 434). This query fully preserves its bite: Do nested hierarchies of successive climates and microclimates have an independent reality — as held by old or new partisans of nonstationarity? Or am I closer to the truth in claiming that there has been thus far no objective proof that they are anything but most convenient *after-the-fact* labels for seven-, seventeen-, or seventy-year-long periods of dry or wet weather?

Even the staunchest partisans of "nonstationarity" would, or so I believe, agree that it is unlikely that the historical swings of the climate should be as severe and ultimately divergent as those of a truly nonstationary process, say, of Brownian motion itself. I tend to believe that in many cases, the urge to involve nonstationarity ought to be satisfied by the degree of variability present in either of two classes of processes that I have been using. They are, so to speak, on the "borderline" between the kin of white noise and the kin of Brownian motion. The first class includes those random functions (some are mentioned earlier in the section) that are mathematically but nonintuitively stationary. The second class includes those random functions that from the mathematical viewpoint are only *conditionally* stationary in the sense of Mandelbrot 1967b. They are related to the fractal events of Chapter IV and are sketched in Chapter XII under STATIONARITY....

Between the classical extremes of intuitive stationarity and total nonstationarity, the categories described above add a third possibility, which we may have no choice but to consider a satisfactory compromise.

CHAPTER XI

Biographical and Historical Sketches

This chapter is mainly devoted to bits of biography. As a prelude, let us note that a life story that is interesting to tell is very rarely the reward (or is it a punishment?) of those who keep to the mainstream of science – even (or should one say, especially?) for those who achieve high fame at an early age, like John William Strutt, Third Baron Rayleigh. The straight flow of his triumphs has made his name recognizable in almost every province of science. Yet, with one exception, his life appears uneventful – in other words, totally subordinated to his natural evolution as a scientist. The unexpected hits when, having been admitted to Trinity as a birthright, being the elder son of a landowning Lord, he decides to become a scholar.

Science does have a few great Romantics, like Evariste Galois, whose story combines within the confines of one day (as demanded in French Court tragedy) his eclosion as scientist and his death in a duel. But most scientists' stories are like Rayleigh's, predictable except that the circumstances of their eclosion and entry into the mainstream are occasionally colorful. The three-year-old Carl Friedrich Gauss corrects the arithmetic of his father. The adolescent Srinivasa Ramanujan reinvents mathematics. Harlow Shapley finds that he must wait out a term before he can register in a school of journalism. To keep busy, he picks up the University Bulletin and selects a department from the alphabetical table of contents. He skips archaeology because he does not know the word's meaning, proceeds to astronomy, and meets his fate. More atypical is the story of Felix Hausdorff. Until the age of 35, he devotes most of his time to philosophy, literature, writing and directing plays, and similar endeavors. Then he settles down to mathematics and ten years later produces his masterpiece, Hausdorff 1914.

Tales according to the above pattern are legion, but the stories selected for this chapter are entirely different. Entry into the mainstream is postponed, and in many cases it is even posthumous. Strong feelings persist of really belonging to other times. The actors are loners. Like certain painters, they might be called *naives* or *visionaries,* but there is a better term for them in American English: *mavericks.* When the curtain falls on the prologue, they are still, by choice or by chance, unbranded.

Their work frequently exhibits a peculiar freshness. Even those who fail to achieve greatness seem to share with the giants the possession of a sharply personal style. The key seems to be time to spare. In the words of the daughter of d'Arcy Thompson about his book *On Growth and Form* (Thompson 1917), "It is a matter of speculation whether [such a work] would ever have been written if [he] had not spent thirty years of his early life in the wilderness." Indeed, he was 57 when he published it, and the ages at which our other mavericks did their best are such that the cliché that science is very largely a young man's game is definitely not true in their case.

I find such stories appealing, and during the preparation of this book they seem to have presented themselves in unusually high numbers. I would like to share the emotions a few of them evoke.

As mavericks should, our heroes differ greatly from one another. Paul Lévy (like d'Arcy Wentworth Thompson, who would not be out of place in this company) lived long enough to set his mark deeply in his province of science, but his admirers think (as I do) that he deserves even better; call it true fame. Lewis F. Richardson also made it (though barely). Louis Bachelier's story was sadder; no one read him through, and his standing as a perennially unsuccessful applicant lasted until all his work had been duplicated by others. Hurst had better luck, and his story is intriguing. And Fournier d'Albe and Zipf do deserve a footnote. Thus each of the stories in this chapter brings us a new insight into the psychology of a peculiar kind of strong mind.

In cases where standard biographies exist, they are not repeated unless necessary. One, the *Dictionary of Scientific Biography*, Gillispie 1970-1976, is very useful and in particular includes bibliographies. Its omissions are also significant.

It was also found convenient to include in this Chapter historical sketches on Brownian motion, on recent mentions of applications for Cantor sets, and on the Weierstrass function. *Postscript.* And when another entry on Cantor sets and an entry on the Joseph Effect were drafted after the book had been written, it was easy to insert them in this chapter.

LOUIS BACHELIER (1870 - 1946)

The story of the beginnings of the theory of Brownian motion is worth knowing and is touched upon in the next sketch in this chapter. However, were it not for quirks of the way science is practiced, physics might have been preceded in this context by mathematics – and also (a most unusual sequence of events) by economics.

The fact is that a truly incredible proportion of the results of the mathematical theory of Brownian motion had been described in detail five unquestionable years before Einstein et al. in the writings of Louis Bachelier (*Dictionary of Scientific Biography,* **I,** 366-367).

The story begins with his doctoral dissertation in the mathematical sciences, defended in Paris on March 19, 1900. Sixty years later, it was reprinted – a rare honor – in an English translation, with extensive comments. However, the committee that examined it had not been overly impressed and gave it the "mention honorable" at a time when no one stood for the French doctorate unless he foresaw an academic opening and felt sure of a "mention très honorable." It is to be expected therefore that this particular dissertation had no direct influence. Bachelier, in turn, was not influenced by anything written in this century, even though he remained active and published (in the best journals) several other papers (long and full of manipulations). In addition, his popular book, Bachelier 1914, enjoyed several printings and even now bears being read. It is not to be recommended to just anyone, because its subject matter has changed profoundly, and because it is written in aphorisms. In many cases, it is not clear whether these are summaries of established knowledge or outlines of problems yet to be explored. The cumulative effect of such ambiguity is rather disconcerting. Only very late, after repeated failures, was Bachelier finally appointed to a University professorship, in Besançon.

In view of his slow career and the thinness of the personal trace left by him (my search, though fairly diligent, has discovered only some odd scraps of recollections by students and colleagues, and not a single photo), his life seems to have been mediocre, and the posthumous fame of his dissertation makes him an almost romantic personality. Why the sharpness of this contrast? To begin with, his life might have been brighter were it not for a certain mathematical error. The story is told in Lévy 1970 (pp. 97-98) and in greater detail in a letter Paul Lévy wrote me on January 25, 1964.

"I first heard of him a few years after the publication of my *Calcul des Probabilités,* that is, in 1928, give or take a year. He was a candidate for a professorship at the University of Dijon. Gevrey, who was teaching there, came to ask my opinion of a work by Bachelier published in 1913 (*Annales de l'Ecole Normale*). In it, he had defined Wiener's function (prior to Wiener) as follows: In each of the intervals $[n\tau,(n+1)\tau]$, he considered a function $X(t|\tau)$ that has a constant derivative equal to either $+$ or $-$ v, the two values being equiprobable. He then proceeded to the limit (v *constant,* and $\tau \to 0$), and claimed he was obtaining a proper function $X(t)$! Gervey was scandalized by this error and asked my opinion. I agreed with him and, upon his request, confirmed it in a letter which he read to his colleagues in Dijon. Bachelier was blackballed. He found out the part I had played and asked for an explanation, which I gave him and which did not convince him of his error. I shall say no more of the immediate consequences of this incident.

"I had forgotten it when in 1931, reading Kolmogorov's fundamental paper, I came to 'der Bacheliers Fall.' I searched for Bachelier's works, and saw that this error, which is repeated everywhere, does not prevent him from obtaining results that would have been correct if only, instead of v = constant, he had written $v = c\tau^{-1/2}$, and that, prior to Einstein and prior to Wiener, he happens to have seen some important properties of the so-called Wiener or Wiener-Lévy function, namely, the diffusion equation and the distribution of $Max_{0 \le r \le t}X(t)$.

"We became reconciled. I had written him that I regretted that an impression, produced by a single initial error, should have kept me from going on with my reading of a work in which there were so many interesting ideas. He replied with a long letter in which he expressed great enthusiasm for research."

That Lévy should have played this role is tragic, for his own career, as we will see very soon, also nearly floundered due to lack of rigor. It would be cruel to speak of the degree of mathematical rigor present in some physical theories.

We now reach the deeper reason for Bachelier's difficulties. It is revealed by the title of his dissertation, which (on purpose) I have not yet mentioned, "Mathematical theory of speculation." The title did not by any means refer to (philosophical) speculation on the nature of chance, rather to (money-grubbing) speculation on the ups and downs of the market for consolidated state bonds (*"la *rente"*). Thus the function X(t) mentioned by Lévy was the price of these bonds. Bachelier's troubles were announced in the following delicately understated words of Henri Poincaré, who wrote the official report on this dissertation and apparently did not notice the error in question, "The topic is somewhat remote from those our candidates are in the habit of treating." One may argue that Bachelier should have avoided imposing his interests on unwilling mathematicians, but he had nowhere else to go. No one knew where to pigeonhole his results, so no one could retrieve them when necessary.

There is nowhere any indication on how this topic had been chosen (the idea of assigning thesis subjects was totally foreign to French professors of that period – above all, to loners like Poincaré). Although Bachelier used stock-market vocabulary with ease, he does not appear to have been a gambler ("No one ever gets rich through skill"). He was the true creator of the probabilistic concept of "martingale" (this is the proper formulation of the notion of *fair game* or of *efficient market*), and he was well ahead of his time in understanding many specific aspects of uncertainty as related to economics. Nevertheless, the most likely conjecture is that he followed tradition and looked at gambling as providing, in

his own words, the "clearest picture of the effects of chance."

Regardless of its initial trigger, Bachelier's concept that prices follow the Brownian motion process took a long time to be accepted but finally acquired great importance in economics (see the last sections of Chapter X). Bachelier had to feel strongly about the importance of his work. In a *Notice* he wrote in 1921 (while applying for some unspecified academic position), he stated that his principal scholarly contribution had been to provide "images taken from natural phenomena, like the theory of radiation of probability, in which [he] likens an abstraction to energy – a strange and unexpected linkage and a starting point for great progress. It was with this concept in mind that Henri Poincaré had written, 'Mr. Bachelier has evidenced an original and precise mind'."

The preceding sentence is taken from the already-mentioned report on the dissertation, which deserves further excerpting: "The manner in which the candidate obtains the law of Gauss is most original, and all the more interesting as the same reasoning might, with a few changes, be extended to the theory of errors. He develops this in a chapter which might at first seem strange, for he titles it 'Radiation of Probability'. In effect, the author resorts to a comparison with the analytical theory of the propagation of heat. A little [sic] reflection shows that the analogy is real and the comparison legitimate. Fourier's reasoning is applicable almost without change to this problem, which is so different from that for which it had been created. It is regrettable that [the author] did not develop this part of his thesis further."

Poincaré, therefore, had been aware for a moment that Bachelier had advanced to the very threshold of a general theory of diffusion. It is, however, obvious – and unfortunate – that this work had escaped Poincaré's mind when the latter came to take an active part in discussions concerning Brownian diffusion.

Other comments in Bachelier's *Notice*, relative to later papers, are also worth summarizing: "1906: *Théorie des probabilités continues.* This theory has no relation whatsoever with the theory of geometric probability, whose scope is very limited. This is a science of another level of difficulty and generality than the calculus of probability. Conception, analysis, method, everything in it is new. 1913: *Les probabilités cinématiques et dynamiques.* These applications of probability to mechanics are the author's own, absolutely. He took the original idea from no one; no work of the same kind has ever been performed. Conception, method, results, everything is new."

Those who have to write such *Notices* are not called upon to be modest, and Louis Bachelier did exaggerate to some extent, and moreover gave no evidence of having read anything written in the twentieth century. Unfortunately, *everything* he said was discounted by his contemporaries, and he was refused the position he was seeking!

Does anyone have further knowledge of his life and personality?

◁ The quotes from Poincaré were copied, with permission, from a report filed in the Archives of the Pierre and Marie Curie University (Paris VI), heir to those of the former Faculty of Sciences of Paris. The document is fascinating and of course written in the lucid style characteristic of Poincaré's popular writings.

◁ This instance leads me to hope that other more extensive selections from Poincaré's correspondence and his confidential reports to universities and academies will be made available. As of today, a whole aspect of his personality is absent from his *Works*. ►

BROWNIAN MOTION

Natural Brownian motion is "the chief of those fundamental phenomena which the biologists have contributed or helped to contribute to the science of physics'' (Thompson 1917). The first of the biologists' two contributions was to discover the phenomenon (well before 1800) and the second was to find that it was not biological but physical in nature. This last step was taken by due to Robert Brown in 1828, hence the term *Brownian* is not as completely undeserved as it is sometimes made to appear.

Brown had other claims to fame as a botanist, being "Humboldt's *facile princeps botanicorum* and discoverer of the nucleus of the cell" (Thompson 1917). Biographies are found in *Dictionary of Scientific Biography*, **II,** 516-522; *World Who's Who in Science*, p. 255; and *Encyclopaedia Britannica*, including quite early editions. Incidentally, Brownian motion itself is not mentioned in the biography of Brown in the ninth edition, 1878. In the eleventh to thirteenth editions, 1910 to 1926, it receives a few words in passing. But it is of course treated fully in editions that came out since Perrin's 1926 Nobel Prize.

The slow acceptance of the physical nature of Brownian motion is recounted in Brush 1968 and Nye 1972. Outlines are given in recent editions of the *Encyclopaedia Britannica*, Perrin 1909 and 1913, Thompson 1917, and Nelson 1967.

The developments started by Brown cumulated in the period 1905-1909 with theories due to Einstein, Smoluchowski, Langevin, Fokker, and Planck and with the experiments of Perrin. The impact of the latter's findings was due to the fact that they confirmed the predictions of Einstein's theory, but it was noted in Chapter I, by reference to Perrin 1906, that the motivation for said experiments is to be found in their author's general philosophical stance about the importance of irregularity and heterogeneity in the description of Nature.

My attention had been drawn to Perrin 1906 by Borel 1912, which is referenced in Perrin 1913. Borel exemplified a seemingly universal tendency, and restricted the scope of Perrin's philosophy to the realm of molecules.

CANTOR SETS: POSTSCRIPT ON FORGOTTEN EARLY APPLICATIONS

Yes: The fact seems astonishing, but the triadic set that Georg Cantor (1845-1918) introduced in Cantor 1883 did receive early applications within a year of original publication.

No: The earliest applications were not restricted to the new and controversial area of set theory. Quite to the contrary, they involved the theories of analytic functions and of differential equations, which in the nineteenth century were topics of the highest orthodoxy.

No: Those applications did not go unnoticed. The first was by Henri Poincaré (1854-1912) and involved his widely praised theory of automorphic functions, and the next was by Paul Painlevé (1863-1933), a lesser scholar, though famous in his own right and influential. Painlevé was fascinated with engineering (he was the Wright brothers' first passenger), and eventually entered politics, rising to the post of Prime Minister of France.

Incidentally, Jean Perrin (1870-1942) was a close friend of Painlevé at the University of Paris. In this context, the excerpts quoted in Chapter I lose nothing of their strength and validity, but they no longer seem so oddly isolated.

Yes: The facts sketched above – to be amplified in a moment – were completely unknown to me when I wrote this Essay and let myself be carried away in Chapter I to praise and scold mathematicians for their role in creating and hiding the Cantor sets. All objective criteria suggested that the broad idea of associating Cantor sets and physics was viewed not only as preposterous but as unprecedented.

How odd! How could it be that a program of research described in a famous book by the Nobel Prize winner Perrin and endorsed by his most illustrious contemporaries, far from being carried out, became reduced to the single aspect Perrin had carried furthest, Brownian motion. We may well be dealing here with a typical instance of the concept of "prematurity" described by Stent 1972, a concept I may have underrated in Chapter I. Stent would argue that until science had the data and the tools to carry out Perrin's proposal, nothing was done because nothing could be done.

In any case, the early events surrounding Cantor sets are worth detailing through free translations from Hadamard 1912 and Painlevé 1895.

Hadamard: "Poincaré was a precursor of set theory, in the sense that he applied it even before it was born, in one of his most striking and most justly celebrated investigations. He showed indeed that the singularities of the automorphic functions form either a whole circle or a Cantor set. This last category was of a kind which his predecessors' imagination could not even conceive. The set in question is one of the most important achievements of set theory, but Bendixson and Cantor himself did not discover it until later."

Painlevé: "I must insist on the relations that exist between function theory and Cantor sets. The latter kind of research was so new in spirit that a mathematical periodical had to be bold to publish it. Many readers viewed it as philosophical rather than scientific. However, the progress of mathematics soon invalidated this judgment. In the year 1883 (which will remain doubly memorable in the history of mathematics in this century), *Acta Mathematica* alternated between Poincaré's papers on Fuchsian and Kleinian functions, Cantor's papers, and those of Mittag-Leffler."

Cantor's papers are found on pp. 305-414 of Vol. 2 of the *Acta* (the Cantor set on p. 407), but Painlevé did not tell the whole story, since these papers were not originals but French translations sponsored by Mittag-Leffler, the editor of the *Acta*, who was attempting to help Cantor in his fight for recognition. Much later, in recollections published by *Acta*, he confided that one of the anonymous translators had been Poincaré. However, this detail does not deny Hadamard's claim that "Poincaré had applied sets that did not yet exist," because his results had already been sketched in *Comptes Rendus* before Cantor's work had appeared in German. Poincaré adopted Cantor's outlook so promptly that in his first paper he used the German term *Mengen* without taking time to seek a French equivalent.

We now translate freely from 1934 comments by Arnaud Denjoy (1884-1974), reprinted in Denjoy 1964: "Some scientists share the criteria of High Society. They view certain truths as being in good taste, well-educated and properly brought up, while to others the gentleman's door must forever remain closed. I think mostly of set theory, which is a whole new universe, incomparably more vast and less artificial, more simple and more logical, more apt to model the physical universe; in a word, truer than the old universe. The Cantor set shares many properties of continuous matter, and seems to correspond to a very deep reality."

In another statement, also written in 1934 (see Denjoy 1975, p. 23), we find an echo of Perrin's claim that the derivative lacks relevance to physics, and then we read the following: "I think it obvious that discontinuous models account in a much more satisfactory manner and more successfully than the present ones for a host of natural phenomena. Therefore, the laws of the discontinuous being much less well elucidated than those of the continuous, they should be investigated broadly and in depth. Insuring that the degrees of knowledge of the two orders are comparable will enable the physicist to use one or the other approach according to need."

Unfortunately (with a single but very notable exception, to which we shall turn momentarily), Denjoy's repeated assertion that Cantor sets are more apt to model the physical universe became increasingly strident but could not be buttressed by specific developments more recent than the very early work of Poincaré and Painlevé.

As a matter of fact, the Hadamard reference and the first Denjoy reference were eulogies for Poincaré and Painlevé and revived suggestions that the originators had not deemed worthy of being refreshed by repetition. In most cases, material that is only fit for eulogy has no impact. It tends to become lost unless, as with Hadamard and Denjoy, the eulogist's own fame warrants the publication of *Collected Papers* that gather less dust than is customary. Incidentally, it is quite possible that Hadamard's remarks were intended to counterbalance a widespread later image of Poincaré as one who disapproved of "Cantorism," except insofar as it promised him "the joy of a doctor called to follow a fine pathological case."

Devoting so much space to Denjoy may seem odd, since his fame is mostly restricted to very pure mathematicians and seems to rest on achievements totally removed from physics. Not only (to put it mildly) did his personality fail to endow his opinions with authority, but he kept these opinions practically confidential. Nevertheless, there is a point in persisting on following this barely traced track, because I do not know of any other, and because (again) the other quotes

came to my attention through a memorial volume, Denjoy 1975.

Denjoy's fourth achievement, in his own count, centers on a 1932 paper concerning differential equations on the torus. Answering a question raised by Poincaré, he succeeded in showing that the intersection between a solution and a meridian could be the whole meridian or any Cantor set one may wish to prescribe. The former behavior agrees with the physicist's notion of ergodic behavior, and the latter behavior is in conflict with intuitive ergodicity. Perhaps Denjoy was simply trying to say that his result reinforced Poincaré's and Painlevé's hunches and indicated that a physical system need not be ergodic. (Note that an example analogous to Denjoy's had been given by Bohl in 1916.)

Denjoy could not have been alone in assessing in advance the basic nature of the changes in viewpoint that adoption of Cantorlike fractal models would threaten to involve. The customary and perhaps proper response is to demand that the evidence in favor of such models satisfy unusually stringent standards.

CANTOR SETS: RECENT MENTIONS OF POSSIBLE APPLICATIONS

This entry refers to applications other than those reported in this Essay. Indeed, while the events sketched in the preceding entry lay forgotten, Stanislaw

Ulam, beginning in 1950, had made several tantalizing suggestions concerning the possible role for Cantor sets in the theory of gravitational equilibrium of large aggregates of stars. (See his collected papers, Ulam 1974.) My work reported in this Essay was started independently, but I soon became aware of Ulam's papers. In addition to enjoyment, they provided encouragement but, so far, no substantive influence that I could report. I hope an actual link will not be slow in coming.

Recently, again independently of earlier work but in a vein analogous to Ulam though in a different context, Ruelle & Takens 1971 (see also Ruelle 1972) wrote tantalizingly about a possible role for Cantor sets in the theory of turbulence. Their argument was inspired by an example of dynamical attractor due to S. Smale. Since this attractor is nonstandard, it came to be ordinarily called *strange*. Moreover, as soon as one turns one's thoughts to modern analytic dynamics, other nonergodic systems (notably, but far from exclusively, the work of J. Moser and V. Arnold, not to mention Denjoy) are equally inspiring.

Furthermore (see Ruelle 1975), the Ruelle & Takens papers led to a fuller appreciation of the contribution of E. N. Lorenz. Lorenz 1963 obtains numerical solutions of a certain ad hoc simplification of the equations of the motion of the atmosphere, and observes (without being aware of Cantor sets) that these solutions manifest a quite unexpected degree of irregularity and of seeming "unpredictability." This line of work is being pursued extensively, recent references being Marsden & McCracken 1976 and Temam 1976. One may expect it to soon acquire strong links to my organized fractal phenomenology. But it is too early to comment about the relationships between these two independent approaches, and I do not feel qualified to attempt to summarize the present status of the Lorenz approach.

As to the term *strange attractor,* it is to be hoped it will be short-lived; insofar as the approach that invokes such attractors is successful, they will automatically cease to be "strange." I hope of course that those which are fractal will come to be called *fractal attractors.*

Postscript. A Cantor set has now been observed in yet another basic field of application. Hofstadter 1976 shows that the spectrum in a certain quantum mechanical problem is a Cantor set. The basic qualitative structure had already been described in Azbel 1964.

EDMUND EDWARD FOURNIER D'ALBE (1868 - 1933)

A paragraph in *World Who's Who in Science,* p. 593; his books in just a few libraries; comments about his model of a hierarchical universe – few in number and generally sarcastic in tone, except for the description and generalization due to Charlier (incidentally, Charlier does not seem to have wished in any way to appropriate a work that he admired). Those are the only traces left on the record by Fournier d'Albe.

We hear from his son that he chose to make his living as a freelance science journalist and inventor (he constructed a prosthesis to enable the blind to "hear" letters and was the first to transmit a television signal from London). He was a believer in spiritualism and a religious mystic. His name was witness of Huguenot ancestry. Despite his partly German education and his eventual residence in London, where he had obtained his A.B. by attending evening college, a stint in Dublin had transformed him into an Irish patriot and a militant in a Pan-Celtic movement.

Two New Worlds remains his claim to a degree of immortality. It received very good reviews in *Nature*, which called its arguments "simple and reasonable," and *The Times*, which called its speculations "curious and attractive." At the end, however, those around him attached little importance to this book. *Nature* and *The Times* did not mention it in their obituaries.

It is the kind of work in which one is surprised to find anything sensible. One fears attracting attention to it, lest the more disputable bulk of the material be taken seriously. Its peculiar author hardly rose to the level of accomplishment of the others listed in this chapter. And yet the fact remains that (as we saw in Chapter V) he was the first to formalize a very important intuition about galactic clustering, so we are indebted to him for something of lasting value.

HAROLD EDWIN HURST (1880 –)

Hurst, who has been hailed as perhaps the foremost Nilologist of all time, and who is sometimes spoken of as "Abu Nil," the Father of the Nile, spent the bulk of his career in Cairo as a civil servant of the British Crown, then of Egypt. (*Who's Who, 1973*, p. 1625, and *Who's Who of British Scientists 1969/70*, pp. 417-418; not eligible for *Dictionary of Scientific Biography*.)

His early training is worth recounting – mostly in his own words and those of a memorandum kindly prepared by Mrs. Marguerite Brunel Hurst. The son of a village builder of limited means, whose family had lived near Leicester for almost three centuries, he left school at age 15, having been mostly trained in chemistry, and also in carpentry by his father. He then started as a pupil teacher at a school in Leicester, attending evening classes to continue his own education. At age 20, he won a scholarship that enabled him to go to Oxford as a noncollegiate student. After a year, he became an undergraduate at the recently reestablished Hertford College. He switched to a major in physics and worked at Clarendon Laboratory. His lack of preparation in mathematics was a handicap, but thanks to the interest that Professor Glazebrook took in an unusual candidate who was very strong in practical work, he won a first class honors degree, to everyone's surprise, and was asked to stay for three years as a lecturer and demonstrator. In 1906, he went to Egypt for a short stay that lasted 62 years, of which the most fruitful were the years after he had turned 65. His first duties included transmitting standard time to the Citadel of Cairo when it fired a gun at mid-day, but he became increasingly fascinated with the Nile, and it was through his study and exploration of the Nile basin that he became well known internationally. He travelled extensively in Sudan, Uganda, Tanganyika, and what used to be the Belgian Congo by river and on land – on foot with porters, using a bicycle, later by car and later still by plane. The low Aswan Dam had been build in 1903, but he realized how im-

portant it was to Egypt that provision should be made not only for the dry years but for a series of dry years. Irrigation storage schemes should be adequate in dealing with every situation, very much in the way in the Old Testament Joseph stored grain for the lean years. He was one of the first to realize the need for the "Sudd el Aali," the High Dam and Reservoir at Aswan. He was thrice awarded the Telford Premium by the Institution of Civil Engineers and in 1957 received the Telford Gold Medal, their highest award, for his outstanding work in this field.

For Egyptians, his work is primarily notable for its practical and descriptive aspects. However, Hurst's name is likely to survive generally for a related but different reason, thanks to the theoretical impact a certain formula that he devised has had, and doubtless will continue to have, on both geophysics and statistics. At first, it seems surprising that anything of the kind could come from an author so weak in mathematics and working so far from any major center of learning, but at second thought this very distance may explain both the birth of his idea and its survival. Had Hurst sought sound statistical advice, one may well fear he would have been directed toward some technique claiming to universal applicability, for example, toward the most difficult of all, spectral analysis. (As von Békésy said, "dehydrated cats and the application of Fourier analysis to hearing problems become more and more a handicap for research in hearing.")

Instead, Hurst investigated the Nile using a peculiar method of analysis of his own design, therefore completely unexplored. The method might be termed narrow and ad hoc but in fact has turned out to be eminently intrinsic. Not being pressed by time and having exceptionally abundant data at his disposal, he was in a position to compare this data and the standard model directly. He examined their respective effects upon the design likely to be adopted for the High Dam, that is, upon a certain standard engineering procedure used to assess a dam's effectiveness. The core of this procedure is a certain expression $R(d)$, which uses the past records to evaluate the size of an optimal reservoir designed for a horizon of d years. Furthermore, and in order to compare different rivers, Hurst divides $R(d)$ by a scaling factor $S(d)$.

For every one of the existing or other "reasonable" models, it was universally expected that $R(d)$ should be proportional to d^H, with $H = 1/2$, and nearly everyone thought it legitimate to pay scant attention to $S(d)$. In fact Hurst found that in the case of the Nile R/S is proportional to d^H with H much above 0.5 (present estimates lie around 0.9), and that for other rivers the same rule applies with a value of H that varies from case to case but almost always exceeds 0.5. The value of $H \sim 0.7$ may be viewed as typical.

One can imagine the amount of hard work implied in such research before the advent of computers – but of course the Nile is sufficiently important to Egypt to justify comparatively large expenditures (and to preclude forcing Hurst to retire).

Hurst was adamant in his conviction that his finding that $H > 1/2$ was significant, despite the fact that, the ratio R/S being unknown to statisticians, there was no test by which such significance could be assessed. Finally, at the ages of 71 and 75, he read two long papers on his discovery, and the potential importance of his work became recognized.

In the words of the British mathematician E. H. Lloyd (but substituting our own notation), "We are, then, in one of those situations, so salutary for theoreticians, in which empirical discoveries stubbornly refuse to accord with theory. All the researches described above lead to the conclusion that in the long run $R(d)$ [sic] should increase like $d^{0.5}$, whereas Hurst's extraordinarily well-documented empirical law shows an increase like d^H, where H is about 0.7. We are forced to the conclusion that either the theorists' interpretation of their work is inadequate or their theories are falsely based; possibly both conclusions apply."

Quoting from Feller 1951: "We are here confronted with a problem which is interesting from both a statistical and a mathematical point of view."

The fractional Brownian motion models of Chapter IX arose as a direct response to the Hurst phenomenon.

◁ It is hard to quibble with glowing comments such as those quoted above. Unfortunately, both were unwittingly based upon an incorrect reading of Hurst's claims. Lloyd's words show that this author was (at the time) unaware of the importance of the division of R by S, and Feller, basing his work upon a third party's verbal report (as he acknowledged), did not realize that a division by S had been performed. The value of Feller's work was not affected by this unawareness, because its bulk was devoted to a more rigorous proof of the $d^{0.5}$ prediction, and because the above comment came at the very end of his paper. The same misunderstanding, however, led other writers astray. For the importance of the division by S, see Mandelbrot & Wallis 1969c and Mandelbrot 1975w. Oddly enough, while the work in which I participated stresses the importance of S in various areas, my exorcism of the Hurst paradox is the only one that is unaffected by the omission of S.

◁ We see again in this instance that when a result is truly unexpected, it is hard to comprehend even by those best disposed to listen. ▶

JOSEPH EFFECT: POSTSCRIPT ON LITTLE-KNOWN EARLY EXAMPLES

Joseph Effect is the term I suggest to designate all phenomena characterized by seemingly endless persistence. The term originated in hydrology and refers of course to the Biblical story of Nile floods and Joseph the son of Jacob, but it seems right to apply the same term to all phenomena having a similar structure.

Earlier conventional wisdom still prevailing in certain circles did not give a thought to the possibility of endless persistence. Evidence suggesting persistence was ignored, dismissed, or described as suspect.

More recently, helped by various phenomena I have studied (of which examples are recorded and discussed in this Essay), this view was replaced by a revised conventional wisdom, according to which endless persistence must be faced sometimes, but only in underdeveloped sciences like hydrology or economics, or in somewhat marginal subbranches of physics such as the theory of $1/f$ noises. Until this book was otherwise finished, I had come to adopt a version of this view.

Now it turns out that both forms of conventional wisdom had been untenable all along. Clear-cut, unmistakable symptoms of Joseph Effect or syndrome, in the form of seemingly endless aftereffects, had already been recognized *over a hundred years ago* in two areas of physics of the highest orthodoxy: electrostatics and elasticity! The elder Kohlrausch contributed basic papers to both fields – Kohlrausch 1847, 1854 – and the elastic torsion of glass fibers was further studied by Boltzmann, in a paper discussed by Maxwell in the ninth edition of *Encyclopaedia Britannica*.

In each instance, the relevant observations are best summarized by saying that the after effects of a stimulation do not decrease exponentially but hyperbolically. One counter theory occurred to all the authors involved: they argued that the hyperbolas may be nothing but a transient effect that is bound to disappear if the experiment is carried far enough. In response, von Schweidler 1907 performed measurements in electrostatics that began by being about 100 seconds apart but lasted for 16 million seconds (200 days, through summer and winter). The same would-be transient behavior shows no sign of abating, in fact continues on the dot. Another counter theory is that the hyperbolas involved are but sums of many exponen-

tials. For example, an early "rough explanation" of the Joseph Effect in Leyden jars was based by Hopkinson upon the assumption that "glass may be regarded as a mixture of a variety of different silicates that behave differently."

It is remarkable that independent of one another, all the students of the Joseph Effect (old or new) should have reacted in essentially identical fashions. Furthermore, recent surveys of elasticity and electrostatics (to which my search had to be limited in this instance) suggest that both fields still lack good theories. To students of other fields characterized by the same syndrome, they provide no assistance and no challenge.

Those concerned with more recently discovered instances of the same syndrome have no reason for being ashamed of what they have achieved. One may even argue that fractal geometry could be the basis of an alternative broad explanation in all cases.

Anyhow, a philosophical comment is in order. It has become prudent to expect every branch of physics to include its own instances of Joseph Effect. The more successful branches no longer seem characterized by their success in dealing with it, but merely by the fact that they have the choice of centering their efforts on Joseph-free areas. While one might conceivably argue that it continues to be best to leave each field to wrangle with such "problem children" in private, the whole tradition of mathematics cries out for a common effort.

◁ See also the Postscript under FRACTAL TIME in Chapter XII. ▶

PAUL LÉVY (1886 - 1971)

Paul Lévy, whom I consider my Teacher even though he himself acknowledged no true pupil, achieved goals that Louis Bachelier only saw from afar. He lived long enough to gain recognition as possibly the greatest probabilist of all time, and (when nearly 80 years old) he finally came to occupy, at the Académie des Sciences, the seat that had been Poincaré's, then Hadamard's.

And yet, almost to the end of his active life, Lévy had been kept at arm's length by the University. Not only did the chair he coveted elude him to the end, but he was made to feel that the noncredit lectures he wanted to deliver might disrupt the curriculum. (He was not eligible for *Dictionary of Scientific Biography*, because the letter L was prepared when he was alive. See *World Who's Who in Science*, p. 1035.)

His life, thoughts, and opinions are documented at length in Lévy 1970, a book well worth reading because of a lack of self-conscious attempt to appear better or worse than life. Certain developments are best skipped, especially toward the end, but the best passages are splendid. In particular, he described at length and in touching terms both his fear of being "a mere survivor of the last century" and his feeling of being a mathematician "unlike all the others." This feeling was widely shared. In 1954, I had heard it expressed independently by John von Neumann who, knowing I felt close to Lévy, told me, in substance: "His works make me ill at ease; I think I understand how every other mathematician operates, but Lévy is like a visitor from a strange planet. He seems to have his own private methods of arriving at the truth."

He had few obligations to distract him, aside from a score of lectures each year as a professor of mathematical analysis at the Ecole Polytechnique. Working alone, he transformed probability theory from a small collection of odd results into a discipline in which rich and varied results could be obtained through methods so direct as to be truly classical. He became interested in the topic when asked to give one lecture on errors in the firing of guns. He was near 40, a promising man who never quite fulfilled his promise and a professor at Polytechnique at a time when the school favored its alumni. His major books were written at ages 50 and 60, and much of his work on Hilbert space-to-line Brownian functions came later.

Of the countless interesting tales to be found in his autobiography, I shall extract one more, because it relates to the short paper he devoted to the Olbers paradox, as relative to the Newtonian gravitation potential. Having as a 19-year-old student in 1904 independently discovered the Fournier model of the universe, Lévy believed that "the argument was so simple that I would not have thought of publishing it if, 25 years later, chance had not made me overhear a conversation between Jean Perrin and Paul Langevin. These illustrious physicists agreed that one could only escape the paradox by assuming the universe to be finite. I spoke up to point out their error. They did not seem to see my point, but Perrin was shaken by my self-assurance and asked me to write down my ideas, which I did."

◁ Apropos of results being "too simple to publish," the phrase appears often in Lévy's recollections, and the feeling must be widespread among major creators. If it were not, it would be hard to understand why so many achieve immortality as authors of "lemmas." The latter, of course, are propositions felt to be "too simple" in themselves and only justified by their role in proving real theorems, but creators' judgment about the relative importance of different statements is often inverted by History. ►

The remarks that follow paraphrase part of what I said at a ceremony in Lévy's memory: "First, let us mention his teaching at Polytechnique. The trace left in my memory by his spoken lectures has become very blurred, because chance had placed me all the way back in the rear of a large amphitheater, and Lévy's voice was rather weak and not amplified. The most vivid recollection is that of the resemblance some of us noticed, between his figure – long, grey, and well groomed – and the somewhat peculiar way he had of tracing on the blackboard the symbol of integration.

"But his written course notes were quite another matter. They were not the traditional well-ordered procession, beginning with a regiment of definitions and of lemmas followed by theorems, every assumption being clearly stated, this majestic flow being perhaps interrupted by the statement of a few unproven results, clearly emphasized as such. Rather, the recollection I have is of a tumultuous flood of remarks and observations. In his autobiography, Lévy suggests that in order to interest children in geometry, one should proceed as quickly as possible to theorems they are not tempted to consider evident. His method at Polytechnique was not all that different. To give an account of it, we are irresistibly attracted to images borrowed from geography and mountaineering. We are thus reminded of an old review of another great *Cours d'Analyse de l'Ecole Polytechnique.* The course had been taught by C. Jordan, and the reviewer was Henri Lebesgue. Because Lebesgue's disdain for Lévy's work was strong and public, it is ironic that his comments in praise of Jordan apply so well to Lévy. He was unlike 'a person who would attempt to reach the peak of an unknown region, but who would not allow himself to look around before reaching his goal. If led there by someone else, perhaps he may be able to look down upon many things, but he could not know what they are. In fact, one cannot generally see anything from a very high peak; mountaineers climb them only for the sake of the effort.'

"Needless to say, Lévy's course notes were not universally popular. To many excellent Polytechnique students, they were a source of worry when cramming for the general examination. In the ultimate rewrite, which I had to study in 1957 as his Maître de Conférences, all those features had become even more strongly accentuated. For example, the treatment of the theory of integration was frankly no more than an approxima-

tion. No one, he had written, can do a good job by trying to force his talent. It would seem that in his last course notes, his talent had been forced. But my recollection of the course he had taught to the class admitted in 1944 remains extraordinarily positive. Intuition, though it cannot be taught, can only too easily be thwarted. I believe that this is what Lévy was trying above all to avoid, and I think he had mostly succeeded.

"At Polytechnique I had heard many references to his creative work. One would praise it as being very important, then promptly add the comment that it did not contain a single fautless mathematical proof, and included infuriatingly many verbal arguments of uncertain footing. In conclusion, the most urgent thing was to make everything rigorous. This task has been performed, and today the intellectual grandchildren of Lévy rejoice in being accepted as full-fledged mathematicians. As one of them put it a moment ago, they see themselves as 'probabilists turned bourgeois.' I fear that too much may have been paid for this acceptance. In every branch of knowledge, there seem to be many successive levels of precision and generality. Some are unsuited to attack any but the most trivial problems. More and more, however, and in almost every branch of knowledge, one is able to push precision and generality to excess. For example, a hundred pages of preliminaries may be needed (without disclosing a new horizon) to make it possible to prove one theorem in a form that is slightly more general than before. Finally, some branches of knowledge are fortunate in allowing an intermediate level of precision and generality that may be termed classical. Paul Lévy's almost unique greatness lies in the fact that he was, at the same time, a forerunner and the classic in his field.

"Lévy rarely concerned himself with anything but pure mathematics. Also, those who have to solve a problem that has already been well-posed rarely find in his work a formula ready to serve them with no further effort. On the other hand, if I can believe my personal experience, Lévy's approach to more basic issues of formulation of chance makes him stand out more and more as a giant.

"Whether in the diverse topics to which the present book is devoted or in those I examine in other works, a proper mathematical formalization seems to demand very quickly either a conceptual tool that Lévy had already provided or a tool wrought in the same spirit and possessing the same degree of generality. More and more, the inner world which Lévy explored as if he were a geographer reveals itself as sharing with the world which surrounds us a kind of premonitory accord that is, without doubt, a token of his genius."

LEWIS FRY RICHARDSON
(1881 - 1953)

Even by the standards of the present chapter, the life of L. F. Richardson was wrought of strands that fail to become integrated along any predominant direction. (He was, incidentally, the uncle of Sir Ralph Richardson, the actor. See *World Who's Who in Science,* p. 1420. Also E. Gold in *Obituary Notices of Fellows of the Royal Society,* **9,** 1954, 217-235. This entry is summarized in Richardson 1960a and 1960s.)

In the words of his more influential contemporary G. I. Taylor, "Richardson was a very interesting and original character who seldom thought on the same lines as did his contemporaries, and often was not understood by them." To paraphrase Gold, his scientific work was original, sometimes difficult to follow, sometimes illuminated by lucid unexpected illustrations. In his studies of turbulence and in the publication that led to Richardson 1960a and 1960s, he was occasionally, and not unnaturally, groping, perhaps with a little confusion. He was breaking new ground in both these subjects and had to find his way with the assistance of a knowledge of advanced mathematics gained as he went – not drawn from a stock obtained in his university career. In view of his inclination to explore new subjects – or even "bits of subjects" – his achievement might seem surprising if one did not realize his amazing and orderly industry.

He had earned his Cambridge B.A. in physics, mathematics, chemistry, biology, and zoology, for he was uncertain as to the career he should follow. Helmholtz, who had been a physician before becoming a physicist, seemed to Richardson to have partaken of the feast of life in reverse order. At 47 he received a psychology degree in London.

He began his career at the Meteorological Office, where one of his first experiments consisted in measuring wind velocity within clouds by shooting steel marbles, with sizes varying from that of a pea to that of a cherry. A Quaker and conscientious objector in 1914-1918, he resigned when the Meteorological Office was integrated into the new Air Ministry after the War.

His book on weather prediction by numerical process, Richardson 1922-1965, was the work of a practical visionary, unfortunately tainted by a fundamental mistake. Indeed, when he attempted to approximate the differential equations of the evolution of the atmosphere with equations of finite differences, he hit upon the worst possible values of the elementary steps of space and time. Since the need for care had not yet been perceived, his mistake was hardly avoidable. Because of it, however, it took twenty years for the validity of his method to be recognized.

Nevertheless, one aspect of his book has survived very well, the concept of the cascade of energy from low to high frequencies. It is expressed in a parody of Jonathan Swift, which became famous and remains productive. Every forward step in the study of turbulence seems to provide it with a new variant.

Swift 1733, lines 337-340:
So, Nat'ralists observe, a Flea
Hath smaller Fleas that on him prey,
And these have smaller Fleas to bit 'em,
And so proceed ad infinitum.

Richardson 1922, p. 66:
Big whorls have little whorls,
Which feed on their velocity;
And little whorls have lesser whorls,
And so on to viscosity.
(in the molecular sense)

Quite naturally, he continued with the study of turbulence, with work that earned him election to the Royal Society. The second section of Richardson 1926 is titled "Does the Wind Possess a Velocity?" and begins as follows: "The question, at first sight foolish, improves on acquaintance." He then goes on to show how wind diffusion may be studied with no need to mention its velocity. In order to give an idea of the degree of irregularity of the motion of air, a fleeting mention is made of the Weierstrass function (which is continuous but has nowhere a derivative; see Chapter I for a reference to it by J. Perrin; also Chapters VI and XII for more detail). Unfortunately, the matter is dropped immediately. What a pity! It so happens that the Weierstrass function can readily be fitted to have the spectrum that Kolmogorov was to derive in 1941. Also, as pointed out by G. I. Taylor, Richardson defined the law of turbulent mutual dispersion of particles but failed by a hair's breadth to obtain the Kolmogorov spectrum. However, his papers remain interesting; each fresh glance at them seems to show some angle that had passed unnoticed.

More important, at least to me, is that most of Richardson's arguments can very easily be translated into the language of the "fractal" vision of turbulence.

One of his last experiments in turbulent diffusion (Richardson & Stommel 1948) required a large number of buoys, which had to be highly visible, hence preferably whitish in color, while remaining almost totally immersed so as not to catch the wind. His solution was to buy a large sack of parsnips, which were thrown from one bridge on the Cape Cod Canal while he made his observations from another bridge downstream.

He spent many years teaching in colleges off the beaten path. Then an inheritance enabled him to retire early to devote himself fully to the study of the psychology of armed conflicts between states, which he had been pursuing since 1919. Two books appeared after his death, Richardson 1960a and 1960s (Newman 1956, pp. 1238-1263 reprints the author's own summaries of these books). Also, several articles were issued, including Richardson 1961 (which is the work on the length of coastlines that was described in Chapter II).

WEIERSTRASS FUNCTION: HISTORICAL SKETCH

The publication of the Weierstrass function in duBois Reymond 1875 (under circumstances to which we shall return later in this entry) has been used in the Introduction to mark the beginning of the great crisis of mathematics that ended in 1922 with the clarification of the concept of topological dimension. Naturally enough, essentially the same function had occurred to other scholars, but it happens that the usual conflicting claims to first publication were not heard in the present context. All claimants to this discovery had abstained from publishing it! The story may have an interesting moral and is worth telling. The thread that led to the information recorded below started with van Emde Boas 1969, which led to Singh 1935-1953. On the general background, an excellent source is T. Hawkins 1970.

Outside the mainstream, the best known and oldest "predecessor" of the Weierstrass function is one constructed by Bernard Bolzano (1781-1848). His work on this account was written in 1834 but lay dormant until the 1920s. See Manheim 1964, p. 70, and Singh 1935, which gives on page 8 evidence that Bolzano was unaware of the interesting properties of his function.

By contrast, the story of the other claim is hardly ever mentioned. It begins with the death of a professor of mathematics in Geneva, Charles Cellérier (1818-1890). While he had not gained special distinction in his lifetime, the manuscripts he had kept unpublished turned out to be (in the words of a biographer) "a revelation." In one of them, a limit case of the Weierstrass function was used for the familiar purpose: as a counterexample to the long prevailing notion of the necessary differentiability of continuous functions. The text was promptly published, with a footnote (by Cailler) including the following words:

"Entirely written by the author on paper yellowed by time, this text was found in a folder marked 'Very important and I think new. Correct. Can be published as is.' The text is not dated and it will probably be impossible to determine whether or not the essential results it contains had been achieved before those of Weierstrass. ... The arguments are rigorous and the examples are simple: even today ... [it] would constitute at least an excellent *lesson* on the principles of analysis."

Concerning priority, one should note that according to recollections by Raoul Pictet, recorded in 1916, Cellérier had referred to this work in class when Pictet was his student, around 1860.

Furthermore, Riemann's students reported that he mentioned analogous examples in his lectures around 1861, but, again, he did not publish them. (See Manheim 1964, Dugac 1973, pp. 93-94. See also Dugac 1976, p. 306.)

Finally, Weierstrass presented his counterexample to the Berlin Academy on July 18, 1872, but then kept his notes to himself. They only appeared in Weierstrass 1895, his *Collected Works*, long after the main finding had been put in the record by a third party in duBois Reymond 1875, a paper that includes the following comments: "The metaphysics of these functions seems to hide many puzzles, as far as I am concerned, and I cannot get rid of the thought that [they] will lead to the limit of our intellect."

More recently, the "proof" that the Cellérier function is not differentiable anywhere has turned out to be incorrect; it has derivatives at some points. So does Riemann's function $\Sigma n^{-2}\cos(n^2 t)$. (See Gerver 1970; the solution of *any* problem posed by Riemann is an occasion for rejoicing.) Hence, Weierstrass preserves all his claims to fame. However, this does not matter for the purposes of the present sketch. We do not attempt to straighten out the priorities but to document an exceptional historical occurrence. Not only has no one rushed to make such functions public but many distinguished thinkers (some of them famous) experienced less fear of being "scooped" on this account than of being associated with a thought they did not seem to enjoy particularly. Some (like Darboux) came to act as if to forget the role they played in their youth around 1875. They were an uncannily reluctant group of revolutionaries.

◁ Since the Weierstrass function has been used to argue against the relevance of fine mathematics to mechanics and physics, it may be of some slight interest to mention its discoverer's view on this matter. To freely translate from his in-augural lecture for 1857 (see Hilbert 1932-1965, **3**, pp. 337-338), Weierstrass stressed that to him it did make a difference whether or not a mathematical theory was capable of applications. The physicist should not see in mathematics a simple auxiliary discipline and the mathematician should not consider the physicist's questions as a simple collection of examples for his methods. "To the question of whether it is really possible to extract something useful from the abstract theories that present-day mathematics seems to favor, one could answer that it was only on the basis of pure speculation that Greek mathematicians derived the properties of conic sections, long before one could guess that they represent the planets' orbits." ▶

GEORGE KINGSLEY ZIPF
(1902 - 1950)

Zipf was an American scholar who started by being a philologist but little by little came to describe himself as a statistical human ecologist. At the time of his death, after twenty years as Lecturer at Harvard, he had just published, apparently at his own expense, *Human Behavior and the Principle of Least Effort,* Zipf 1949-1965.

I know very few books (Fournier d'Albe 1907 is another) in which so many flashes of genius, projected in so many directions, are lost in so thick a gangue of wild notions and extravagance. On the one hand it includes a chapter dealing with the shape of sexual organs and another in which the Anschluss of Austria into Germany is justified by means of a mathematical formula. But on the other hand, the book has an array of figures and tables that hammer away ceaselessly at the empirical law that can be summarized by stating that social science statistics tend to be ruled by the hyperbolic distribution. Two typical applications of that law were cited in Chapter X of this Essay, and Zipf listed innumerable other instances where it provides the best combination of mathematical convenience and empirical fit.

The natural scientists reading this Essay will have recognized the power-law expressions, scaling laws, and so on, which they have accepted with no extraordinary emotion. Natural scientists see more than jest in the remark found in von Kármán's autobiography: "When a straight line, the mathematical evidence of a law, failed to appear on the paper, I suggested different kinds of logarithmic paper. If you cannot simplify the curve on one kind of paper, simplify the paper." Even the less bold among physicists would find it hard to imagine the fierceness of the opposition that the very same procedure and outcome experienced in the social sciences. The procedure was unforgivable because its outcome flew in the face of the Gaussian dogma, which has long ruled uncontested among professional statisticians and hence among social scientists. (The lognormal distribution – a refuge when the Gaussian is untenable – helped the opposition and delayed progress in the social sciences but not in physics.) Because of the struggle he went through, Zipf's work had been of considerable historical importance. Also, it has not yet been exhausted.

It is somewhat irrelevant therefore, yet worth noting, that he was an encyclopedist and a fighter more than a genuine innovator. Much of his writing is notable primarily for having kept alive the awareness of works of others, including Vilfredo Pareto. The laws of which he was the first or the only author were few and often were the less interesting ones.

We like to see sad stories have a happy ending, even a posthumous one when death cuts them short. Zipf, however, remains in purgatory. While struggling against a statistical dogma, he had forged his own conceptual dogma, which – if it had spread – might have been equally harmful.

One sees in him, in the clearest fashion – even in caricature – the extraordinary difficulties that surround any interdisciplinary approach.

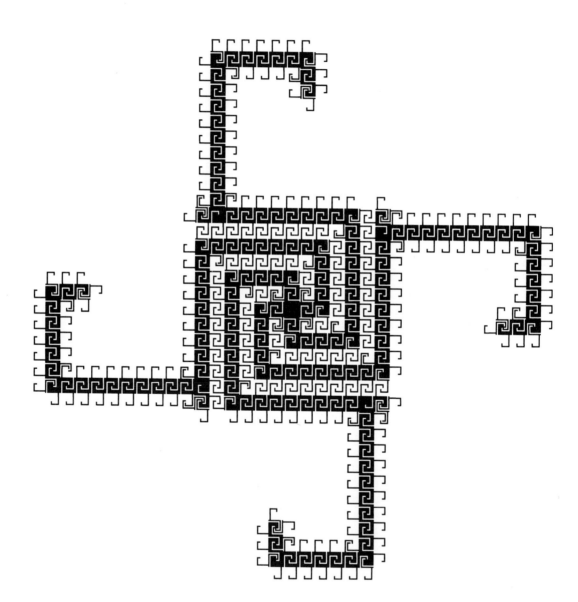

CHAPTER XII

Mathematical Lexicon and Addenda

Deliberately the text of this Essay, especially in early Chapters, is kept as free as possible of mathematical formulas, the definitions being informal and split between sections. I hope, however, that many readers will want to find out about the technical aspects of fractals. This lexicon is designed to help them and to direct them to the source material.

The entries are ordered alphabetically. A logical ordering would have started with HAUSDORFF MEASURE. No attempt is made to avoid repeating the earlier chapters' substance in different words. Most entries are organized in more or less logical order into several numbered subentries. A few entries (preceded by a black square) concern terminology.

Further, this lexicon includes various last minute complements.

This chapter makes free use of spectral analysis, also called harmonic or Fourier analysis. Indeed, spectral and fractal structures are intimately related, and neither can be followed far without the other. Nevertheless, they are definitely distinct mathematical structures, and each can be defined by itself. This is why, wishing to simplify and to stress the specificity of fractal structures, it seemed best in the body of the Essay to avoid completely a reference to spectra. Now we encounter the opposite pedagogic constraint, but of course we have neither the motivation nor the space for a tutorial on harmonic analysis.

We shall repeatedly encounter a well-known feature of mathematical terminology, one that is both endearing and infuriating: extensive reliance upon simple terms (stable, distribution, characteristic, structure, measure, and the like, not to mention dimension), each of which is given multiple conflicting meanings depending upon the context. As soon as several techniques are combined, one deals with phrases like "the proability distribution of the Schwartz distribution relative to the distribution of stellar matter!"

The terms coined in this Essay seek to avoid this pitfall – with the help of other rich vocabularies, especially that of the kitchen.

AFFINITY AND SIMILARITY

The terms *self similar* and *self affine* were applied in the text to either bounded or unbounded sets (without, I hope, introducing ambiguity). These meanings will now be distinguished, while keeping to Euclidean space.

Scaling. Many discussions of turbulence use self similar in a "generic" sense that includes the meaning of self affine. (In fact, the latter is a neologism.) Such casual usage is encountered in some of my own papers, but I hope not in this Essay. The generic meaning is best implemented by a term familiar to physicists and some mathematicians, *scaling* – which can be used both as a noun and as an adjective. See also AUTOMODEL.

1. SELF SIMILARITY

In the Euclidean space of dimension E, a real number r determines a transformation called *similarity*. It transforms the point $\mathbf{x} = (x_1,...x_\delta,...x_E)$ into the point $r(\mathbf{x}) = (rx_1,...rx_\delta,...rx_E)$, and hence transforms a set S into the set r(S).

BOUNDED SETS. One says that the bounded set S is *self similar* with respect to the ratio r and an integer N when S is the union of N nonoverlapping subsets, each of which is identical to r(S) except possibly for a displacement.

One says that the bounded set S is *self similar* with respect to the array of ratios $r^{(1)}...r^{(N)}$ when S is the union of N nonoverlapping subsets, respectively identical to $r^{(n)}(S)$ except possibly for a displacement.

One says that a bounded random set S is *statistically self similar* with respect to the ratio r and an integer N when S is the union of N non-overlapping subsets, each of which is of the form $r(S_n)$ where the N sets S_n are identical in distribution to S, except possibly for a displacement.

UNBOUNDED SETS. One says that the unbounded set S is *self similar* with respect to the ratio r and the focus F in S when the set r(S) is identical to S except possibly for a displacement.

One says that the unbounded set S is *self similar* with respect to the ratio r and an arbitrary focus if the preceding definition holds with F an arbitrary point in S.

2. SELF AFFINITY

In the Euclidean space of dimension E, a collection of positive ratios $\mathbf{r} = (r_1...r_\delta...r_E)$ determines a transformation called *affinity*. It transforms each point $\mathbf{x} = (x_1...x_\delta...x_E)$ into the point

$$\mathbf{r}(\mathbf{x}) = \mathbf{r}(x_1...x_\delta...x_E) = (x_1r_1...x_\delta r_\delta...x_E r_E),$$

hence a set S into the set $\mathbf{r}(S)$.

THE BOUNDED CASE. One says that the bounded set S is *self affine* with respect to the ratio vector **r** and an integer N when S is the union of N nonoverlapping subsets, each of which is identical to **r**(S) except for translation. (When two or more of the r_δ are identical, translation may be followed by rotation in the space defined by these coordinates.)

When a set S is *self affine* with respect to the ratio vector ·**r**, its projection on each coordinate axis is self similar. When several of the r_δ coordinates of **r** are identical, the projection of S on the corresponding subspace is self similar.

THE UNBOUNDED CASE. One says that the unbounded set S is *self affine* with respect to the ratio vector **r** and the focus F in S when the set **r**(S) translated by F−**r**(F) is identical to S (except, when two or more r_δ are identical, for possible rotation).

LAMPERTI SEMI-STABILITY. The preceding definition is often applied under the following conditions: (a) S is the graph of a function X(t) from scalar time t to a E−1 dimensional Euclidean vector; (b) one has $r_1 = \ldots r_\delta \ldots = r_{E-1} = r$, where r is a positive real number; (c) the remaining ratio r_E is different from r. In this case, a direct definition of self affinity runs as follows: A time-to-vector function X(t) is *self affine* with respect to the exponent α and the focal time t_0 if there exists an exponent $\log r_E / \log r = \alpha > 0$ such that for every h>0 the function $h^{-\alpha} X[h(t-t_0)]$ is independent of h. The random version of the concept is due to Lamperti 1962, who uses the term *semi-stability* for it. (Among the many ambiguous and misleading terms of mathematics, this must be one of the most unfortunate.)

3. AFFINE OR "JOSEPH" NOISES; NOISES WITH AN 1/f SPECTRUM

By definition, these noises are natural fluctuations with a measured spectral density approximately proportional to $f^{-\beta}$, $\lambda^{-\beta}$, $\omega^{-\beta}$ or $k^{-\beta}$, where β is a constant, and the letters f, λ, ω, and k are common alternative notations for the frequency.

The spectral power in frequencies above f is $f^{-\beta+1}$. Hence, for the noise X(ht), it is $f^{-\beta+1}h^{\beta-1}$, and for $h^{-\alpha}X(ht)$, it is $h^{-\alpha+\beta-1}f^{-\beta+1}$. In other words, such noises are self affine with the exponent $\alpha = \beta-1$. ◁ The preceding argument is heuristic. Though its final result is correct, a fuller justification often has to go beyond the Wiener-Khinchin form of spectrum, for example, to the conditional spectrum of Mandelbrot 1967b. See STATIONARITY AND KIN, 3. ▶

The above noises are by far less widely known than they deserve, and scholars or engineers acquainted with one among them tend to ignore the existence of others and therefore tend to underestimate the importance of the phenomenon they represent. The best-known examples are the $1/f$ noises in electronic contacts or semiconductors (a good general reference is van der Ziel 1954) and their possible cousins in contact noises of neurophysiology (Verveen & DeFelice 1974). One striking example concerns uncertainties in time as measured by an atomic clock: because they are $1/f$ noises, the error of measurement on one second (minute, hour) is 10^{-12} seconds (minutes, hours). Other $1/f$ noises also occur in the wobble of Earth's pole (Mandelbrot & McCamy 1970) and in music (Voss & Clarke 1975).

Much of this Essay springs from the recognition that yet another example of $1/f$ noise is involved in hydrology (Mandelbrot 1965h). This last instance is related to the Biblical story of Joseph the son of Jacob, hence the neologism "Joseph noises." (See Chapter XI, JOSEPH EFFECT.)

4. INFRARED AND ULTRAVIOLET CATASTROPHES

When $\beta > 0$, the spectral density of an affine noise is infinite at the origin and may be either nonintegrable (when $\beta \geq 1$) or integrable (when $\beta < 1$). The former symptom is the *infrared catastrophe*, and the second may therefore be called an *infrared crisis* (a neologism). In either case, the predominant factor lies in the low frequencies (hence these phenomena could be called "hyponoises"). When $\beta \leq 1$, the spectral density is nonintegrable at infinity, which is a different symptom, the *ultraviolet catastrophe*.

Mathematicians have worked hard at the ultraviolet catastrophe, since it is related to the problems of differentiability discussed by Weierstrass and Cantor. By self similarity, I have also made use of analogous considerations in the study of low-frequency phenomena.

Both for the Weierstrass function and the Cantor staircase, nondifferentiability is associated with an $f^{-\beta}$ spectrum (see SPECTRAL PROPERTIES under CANTOR AND LEVY STAIRCASES and under WEIERSTRASS FUNCTION AND KIN). Nevertheless, they are profoundly different functions. They are examples on opposite sides of a basic dichotomy between affine noises in Euclidean time – for example, the increments of a Gauss-Weierstrass or a fractional Brownian function – and affine noises in fractal time – for example, the increments of a Cantor staircase or the fractal events examined in Chapter VI. See FRACTAL TIMES and STATIONARITY AND KIN.

■ANOMALOUS DIMENSION – A term used in certain fields of physics. In

E-dimensional space one tends to expect correlation to be a solution of the Poisson equation, hence proportional to d^{-E+2}. If they actually turn out to be of the form d^{-F}, with $F \neq E-2$, the quantity $F+2$ is described as an *anomalous dimension*. See POTENTIALS, 3.

■ AUTOMODEL – By definition, an automodel process is a model (in the sense used in "model airplane") of itself. This term seems a mistranslation of a Russian word meaning "scaling" (see AFFINITY AND SIMILARITY, opening paragraph). *Automodel* should not compound the present already excessive terminological confusion.

The process used in Sinai 1976 is simply the discrete increment process of a fractional Brownian function as defined under BROWNIAN SETS, 5.

■ BROWNIAN MOTION – An ambiguous term. Except when ambiguity is tolerated or desired, in this Essay this term is restricted to the physical phenomenon. Otherwise, it is replaced by either of the Brownian sets listed below.

(Physicists compound the ambiguity by also using the term to designate the Gauss-Markov process of Uhlenbeck & Ornstein; see Chandrasekhar 1943.)

BROWNIAN SETS

1. THE LINE-TO-LINE FUNCTION

This is the (necessarily pedantic) term chosen in this Essay to denote the *classical ordinary Brownian motion*, also called *Wiener function*, *Bachelier function*, or *Bachelier-Wiener-Levy function*. The following definition is more cumbersome than those given in the text but allows an easy classification of the various generalizations to come.

ASSUMPTIONS. (a) The time t is a real number. (b) The variable x is a real number. (c) The parameter H is equal to $H = 1/2$. (d) The probability distribution $F(x) = Pr(X < x)$ is the error function defined as the distribution of the reduced Gaussian random variable with $\langle X \rangle = 0$ and $\langle X^2 \rangle = 1$.

DEFINITION. A random function is an ordinary scalar Brownian function $B(t)$ if, for all t and Δt,

$$Pr([B(t+\Delta t)-B(t)]/|\Delta t|^H < x) = F(x) .$$

In spectral terms, the spectral density of $B(t)$ is proportional to f^{-2H}, that is, to f^{-2}. The function $B(t)$ is continuous but is not differentiable.

REFERENCES. Lévy 1937-1954 and 1948-1965 have a well-deserved reputation for cryptic elegance and very personal style. However, they are unmatched for intuitive depth and simplicity. Businesslike recent references are too numerous to list.

2. METHODS OF GENERALIZING THE BROWNIAN FUNCTION

Of all continuous random processes, Brownian motion is among the easiest to define (which is why it was among the first) and among the easiest to study. In addition, it is the easiest to generalize, which gives it an especially central position. Indeed, each of the assumptions (a, b, c, and d) listed above in BROWNIAN SETS, 1 has a natural generalization, and essentially every process obtained by generalizing one or more of the statements "happens" to be significantly different from the original Brownian function and to have significant applications.

(a) The real time t may be replaced by a point in Euclidean space, alternatively by a point on a circle or a sphere.

(b) The real (scalar) X may be replaced by a point P in E-dimensional Euclidean space, alternatively by a point on a circle or a sphere.

(c) The parameter H may be given a value different from $1/2$. When $F(x)$ is Gaussian, H can be allowed to lie anywhere in the range $0 < H < 1$.

(d) The Gaussian distribution may be replaced by a non-Gaussian distribution

discussed in STABLE RANDOM ELEMENTS IN THE SENSE OF LEVY.

3. BROWNIAN LINE-TO-CIRCLE FUNCTION

A *Brownian line-to-line* function having been expressed as a real number, one can drop its integer part and multiply the fractional remainder by 2π. The result determines a point's position on the unit circle, and serves to define a Brownian line-to-circle function. This function is mentioned primarily to warn against confusing it with the very different function that follows.

4. BROWNIAN CIRCLE-TO-LINE FUNCTION

This is also called *periodic Brownian function*, *Brownian Fourier series*, or *Brown Wiener series*. The variation of the Brownian line-to-line function $B(t)$ between $t=0$ and $t=2\pi$ can be decomposed into two terms: (a) the trend, as defined by $B(0) + (t/2\pi)[B(2\pi)-B(0)]$, and (b) an oscillatory remainder. In this special case (but not in general!), these two terms happen to be statistically independent. The first is merely a straight line with a random Gaussian slope. The oscillatory term is identical in distribution to a *Brownian bridge*, defined as a

Brownian line-to-line function that has been pinned down or constrained to satisfy $B(2\pi) = B(0)$. This decomposition suggests the following concept.

The *Brownian circle-to-line* function is by definition that periodic function of t which coincides over the time span $0 < t \le 2\pi$ with the Brownian bridge adjusted so its average vanishes. This process turns out to be stationary; that is, a change in origin leaves the statistical rules of the process unchanged. It can therefore be represented by a random Fourier series. The coefficients turn out to be independent Gaussian random variables, having wholly random phases and moduli proportional to n^{-1}. In other words, the discrete spectrum is proportional to n^{-2}, that is, to f^{-2}.

PRACTICAL CONSEQUENCE. The simulation of $B(t)$ is necessarily carried out over a finite time span, viewed as $[0, 2\pi[$. We see that this simulation can rely on discrete finite Fourier methods. One computes a Brownian bridge using a fast Fourier transform, and the required random trend is then added separately.

REFERENCES. Paley & Wiener 1934 deserves its reputation for relentlessly hard algebra. However, the profound expository paragraphs of Chapters IX and X still deserve to be read. Kahane 1968 is recommended but results are stated only in the most general form known, and the original context in which they are simplest is not mentioned.

5. FRACTIONAL BROWNIAN LINE-TO-LINE FUNCTION

The definition of this function, denoted by $B_H(t)$, is best achieved by taking the definition of the ordinary Brownian line-to-line function and replacing the exponent $H = 1/2$ by a real number satisfying $0 < H < 1$. Cases where $H \ne 1/2$ are called *properly fractional*.

All the $B_H(t)$ are continuous and nondifferentiable. The earliest mention of them I could locate is in Kolmogorov 1940. Other scattered references and various properties are listed in Mandelbrot & Van Ness 1968. See also Lawrance & Kottegoda 1977.

In spectral terms, the spectral density of $B_H(t)$ is proportional to f^{-2H-1}. The exponent is not an integer, hence the term *fractional* as applied to this process. (See the entry below on FRACTIONAL INTEGRO-DIFFERENTIATION.)

The increments of $B_H(t)$ over successive unit time spaces are called a *discrete fractional Gaussian noise*. They have the correlation

$$2^{-1}[\,|d+1|^{2H} - 2\,|d|^{2H} + |d-1|^{2H}\,].$$

Setting $B_H(0) = 0$, the most characteristic property of $B_H(t)$ concerns the correlation between the past increment defined as $-B_H(-t)$ and the future increment defined as $B_H(t)$. The covariance is

$$\langle -B_H(-t)B_H(t)\rangle$$
$$= 2^{-1}\{\langle [B_H(t)-B_H(-t)]^2\rangle - 2\langle [B_H(t)]^2\rangle\}$$
$$= 2^{-1}(2t)^{2H}-t^{2H}.$$

To obtain the correlation, one divides this result by $\langle B_H(t)^2\rangle = t^{2H}$, thus obtaining $2^{2H-1}-1$. In the classical case $H=1/2$, this quantity vanishes, as expected. For $H>1/2$, the correlation is positive, expressing persistence, and it becomes 1 when $H=1$. For $H<1/2$, the correlation is negative, expressing antipersistence, and it becomes $-1/2$ when $H=0$.

The fact that this correlation should be independent of t even in cases when it does not vanish is merely a corollary of the self affinity of $B_H(t)$. To an educated intuition, the lack of dependence of said correlation on t is obvious, but observation suggests that many students of randomness expect independence only when the correlation vanishes. In all other cases, they begin by viewing it as surprising and/or disturbing.

PRACTICAL CONSEQUENCES. To generate a random function for all integer times between $t=0$ and $t=T$, it is customary to select an algorithm in advance with no regard to T, and then to let it run for a time dependent upon T. From this viewpoint, the algorithms needed to generate the persistent fractional Brownian functions are atypical: they necessarily depend on T. An example of dependence is described under GAUSS WEIERSTRASS FUNCTION AND KIN. Another example is encountered in the "fast" generator of Mandelbrot 1971f ◄which uses a procedure reminiscent of the physicists' group normalization techniques. ►

◄WARNING. Mandelbrot 1971f is marred by a potentially very disturbing misprint: in the first fraction on p. 545, 1 must be subtracted from the numerator and added to the whole fraction. ►

FRACTAL DIMENSION OF THE GRAPH. The proof that almost surely $D=2-H$ was provided by Orey 1970.

FRACTAL DIMENSION OF LEVEL SETS. Almost surely $D=1-H$. A slightly weaker result (nevertheless sufficient for practical purposes) was proven by Berman 1970. The result as claimed follows from the arguments in Marcus 1976.

6. FRACTIONAL BROWNIAN CIRCLE-TO-LINE FUNCTION

As a generalization of the Brownian circle-to-line function, it is quite natural to consider the sum of a Fourier series with independent Gaussian coefficients, wholly random phases and moduli proportional to $n^{-H-1/2}$.

The mathematical transformation

leading from Brownian to fractional Brownian circle-to-line functions is known as the fractional integro-differentiation (an entry is devoted to it later in this chapter). The same transformation also leads from Brownian to fractional Brownian line-to-line functions. (See, for the corresponding nonstochastic problem, Zygmund 1959, **II,** p. 133.)

WARNING. A superficial analogy would suggest that the fractional Brownian circle-to-line function might be obtained by the same process as in the nonfractional case: form the bridge of a fractional Brownian line-to-line function, then adjust it, and finally repeat it. Recall that a bridge is constrained to satisfy $B_H(2\pi) = B_H(0)$, and that adjustment consists in setting the average to vanish.

Unfortunately, the periodic function obtained in this fashion and the sum of the Fourier series with coefficients $n^{-H-1/2}$ are different random functions. In particular, the Fourier series is stationary, while the repeated adjusted bridge is not. For example, over a small interval on both sides of t=0, the repeated bridge is constructed by joining together two nonconsecutive subpieces of $B_H(t)$. The pinning down involved in the definition of the bridge is sufficient to make the combined piece continuous but is not sufficient to make it identical in distribution to a small piece made of two consecutive subpieces on both sides of, say, t=π.

Therefore, the adjusted bridge cannot conceivably be represented by any Fourier series with statistically independent coefficients.

As a practical consequence, the computation of fractional Brownian line-to-line functions by finite discrete Fourier methods is theoretically impossible and in practice is workable but very tricky. The most straightforward procedure is to (a) compute the appropriate circle-to-line function, (b) discard it except for a limited portion corresponding to the subinterval of time, from 0<t<t∗, and (c) add a separately computed very low frequency component. As H→1, this length t∗ must tend to 0.

FRACTAL DIMENSION OF THE GRAPH. It is almost surely 2–H (Orey 1970).

FRACTAL DIMENSION OF THE LEVEL SET. When the level set is nonempty, it is almost surely of dimension 1–H. This result is due to Marcus 1976 (who strengthened Theorem 5, p. 146, in Kahane 1968).

ABSENCE OF TRANSITION. It may be observed that the random Fourier series with independent Gaussian coefficients proportional to $n^{-1/2-H}$ converges for all H>0. When H crosses the value H=1, no singularity is encountered, merely an increase in smoothness, since the sum becomes not only continuous but also differentiable. By contrast, the fractional Brownian process is only defined up to the bound $H=1$. This difference in the range of admissible values of H confirms that these two processes are quite different. It also suggests that physical transition phenomena are more likely to be modeled by the line-to-line than by the circle-to-line function.

7. FRACTIONAL BROWNIAN LINE (OR CIRCLE)-TO-SPACE TRAILS

The most worthwhile comment concerns the dimension in the circle-to-space case. Assuming $H<1$, the trail's dimension is $\min(E, 1/H)$. This is part of Theorem 1, p. 143, in Kahane 1968. The remainder of said theorem involves that limited portion of a trail which corresponds to time instants forming a subset of dimension δ. The dimension of the partial trail is the smaller of the two quantities E and δ/H.

The claim made in Chapter IX, that the same result also holds for the line-to-space function, has not been proved in full detail, to my knowledge, but it is difficult to entertain any doubt.

8. SPACE-TO-LINE FUNCTIONS AND TRAILS

Lévy 1948 has opened up the study of Brownian functions from a space Ω to the real line. He considered primarily the case when Ω is *either* $\mathbb{R}^E$ with the ordinary distance, *or* a sphere in $\mathbb{R}^{E+1}$ with distance defined along a geodesic, *or* a Hilbert space. The Brownian function requires $B(P)-B(P_0)$ to be a Gaussian random variable of zero mean and variance $G(|PP_0|)$ with $G(x)=x$.

The early literature on the subject includes Chentsov 1957; Lévy 1957, 1959, 1963, 1965; and McKean 1963. See also Cartier 1969.

The generalization to functions $G(x)$ other than x is investigated by Gangolli 1967. He was anticipated on certain specific points by Yaglom 1957 (with scant credit to Lévy).

For calculations of dimensions, see Yoder 1974, 1975.

■ **CANTOR SETS** – This is a mathematical term of average degree of ambiguity. The original and most specific meaning (covered in this Essay by the more explicit term *triadic Cantor set*) is a "set of real numbers having a ternary representation of the form $t=0, t_1, t_2, \ldots t_n, \ldots$ in which the t_n can equal 0 or 2 but never 1" (Cantor 1883).

The same term is also applied generically and loosely to more general perfect nowhere dense fractals when they are self similar, or at least sufficiently regular to be viewed as nearly self similar.

CANTOR AND LEVY STAIRCASES

CANTOR STAIRCASE. We know that the nth approximation to the triadic Cantor set is made of 2^n segments of length 3^{-n}, scattered over $[0,1]$ and separated by cutouts. By cutting out the middle thirds of these segments, one advances from the nth to the $(n+1)$st approximation.

With each successive approximation, Cantor associated a polygon, namely, the graph of a continuous function $C_n(x)$ such that $C_n'(x)=0$ if x lies in a cutout, and $C_n'(x) = (3/2)^n$ otherwise, which implies that $C_n-C_n(0)\equiv 1$. As $n\to\infty$, the sequence $C_n(x)$ has a limit $C(x)$, which is called *Cantor staircase.*

Note that the inverse function $x(C)$ is *not* one-to-one. Whenever C is an integer multiple of 2^{-k}, $x(C)$ is indetermined over a whole interval that results by translation from a cutout of the Cantor set. However, to force $x(C)$ to become one-to-one, it suffices to make it left continuous.

LEVY STABLE SUBORDINATOR AND STAIRCASE. A stable subordinator $x\star(L)$ is the left continuous determination of a nondecreasing Lévy stable random process of exponent $D<1$. It is a one-to-one function. However, by inverting the procedure suggested above to construct $x(C)$, it is easy to make $x\star(L)$ into a a one-to-many function $x(L)$. It suffices to let $x(L)$ lie anywhere between $x\star(L)$ and $x\star(L+0)$. The inverse function of $x(L)$ is a one-to-one continuous nondecreasing function, which will be called *Lévy staircase.*

SINGULAR FUNCTIONS. $C(x)$ is the first and best example of a noted monster: a nondecreasing and nonconstant function that is singular in the sense that it is continuous but not absolutely continuous and hence nondifferentiable. It has a vanishing derivative almost everywhere, and its continuous variation manages to occur over a set of vanishing linear measure. $L(x)$ is a newer example of a singular function; its definition is more cumbersome but its behavior is in many ways simpler because it lacks any of the arithmetic properties which complicate the study of $C(x)$.

Any nondecreasing continuous function can be written as the sum of a singular function and of two other components, namely, a function made of discrete jumps and a differentiable function. These last two components are classical in mathematics and of wide use in physics. On the other hand, the singular component is still sometimes viewed as nonclassical in mathematics and is widely regarded in physics as pathological and useless. But of course this last opinion must be wrong, since in Chapters IV and V singular functions were all but unavoidable.

◁ Perrin seems aware of the possible usefulness of this function. In *Les atomes*, he asserts that the "true density vanishes almost everywhere, except at an infinite number of isolated points, where it reaches an infinite value." Then he adds: "The function that represents a [density] will form a *continuum* that presents an infinite number of singular points." In this second sentence, the word "isolated" is missing, as mathematics says it should be. ▶

SPECTRAL PROPERTIES. The spectrum of L(x) was obtained by Kahane & Mandelbrot 1965 (see DIMENSION (FOURIER).) It happens to be near identical on the average to the spectrum of the fractional Brownian line-to-line function and to be a smoothed form of the spectrum of the Gauss-Weierstrass function. The spectrum of C(x) has the same overall shape as that of L(x) with one difference, occasional sharp peaks of nondecreasing size. See Hille & Tamarkin 1929. (This difference plays a vital role in the theory of sets of unicity; Kahane & Salem 1963 and DIMENSION (FOURIER).)

FRACTAL TIME. The fact that the spectra of the function L(x) and of a fractional Brownian function are near identical on the average may seem paradoxical, as it seems to imply an identity between the functions themselves. However, this last inference would be unwarranted, as it holds only among Gaussian random functions. Since L(x) is drastically non-Gaussian, there is no paradox whatsoever.

Physicists and engineers like the Gaussian distribution too much. They consider it as being at least a representative first approximation. But such need not be the case. The above juxtaposition of examples illustrates that under the placid cover of the same spectrum, drastic differences of behavior are conceivable. Furthermore, both kinds of behavior are actually observed for different actual affine $1/f$ noises. An affine noise that is related to fractional Brownian motion varies all the time and can be said to be "on" in Euclidean time. An affine noise related to the Lévy staircase, on the contrary, is "on" solely for instants that belong to a fractal subset of the time scale.

The preceding distinction was alluded to in AFFINITY AND SIMILARITY, 3. It demonstrates that the recurrent attempts by theoretical physicists to find a model that would explain *all* $1/f$ noises in one swoop are predoomed, and that the experimental physicists content with measuring spectra had been indoctrinated into discarding much of the important information. Different geometric aspects of both kinds of $1/f$ noise have been studied in Mandelbrot 1965c, 1967i, 1969e.

COVERING BY SPHERES
OF A SET OR ITS COMPLEMENT

1. BRUTE FORCE METHOD
(CANTOR-MINKOWSKI-BOULIGAND)

Consider a metric space Ω, that is, a space in which one has defined the distance between two points. As a result, a ball of center ω can be defined as the set of all points such that their distance to ω is at most ρ. (Balls are "solids" and spheres are their "surfaces.") Given a bounded set S in Ω, there are many ways of covering it with balls of radius ρ.

The crudest method goes back to G. Cantor and consists in centering a ball on every point of S. It yields a smoothed-out version of S, to be called $S(\rho)$. Minkowski 1901 used this method to relate the concepts of length and area to the concept of volume. He had therefore to assume from the outset not only that Ω is a metric space but in addition that it is an E-dimensional Euclidean space in which the volume is defined. Write

$$\gamma(d) = [\Gamma(1/2)]^d / \Gamma(1+d/2).$$

Then the quantities

$$\lim \sup_{\rho \downarrow 0} \text{volume}(S(\rho))/\gamma(E-d)\rho^{E-d}$$

and

$$\lim \inf_{\rho \downarrow 0} \text{volume}[S(\rho)]/\gamma(E-d)\rho^{E-d}$$

are the upper and the lower Minkowski d-contents of S. When they are equal, their common value is the content of S. Minkowski observes that if $d>D$, the upper content of S vanishes and if $d<D$, the lower content of S vanishes.

The extension of this definition to nonstandard cases where d is a fraction is due to G. Bouligand 1928, 1929. The number it yields may be called the Minkowski Bouligand dimension D_{MB}. Bouligand recognizes that the value of D_{MB} is often different from what is expected on intuitive grounds, and more generally that D_{MB} is less desirable than D but often easier to evaluate. The case $E=1$ is discussed in Kahane & Salem 1963, p. 29, who confirm that D_{MB} is often equal to the Hausdorff Besicovitch dimension, cannot be smaller, and can be greater. See DIMENSION OF A MEASURE'S CONCENTRATE.

2. AN ECONOMICAL METHOD
(PONTRJAGIN & SCHNIRELMAN)

Among all collections of balls that cover a set S in the metric space Ω, the most economical is by definition the one that requires the smallest number of balls. When S is bounded, this smallest number is finite and can be denoted by $N(\rho)$. Pontrjagin & Schnirelman 1932 have advanced the expression

$$\lim \inf_{\rho \to 0} \log N(\rho)/\log (1/\rho)$$

as an alternative definition of dimension.

This approach was later taken up by Kolmogorov & Tihomirov 1959, who (inspired by Shannon's information theory) labeled $\log N(\rho)$ the ρ-entropy of S and studied it extensively. Their work has since been further extended by Hawkes 1974, who calls the corresponding dimension the lower entropy dimension, and the variant obtained by replacing $\lim \inf$ by $\lim \sup$ the upper entropy dimension. He shows that the Hausdorff Besicovitch dimension is at most equal to the lower entropy dimension; they often coincide but sometimes fail to do so.

Kolmogorov & Tihomirov 1959 also study $M\rho$, the largest number of points of S such that their mutual distances exceed 2ρ. Clearly,

$$\lim \inf_{\rho \to 0} \log M(\rho)/\log (1/\rho)$$

is still another form of fractal dimension. (For sets on the line, $N(\rho) = M(\rho)$.)

◁ They use the term *capacity* for $\log M(\rho)$, which is most unfortunate, because this term has an entirely different and better justified meaning in potential theory. In particular, one must avoid the temptation of designating the dimension in the preceding paragraph as a capacity dimension. See POTENTIALS, CAPACITIES, AND THE DIMENSION. ▶

3. COVERING OF THE COMPLEMENT OF A SET ON THE LINE (BESICOVITCH & TAYLOR)

When Ω is $[0,1]$ or the real line, a Cantor-like set S (as we saw in Chapter IV) is fully determined by its complement, which is the union of the maximal open intervals we called intermissions. Working within $[0,1]$, the lengths of the intermissions add up to 1 and (Chapter IV) were found to follow the hyperbolic distribution $\Pr(U>u)=\sigma u^{-D}$. As a result, the length λ_n of the nth intermission by decreasing size has an order of magnitude of $n^{-1/D}$.

For general line sets of zero Lebesgue measure, the question of the behavior of the λ_n is raised by Besicovitch & Taylor 1954. They start with the observation that the series $\Sigma \lambda_n^d$ converges if $d>D$ (and in particular converges to 1 when $d=1$). On the other hand, it diverges when $d<D$, from which it follows that D is the infimum of the real numbers d such that $\Sigma \lambda_n^d < \infty$. The virtue of this last definition lies in its potential generality: it applies to all sets for which the complement is a union of open intervals and defines what may be called the Besicovitch & Taylor exponent D_{BT}. It does not, however, warrant being made into an independent dimension, since Hawkes 1974 (p. 707) proves that D_{BT} coincides with the upper entropy dimension. One has in every case $D_{BT} \geq D$. See also PARADIMENSION.

For zero measure sets, the Fourier dimension and the Besicovitch Taylor exponent together bound the Hausdorff Besicovitch D. Such multiplicity of dimensions may seem tiresome, but it includes an important message. Like the Fourier dimension, the Besicovitch & Taylor exponent turns out to have arithmetic properties (see Kahane 1971, p. 89), but the Hausdorff Besicovitch dimension has absolutely no link to arithmetic; it alone is a purely fractal concept.

■ CURD AND WHEY – Curdling, curd, and whey are given new meanings.

Curdling will designate any cascade of instabilities resulting in contraction.

Curd will be used to designate a volume within which a physical characteristic becomes increasingly concentrated as a result of curdling.

Whey will therefore be a natural choice ◁incidentally, a choice Miss Muffet should not mind▶ to designate the space not occupied by curds.

Etymology. The term *curd* is derived from the old English *crudan,* 'to press, to push hard.' It is related etymologically to *clot* (of blood), *clod* (of earth), and *cloud.* In old English, *clud* designated both a cloud and a rock or a hill. *Clot* is also related to *cluster.* All this erudition – straight from Partridge 1958 – is not necessarily irrelevant, since the etymological kin of curd doubtless include fractal kin of interest.

Before settling on the pairing curd-whey, I had thought of clot-serum, which is less appetizing and not related to the various Swiss cheese structures.

DIMENSION (FOURIER)

S being a closed set on $[0,1]$, consider all the nondecreasing functions $\mu(x)$ that are constant outside of S, meaning that $d\mu(x)$ is supported by S, and denote their Fourier transforms by

$$\mu \star (f) = \int \exp(ifx) d\mu(x).$$

The greatest degree of smoothness of μ corresponds to the fastest possible rate of decrease of $\mu\star$. Let D_F be the largest exponent such that at least one function μ supported by S satisfies

$$\mu \star (f) = o(|f|^{-D_F/2+\epsilon})$$

whichever $\epsilon > 0$, but no μ satisfies

$$\mu \star (f) = o(|f|^{-D_F/2-\epsilon}) \text{ for some } \epsilon > 0.$$

When S is the whole interval $[0,1]$, D_F is infinite, therefore devoid of interest. On the contrary, when S is a single point, it is clear that $D_F = 0$. More generally, whenever S is of zero Lebesgue measure, one can prove that D_F is finite and at most equal to the Hausdorff Besicovitch dimension D of S. The inequality $D_F \leq D$ shows that the fractal and harmonic properties of a fractal set are related but not necessarily identical.

◁ To prove that these dimensions can differ, suppose that S is a set on a line for which $D = D_F$. When the same S is viewed as a set in the plane, D is unchanged but D_F falls down to 0. ►

A convenient way of summarizing the harmonic analysis properties of S is to call D_F the Fourier dimension of S. The equality between D_F and D characterizes a certain category of sets called sets of unicity or Salem sets (Kahane & Salem 1963, Kahane 1968).

This equality fails to hold for the classic triadic Cantor set, for which the spectral peaks that were mentioned in CANTOR AND LEVY STAIRCASES imply that $D_F = 0$. This finding is significant because the Cantor set had originally emerged out of G. Cantor's unsuccessful search for a set of unicity. (See Zygmund 1959, I, p. 196.) However, $D_F = D$ does hold (a) for special nonrandom Cantor sets such that r satisfies certain number theoretic properties and (b) for various random Cantor sets, merely because randomness can be made to break up every arithmetic regularity.

The original example of this second kind, due to R. Salem, is very complex. The next example is the Lévy set (Kahane & Mandelbrot 1965 and CANTOR AND LEVY STAIRCASES). Further examples are given in Kahane 1968 (Theorem 1, p. 165; see also his Theorem 5, p. 173), where it is shown that the image of a compact set S of dimension δ by a fractional Brownian line-to-line function of exponent H is a Salem set of dimension $(\min(1, \delta/H)$. (Here, it makes no difference whether S is random or as regular as Cantor's triadic set.)

DIMENSION (TOPOLOGICAL)

We did not dwell on topological dimension in the body of this Essay because (a) its value was intuitively obvious for every fractal actually studied thus far and (b) the topic is treated in many readily available references (e.g., Tietze 1965, Menger 1943) and in an exceptionally good book, Hurewicz & Wallman 1941. This book begins with the intuitive description given in 1912 by Poincaré. Here is a free translation:

"When we say that space has the dimension three, what do we mean? If to divide a continuum C it suffices to consider as cuts a certain number of distinguishable elements, we say that this continuum is of *dimension one*. If, on the contrary, to divide a continuum it suffices to use cuts which form one or several continua of dimension one, we shall say that C is a continuum of *dimension two*. If cuts which form one or several continua of at most dimension two suffice, we shall say that C is a continuum of *dimension three*; and so on.

"To justify this definition it is necessary to check how geometers introduce the notion of dimension three at the beginning of their works. Now, what do we see? Usually they begin by defining surfaces as the boundaries of solids or pieces of space, lines as the boundaries of surfaces, points as the boundaries of lines, and they state that the same procedure can not be carried further.

"This is just the idea given above: to divide space, cuts that are called surfaces are necessary; to divide surfaces, cuts that are called lines are necessary; we can go no further and a point can not be divided, a point not being a continuum. Therefore the lines, which can be divided by cuts which are not continua, will be continua of dimension one; the surfaces, which can be divided by continuous cuts of dimension one, will be continua of dimension two; and finally space, which can be divided by continuous cuts of two dimensions, will be a continuum of dimension three."

◁ The preceding words are patently inappropriate to fractal (Hausdorff Besicovitch) dimension. In the case of the interiors of the various islands drawn in the book, the topological and fractal dimensions coincide and both equal two, but the coastlines are an entirely different matter: topologically of dimension 1, as mentioned by Poincaré, but fractally of dimension above 1. ▶

Now to freely quote from Hurewicz & Wallman: In 1913 "Brouwer constructed on Poincaré's intuitive foundation a precise and topologically invariant definition of dimension, which for a very wide class of spaces is equivalent to the one we use today. Brouwer's paper remained unnoticed for several years. Then in 1922, independently of Brouwer and of each other, Menger and Urysohn recreated Brouwer's concept, with important improvements.

"Before then mathematicians used the term dimension in a vague sense. A configuration was called E-dimensional if the least number of real parameters needed to describe its points, in some unspecified way, was E. The dangers and inconsistencies in this approach were brought into clear view by two celebrated discoveries of the last part of the 19th century: Cantor's one-to-one correspondence between the points of a line and the points of a plane, and Peano's continuous mapping of an interval on the whole of a square. The first exploded the feeling that a plane is richer in points than a line, and showed that dimension can be changed by a one-to-one transformation. The second contradicted the belief that the dimension of a space could be defined as the least number of continuous real parameters required to describe the space, and showed that dimension can be raised by a one-valued continuous transformation.

"An extremely important question was left open: Is it possible to establish a correspondence between Euclidean E-space and Euclidean E_0-space combining the features of both Cantor's and Peano's constructions, that is, a correspondence which is both one-to-one *and* continuous? The question is crucial since the existence of a transformation of the stated type between Euclidean E-space and Euclidean E_0-space would signify that dimension (in the natural sense that Euclidean E-space has dimension E) has no topological meaning whatsoever! The class of topological transformations would in consequence be much too wide to be of any real geometric use.

"The first proof that Euclidean E-space and Euclidean E_0-space are not homeomorphic unless E equals E_0 was given by Brouwer in 1911. However, this proof did not explicitly reveal any simple topological property of Euclidean E-space distinguishing it from Euclidean

E_0-space and responsible for the nonexistence of a homeomorphism between the two. More penetrating, therefore, was Brouwer's procedure in 1913 when he introduced an integer-valued function of a space which was topologically invariant by its very definition. In Euclidean space, it is precisely E (and therefore deserved its name).

"Meanwhile Lebesgue had approached in another way the proof that the dimension of a Euclidean space is topologically invariant. He had observed in 1911 (Lebesgue 1972–, **4**, 169-210) that a square can be covered by arbitrarily small 'bricks' in such a way that no point of the square is contained in more than three of these bricks; but that if the bricks are sufficiently small at least three have a point in common. In a similar way a cube in Euclidean E-space can be decomposed into arbitrarily small bricks so that not more than E+1 of these bricks meet. Lebesgue conjectured that this number E+1 could not be reduced further; that is, for any decomposition in sufficiently small bricks there must be a point common to at least E+1 of the bricks. (The proof of this theorem was given by Brouwer in 1913.) Lebesgue's theorem also displays a topological property of Euclidean E-space distinguishing it from Euclidean E_0-space and therefore it also implies the topological invariance of the dimension of Euclidean spaces."

A further reference to the above-mentioned paper by Lebesgue is found in PEANO CURVES, 1.

The correspondence between Cantor and Dedekind and other aspects of the early history of the notion of topologically invariant dimension are discussed in Dauben 1974, 1975. Cantor's work in 1877 was of vital importance, but selecting it as the point of departure of the great crisis of mathematics would hardly change the dates 1875-1922. See also Dugac 1976, p. 335.

DIMENSION OF A MEASURE'S CONCENTRATE

The interrelationship between Hausdorff Besicovitch and similarity dimensions is sufficiently important to be stated once again. The basic concept is the former one but, (a), it seldom yields itself to intuitive argument and, (b), there is in many cases a natural easy standby: similarity dimension. I suspect that every mathematician seeking the value of some D begins with the corresponding similarity dimension, but of course few confess to this step in writing.

These two definitions of dimension, however, are not only different in princi-ple but simple sets exist for which they yield different values. Such is, for exam-ple, the case (see FRACTALS, 4, 5) for the open everywhere dense fractals that are the core of the study of relative intermittency. Their similarity dimension is $E -$ as is their Minkowski Bouligand dimen-sion. Since the Hausdorff Besicovitch dimension becomes even less intuitive for open than for closed sets, I was led in the study of relative intermittency to de-velop yet another set of definitions of dimension. They were not needed in this Essay, but I would like to describe them in order to solicit comment.

Like in the Hausdorff theory, we be-gin with a metric space Ω, but we do not consider a set S in Ω, rather a measure μ with the following properties. (a) $\mu(S)$ is defined when S is a ball of radius δ in Ω (and also for other suitably restricted sets). (b) It satisfies $\mu(S) > 0$ and also $\mu(\Omega) = 1$. It follows that "the set in which $\mu > 0$" is strictly speaking identical to Ω. In the interesting cases, however, intuition suggests that μ is overwhelm-ingly "concentrated" over a very small portion of Ω. In order to transform this intuitive feeling into a definition, the fol-lowing approach (though little explored as yet) seems promising.

Given $\delta > 0$ and $0 < \lambda < 1$, consider all the sets $\Sigma_{\delta\lambda}$, each of which is the union of

$N(\delta,\lambda,\Sigma_{\delta,\lambda})$ balls of radius δ and is such that $\mu(\Omega - \Sigma_{\delta,\lambda}) < \lambda$. Define

$$N(\delta,\lambda) = \inf N(\delta,\lambda,\Sigma_{\delta,\lambda}).$$

The following expressions are dimensionlike.

$$\lim \inf_{\rho \downarrow 0} \log N(\rho,\rho)/\log (1/\rho)$$
$$\lim \inf_{\delta \downarrow 0} \log N(\delta,\lambda)/\log (1/\delta)$$
$$\lim \inf_{\lambda \downarrow 0} \lim \inf_{\delta \downarrow 0} \log N(\delta,\lambda)/\log (1/\delta)$$

The second one is only recommended when its value is independent of λ (so that the third definition coincides with the second). In intuitive shortcuts, of course, one replaces $\inf N(\delta,\lambda)$ by the actual $N(\delta,\lambda,\Sigma_{\delta,\lambda})$ relative to some sensible economical covering $\Sigma_{\delta,\lambda}$.

FRACTAL

1. ROUGH DEFINITION

Fractal is a neologism coined by the author, in 1975, from the Latin adjective *fractus,* which is related to the verb *frangere*, "to break." This coinage responds to the need for a term to denote a mathematical set or a concrete object whose form is extremely irregular and/or fragmented at all scales.

The concept itself had been studied for a long time by mathematicians, but somehow they had not felt the need of a term to designate it – other than *strange, exceptional,* and the like.

2. MATHEMATICAL DEFINITION

A *fractal set* was defined in the Introduction as a set for which one has

Hausdorff Besicovitch dimension > topological dimension.

All fractals studied in this book are sets in a Euclidean space of dimension $E < \infty$. They may be specified more fully by being called Euclidean fractals.

3. SELF CRITIQUE OF THE DEFINITION

I am fully aware of the arbitrariness present in the above mathematical definition, and one should not be surprised if the boundary between fractals and the nonfractal "standard" sets comes eventually to be moved elsewhere.

◁ As a matter of fact, the French version deliberately avoided advancing a definition. ▶

A first element of arbitrariness involves nonrectifiable curves for which $D = 1$ and other sets for which $D = D_T$ but the measure using $h(\rho) = \gamma(D)\rho^D$ is infinite (it cannot vanish). Since calling such sets either fractal or nonfractal would be equally arbitrary, I settled on the shorter definition. If a good reason arises, this definition may be changed. See HAUSDORFF MEASURE, 7.

A second element of arbitrariness involves the choice of the basic fractal dimension. I settled on the Hausdorff Besicovitch dimension because it is of the widest applicability, is not related to arithmetic, and has been studied most carefully.

4. NONCOMPACT FRACTALS

All the fractal sets used in this Essay are closed and compact sets in a Euclidean space, but the above definition (and many of its possible variants) also allow a bounded fractal to be noncompact.

Consider, for example, with Besicovitch 1935 (see also Billingsley 1967), the points of $[0,1]$ for which there exist a binary development such that the ratio

n^{-1} (number of 1 ' s in the first n bits)

converges to a limit p_1 other than $1/2$. Such points form an open set; the limit of a sequence of such points need not be in the set. Also, they are everywhere dense, since any point of $[0,1]$ is the limit of many sequences or points in that set. Let us call it the Besicovitch set. Its Hausdorff Besicovitch dimension is

$$-p_1 \log p_1 - (1-p_1) \log(1-p_1),$$

while its topological dimension is 0. It is therefore a fractal according to the above definition.

However, both mathematically and intuitively, the open Besicovitch fractal "feels" very different from the Cantor fractal, which is closed. The same is true of a random variant of the Besicovitch fractal encountered in the model of turbulent intermittency developed in Mandelbrot 1974f, 1974c and described as "weighted curdling." See also Kahane 1974, Kahane & Peyrière 1976.

5. RELATIVE INTERMITTENCY

Replacement of absolute by weighted curdling leads to replacement of absolute intermittency by the more general notion of relative intermittency. To assess the latter concept, it is good to remember that many of the studies of natural fractals in this Essay had negated some unquestionable knowledge about Nature.

We tried to forget in Chapter IV that between fractal errors the noise that causes them weakens but does not desist.

We neglected in Chapter V our knowledge of the existence of interstellar matter, the distribution of which is doubtless *at least* as irregular as that of the stars. To quote deVaucouleurs 1970, "It seems difficult to believe that, whereas visible matter is conspicuously clumpy and clustered on all scales, the invisible intergalactic gas is uniform and homogeneous ... [its] distribution must be closely related to ... the distribution of galaxies."

Other astronomers describe intergalactic matter by the terms *wisps* and *cobwebs*. (In fact, the notion that it is impossible to define a density is stronger and more widespread for interstellar than stellar matter!)

And when we studied in Chapter VI the pastrylike sheets of turbulent dissipation, we used an obviously oversimplified view of laminar flow.

Had we taken the time to write about minerals (see Mandelbrot 1975m), the obvious model using closed fractals would have implied, for example, that between the regions where copper can be mined the concentration of copper vanishes, while in fact it is in most places very small but vanishes nowhere.

In each case, some areas of less immediate interest were artificially emptied to make it possible to use *closed* fractals, but eventually these areas must be filled. Luckily, this last task may in many cases be performed using *open* fractals, such as, for example, the above-mentioned randomized form of the Besicovitch set.

The fact that these sets are everywhere dense leads to new themes. For example, the Hausdorff Besicovitch dimension remains fully workable, but the similarity and Minkowski Bouligand dimensions are equal to E rather than to the Hausdorff Besicovitch D.

As a result, if one is to do justice to noncompact fractals, a careful study is required. The brief reference to them that formed Chapter IX of the French version was too allusive, and a detailed study would take us too far afield. The topic had better be put aside until a more appropriate occasion.

FRACTAL TIMES, INTRINSIC & LOCAL

In the study of many natural phenomena, it is useful, aside from the "clock time," to consider a second time scale intrinsic to the phenomenon in question. In the case of a stationary point process, such as a Poisson process of independent events, the intrinsic time θ is discrete and changes by unity at each instant when an event occurs. Whichever the origins of their scales, the two times are asymptotically proportional. However, there also exist point processes such that when the common origin of their scales belongs to the process, one has

$$(\theta = \text{intrinsic time}) \sim (t = \text{clock time})^D,$$

with $D < 1$. A mathematical study of recurrent (also called renewal) processes of this type was performed in Feller 1949 and the first concrete application (or so I believe) concerned the problem described in Chapter IV, starting with Berger & Mandelbrot 1963.

Now let us generalize beyond discrete events, for example by interpolating the Berger & Mandelbrot model as done in Chapter IV (following Mandelbrot 1965c, 1967b). Then the intrinsic time becomes a continuous and monotone function $\theta(t)$ of the clock time t. The key assumption of Chapter IV is that this function can be singular; that is, it can be constant except over a fractal subset of clock time, the fractal dimension of which is $D < 1$. The prototypes of $\theta(t)$ are the Cantor and Lévy staircases, and $\theta(t)$ is the uniformizing transformation mentioned in the caption of Plate 101.

CONDITIONAL STATIONARITY AND "BIG BANGS." A priori, it seems that the relationship $\theta \sim t^D$ can only hold with some unique specified origin. Using a terminology familiar to cosmologists, it seems related to a "big bang" at $t = \theta = 0$. However, the discussion of conditional versus absolute stationarity in Chapter V shows that such is definitely *not* the case. With discrete time events, we have a sequence of equal bangs very irregularly spaced in time, and with continuous events, we deal with a sequence of infinitely many infinitely minute bangs. The same relation $t = \theta^D$ would hold if the origin is placed at any of the bangs, while a randomly selected origin almost surely yields θ = constant.

◁ Formally, the preceding comments could be fitted to the Einstein deSitter relation $R \sim t^{2/3}$. Is it conceivable that this observation should go beyond mere use of cosmological terminology? ▶

LOCAL TIME. The above continuous form of intrinsic time is intimately related with the notion of "local time," which arose in the study by Lévy 1948 (pp. 239-241) of the returns of a Brownian one-to-one function to a specified level, say to zero. Since these returns have a vanishing (Lebesgue) measure in clock time, Lévy introduced an ad hoc time in which their sojourns at this level can be compared (a recent reference is McKean 1975). Since the models of Chapter IV (see p. 86) were stimulated by the resemblance I saw between clusters of errors and of Brownian zero crossings, it is natural that the errors' intrinsic time and the crossings' local time should both be fractal and have identical mathematical descriptions.

In Chapter V, when we generated clusters through a Lévy motion, the equivalent of intrinsic time was the mass scattered by the motion.

FRACTAL TRANSFORMATIONS AND RESTRICTIONS OF TIME. There are many ways of combining a function X(t) with a fractal time transformation $\theta(t)$ independent of $X(t)$. For example, $X[\theta(t)]$ is constant except where fractal time varies. When $\theta(t)$ and $X(\theta)$ are scaling (self affine or self similar), so is $X[\theta(t)]$.

One may also construct a function $X(*t)$ by restricting $X(t)$ so it is defined only for the instants t where θ varies. We did so when in Chapter V we introduced the Cauchy process by considering a Brownian function $B^{(1)}(t)$ at those instants where another statistically independent Brownian function $B^{(2)}(t)$ happens to vanish. Suppose that X is a vector in $\mathbb{R}^E$ and the trail of the unrestricted $X(t)$ is of dimension $D_0 < E$, and that the instants where θ varies are of dimension D_2. Thus the rule of thumb is that the trail of $X(*t)$ is of dimension $D = D_0 D_2$. In the example in Chapter V, we had $D_0 = 2$, $D_2 = 1/2$ and $D = 1$. For other examples, see Blumenthal & Getoor 1960m, p. 267; Kahane 1968, pp. 165 and 173. (The latter reference also occurs at the end of DIMENSION (FOURIER).)

NEW APPLICATION TO TRANSIENT PHOTOCONDUCTIVITY. The procedure used in Chapter IV is also of great help in the study of transient photoconductivity in amorphous materials. The main references (of which I became aware on November 5, 1976) are Montroll & Scher 1973 and Scher & Montroll 1975. It is assumed that electrons and holes are trapped in their sites for random time spans that follow the hyperbolic distribution. Therefore the motion of an electron or a hole is extremely intermittent, and is restricted to instants that belong to a finite approximation of a fractal set, namely the set considered in Berger & Mandelbrot 1963. In particular, the spatial displacement of an electron during the clock time t is overall proportional to t^D, assuming it is measured with a yardstick equal to a lattice scale. Very interesting effects are induced by the fact that this phenomenon has a finite outer scale equal to the material's thickness.

Another feature of this theory of amorphous materials has a familiar ring to readers of this Essay. The basic postulate is that the distribution of trapping time is hyperbolic. This postulate was inspired by the discovery of an empirical scaling law, and it has not yet been reduced to any basic law of physics.

FRACTIONAL INTEGRO-DIFFERENTIATION

In pure mathematics, the transformation that led from the Brownian line-to-line function $B(t)$ to its fractional generalization $B_H(t)$ is classical but somewhat obscure. It is the Riemann Liouville fractional integro-differentiation of order $H-1/2$. The underlying idea is that the order of integration and/or differentiation is not necessarily an integer.

Fractional integro-differentiation is important in the Riesz theory of hyperbolic partial differential equations, and it has other numerous but scattered applications (see Ross 1975; Oldham & Spanier 1974; Lavoie, Osler & Tremblay 1976). The application to probability (with references back to the little-known paper Kolmogorov 1940) is newer and less well known. It is discussed in Mandelbrot & Van Ness 1968.

When its order $H-1/2$ is positive, the Riemann Liouville operation is a form of integration, because it increases a function's smoothness. Smoothness equals local persistence, but smoothness obtained by integrationlike operations hinges upon a function's global properties. Therefore, if there are to be limits to a self similar function's susceptibility to smoothing, these limits must be related to its long-range behavior.

A periodic function has a trivial long-range behavior, which fits with the fact that, as we have already noted, the value of H for a fractional Brownian circle-to-line function has no upper bound. Fractional integration of order $H-1/2$ exceeding $1/2$ smoothes a Brownian function enough to make it differentiable. On the contrary, line-to-line functions have a very significant long-range behavior, which is why the corresponding $H-1/2$ can at most equal $1/2$ and $B_H(t)$ is never differentiable.

When $H-1/2 < 0$, the Riemann Liouville operation is a form of differentiation, because it enhances irregularity that depends on local behavior. Hence the Brownian circle-to-line and line-to-line functions exhibit the same limitations with respect to the range of acceptable exponents of fractional differentiation. In either case, local irregularity prohibits differentiation beyond $H=0$, hence the order $0-1/2 = -1/2$.

◁ The idea of fractional integro-differentiation occurred to Leibniz as soon as he developed his version of calculus and invented the notation $d^k F/dx^k$. Writing to de l'Hospital on September 30, 1695, he included the following remarks. (I translate freely from Leibniz 1849, **II, XXIV,** 197ff.): "Jean Bernoulli seems to have told you of my having mentioned to him a marvellous analogy which makes it possible to say in a way that successive differentials are in geometric progression. One can ask what would be a differential having as its exponent a fraction. You see that the result can be expressed by an infinite series, although this seems removed from Geometry, which does not yet know of such fractional exponents. It appears that one day these paradoxes will yield useful consequences, since there is hardly a paradox without utility. Thoughts that mattered little in themselves may give occasion to more beautiful ones." Responding to a December 17/27, 1965, letter from Joh. Bernoulli (Leibniz 1849, **III.1**, XX, 222ff.), Leibniz further elaborated on the above thoughts in a letter dated December 28, 1695 (Leibniz 1849, **III.1**,XXX, pp. 226ff.). ▶

HAUSDORFF MEASURE AND HAUSDORFF BESICOVITCH DIMENSION

Convenient references on this topic, in addition to a chapter in Hurewicz & Wallman 1941, are Rogers 1970 and Federer 1969.

1. HAUSDORFF MEASURE

The thought that "the general notion of volume or magnitude is indispensable in investigations on the dimensions of continuous sets" had occurred to Cantor but he published nothing on this account. (Around 1900, the problem's difficulty had become apparent and Lebesgue expressed doubt that Cantor could have reached any significant result.) The idea was furthered in Carathéodory 1914 and implemented in Hausdorff 1919.

The classical method for evaluating a planar shape's area begins by approximating this shape S by a collection of very small squares and by adding these squares' sides raised to the power $D = 2$. Carathéodory 1914 had extended this traditional approach by avoiding reliance on coordinate axes. This can be done by replacing squares by circles. In addition, it is convenient to postpone as much as possible any explicit reference to the fact that one deals with shapes of known dimension imbedded in a space of known dimension. Observe therefore that when a planar shape is covered by circles, it is a fortiori covered by balls of which these circles are meridians, so they have the same value of the radius. Hence, to avoid prejudging the planar character of S, it suffices to cover it by balls instead of circles.

Furthermore, a ball is merely the set of points such that their distance from an origin ω does not exceed a prescribed radius ρ. This definition continues to hold when the space Ω in which one works is not Euclidean, as long as a distance is defined. Such spaces are called metric (which is why the Hausdorff measure is a metric concept). If one believes S to be a surface, one obtains its approximate contents by adding expressions of the form $\pi\rho^2$ corresponding to all the covering balls. More generally, if one deals with a standard d-dimensional shape, one will add expressions of the form $\gamma(d)\rho^d$, where

$$\gamma(d) = [\Gamma(1/2)]^d / \Gamma(1+d/2).$$

The argument up to this point is due to Carathéodory 1914, who pursues it further to define a "length" or an "area" for nonstandard shapes. The next step was taken by Hausdorff 1919, who allows d to be fractional (the function $\gamma(d)$ had purposely been written in a fashion that is readily interpolated) and notes that instead of limiting oneself to elementary measures that are powers of ρ, one can use any positive gauge function $h(\rho)$ that tends to 0 with ρ.

The function $h(\rho)$ having been prescribed, a finite covering of the set S by balls of radii ρ_i can be said to have the measure $\Sigma h(\rho_i)$. To achieve economy in covering, one will consider all the coverings that involve the same $\rho = \sup \rho_i$, and one will form the corresponding infimum

$$\inf_{\rho_m < \rho} \Sigma h(\rho_i).$$

In a final step, one lets $\rho \to 0$. Since the constraint $\rho_m < \rho$ becomes increasingly stringent, the expression $\inf \Sigma h(\rho_m)$ can only decrease, and since it is positive, it has a limit

$$\lim_{\rho \to 0} \inf_{\rho_m < \rho} \Sigma h(\rho_m).$$

This limit defines the h-measure of E.

When $h(\rho) = \gamma(d)\rho^d$, h-measure is called d-dimensional.

When $h(\rho) = 1/\log|\rho|$, h-measure is called logarithmic.

A gauge function $h(\rho)$ is called intrinsic for S and denoted by $h_S(\rho)$ if the h_S-measure of S is positive and finite. For the standard shapes in Euclid, the intrinsic gauge function is always of the form $h_S(\rho) = \rho^D$ with some integer value of D. The first step in the generalization to fractals allows an intrinsic $h_S(\rho) = \rho^D$ with noninteger D. Such is the case for the simple nonrandom self similar fractals (Cantor sets and Koch curves).

On the other hand, in the case of random fractals, even self similar ones, the intrinsic $h_S(\rho)$ is often more complicated. If, for example, $h_S(\rho) = \rho^D |\log \rho|$, the h-measure of S with respect to $h(\rho) = \rho^D$ vanishes; the shape has a shade less substance than if it were D-dimensional but more than if it were $(D-\varepsilon)$-dimensional. An illustrative example is provided by Brownian motion in the plane, for which Lévy finds $h_S(\rho) = h^2 \log \log (1/\rho)$. (See S. J. Taylor 1964.)

The 2-dimensional measure of any bounded set in the plane being finite, functions such as $h^2/\log(1/\rho)$ need not even be considered.

Much of the work on determining $h_S(\rho)$ for random sets is coauthored or authored by S. J. Taylor; a recent reference is Pruitt & Taylor 1969.

2. HAUSDORFF BESICOVITCH DIMENSION: DEFINITION

The definition of Hausdorff measure as formulated does not require any advance knowledge of a shape's dimension. If one knows ahead of time that S is two-dimensional, it suffices to evaluate the measure for $h(\rho) = \pi \rho^2$. If one knows nothing of a shape except that it is a standard one, one will evaluate the measure for all $h(\rho) = \gamma(d) \rho^d$ with d an integer. If one finds its length to be infinite and its volume to be zero, the shape is surely two-dimensional.

Besicovitch has shown that the core of this last conclusion continues to be valid when d is not an integer and/or when S is not a standard shape. For every set S, there exists a real value D such that the d-measure is infinite for $d < D$ and vanishes for $d > D$. This D is called Hausdorff Besicovitch dimension.

A set's D-dimensional Hausdorff measure may be either zero, or infinite, or positive and finite. Hausdorff had already considered this third and simplest category. It includes the Cantor sets and the Koch curves. If, in addition, the set X is self similar, it is easy to see that its self similarity dimension must equal D. On the other hand, the typical random sets have zero measure in their intrinsic dimension.

◁For a long time, Besicovitch was the author or joint author of nearly every paper on this subject. To paraphrase a well-known witticism, if Hausdorff can be called the father of nonstandard dimension, Besicovitch made himself the mother. ►

3. HAUSDORFF CODIMENSION

When Ω is a Euclidean E-space, $D \leq E$ and $E-D$ is called the codimension.

4. DIRECT PRODUCTS (ADDITIVITY OF THE DIMENSIONS)

Let S_1 and S_2 belong respectively to an E_1-space and an E_2-space, and let S be the set in $(E = E_1 + E_2)$-space obtained as the product of S_1 and S_2. (If $E_1 = E_2 = 1$, it is the set of points (x,y) such that $x \in S_1$ and $y \in S_2$.)

The rule of thumb is that if S_1 and S_2 are "independent," the dimension of S is the sum of the dimensions of S_1 and S_2.

5. INTERSECTIONS (ADDITIVITY OF THE CODIMENSIONS)

The rule of thumb is the following. When S_1 and S_2 are independent sets such that

$E >$ codimension $S_1 +$ codimension (S_2),

this last sum is almost surely equal to

codimension $(S_1 \cap S_2)$.

In particular, if $E = 3$, two sets of the same dimension miss one another if $D \leq 3/2$. Since Brownian trails have the dimension $D = 2$, two trails hit one another when $E < 4$ and miss when $E \geq 4$.

The rule extends in an obvious fashion to the intersection of more than two sets.

The notion of "independence" embodied in this rule has been given a proper meaning in all cases that have been studied (of which a few simple ones are discussed in the text), but it has proven unexpectedly difficult to state and prove generally. See Marstrand 1954a, 1954b; Hawkes 1974; Mattila 1975.

SELF INTERSECTIONS. The set of $M_k(S)$ of k-multiple points of S can be viewed as the intersection of k replicas of S. One is tempted to try the grossly simplifying assumption that from the viewpoint of the intersection's dimension, said k replicas can be viewed as independent. In at least one example, this guess turns out to be correct. S. J. Taylor 1966 (generalizing upon results by Dvoretzky, Erdös & Kakutani) studies the trails of Brownian and Lévy motion in $\mathbb{R}^1$ and $\mathbb{R}^2$. The trails are of dimension D and the sets of k-multiple points are of dimension $\{\max 0, E - k(E - D)\}$. Taylor's guess is that the result holds in $\mathbb{R}^E$ for all values of k, including $k = \infty$.

6. PROJECTIONS

The rule of thumb here is that when a fractal of dimension D is projected as a Euclidean subspace of dimension E_0, the projection's dimension is almost surely the lesser of the two quantities E_0 and D. The rule holds indeed for the fractals constructed by curdling, but it suffers many exceptions. Several will be examined in LIPSCHITZ-HOLDER HEURISTICS.

APPLICATION. Let $x_1 \in S_1$ and $x_2 \in S_2$, where S_1 and S_2 are two fractals in $\mathbb{R}^E$ of dimensions D_1 and D_2, and let a_1 and a_2 be nonnegative real numbers. Then the set S made up of the points of the form $x = a_1 x_1 + a_2 x_2$ has a D satisfying

$$\max(D_1, D_2) \leq D \leq \min(E, D_1 + D_2).$$

The proof consists in taking a direct product of $\mathbb{R}^E$ by $\mathbb{R}^E$, then projecting.

In case of "independence," the upper bound tends to apply. When $D = E = 1$, S may be either a fractal or a set that includes intervals.

EXAMPLE. Let both x_1 and x_2 be of the form $\Sigma V_m 4^{-m}$ with $V_m = 0$ or 2. When $a_1 = 1$ and $a_2 = 1/2$, one finds $D = 1$, S is the interval $[0,1]$, and each $x \in S$ has a single representation $a_1 x_1 + a_2 x_2$. When

$a_1 = a_2 = 1$, $D = \log 3/\log 4$, S is a fractal, and many $x \epsilon S$ have two different representations.

The preceding result plays a role in the nonstandard number representations examined in PEANO CURVES, 4.

7. SUBDIMENSIONAL SEQUENCE

When the set S is such that its intrinsic gauge function is simply $h_S(\rho) = \rho^D$, the set's fractal properties are fully described by its fractal dimension D. Such is the case for all the standard sets in "Euclid" (with D an integer) and for all the sets considered in Hausdorff 1919. In the general case, however, as pointed out by Besicovitch, one can also have, for example,

$$h_S(\rho) = \rho^D [\log(1/\rho)]^{\Delta_1}[\log\log(1/\rho)]^{\Delta_2} ...$$

In such cases, the description of the fractal properties of S requires the whole sequence D, Δ_1, Δ_2. Its members beyond D may be called *subordinate dimensions* or *subdimensions*.

The subdimensions may have a bearing on the very definition of the notion of fractal set. This book calls S fractal if and only if $D > D_T$, but it may become useful to also include the S such that $D = D_T$ but at least one Δ is nonzero.

■ **HYPERBOLIC DISTRIBUTION** – The random scalar U is said to be hyperbolically distributed if $\Pr(U > u) = (u/u_0)^{-D}$. See POWER LAWS.

LIPSCHITZ HOLDER HEURISTICS

Since fractal dimension is a local property, its value in the case of the graph of a continuous function $X(t)$ must be related to such other local properties as the Lipschitz exponent α. The local Lipschitz condition is a way of expressing that near $t = t_0$ the function X satisfies

$$X(t) - X(t_0) \sim |t - t_0|^\alpha .$$

Heuristically, it follows that the number of square boxes of side r necessary to cover the graph of X locally between times t and $t+r$ is roughly equal to $r^{\alpha-1}$.

When X is differentiable for every t between 0 and 1, one has $\alpha = 1$ throughout and the total number of boxes needed to cover the graph is $N \sim r^{\alpha-1}(1/r) = r^{-1}$. In a further heuristic step, let us act as if the curve within each box were a reduced scale image of the whole. It follows that $D = \log N/\log(1/r) = 1$, as is indeed classically the case.

When $X(t)$ is a Brownian function, ordinary or fractional, one has $\alpha = H$ for all t's, and we obtain the heuristic result $N \sim r^{H-1-1}$, hence $D = 2 - H$, which again agrees with the known true value. ◁For the Weierstrass function, Hardy 1916 showed that $\alpha = H$. It would be interesting to check the validity of the natural conjecture that the Hausdorff Besicovitch dimension is $2-H$. ▶

The case of the Cantor staircase (Plate 101) is quite different. Here X varies only for t's that belong to a fractal set having a fractal dimension of $\delta < 1$, and the value of α is no longer independent of t. The $1/r$ equal time spans between 0 and 1 fall into two classes. In $r^{-\delta}$ of them, $\alpha = \delta$ and in the other spans, $\alpha = 1$. Hence the heuristic value of the total number of boxes is $r^{\delta-1}r^{-\delta} = r^{-1}$ and the heuristic dimension is $D = 1$. Such is indeed the case since we observed in the caption of Plate 101 that the staircase is rectifiable (as it must be since its coordinate functions are of bounded variation).

In summary, the heuristic suggests that D cannot exceed $2-\lambda$, with $\lambda = \inf\alpha$. Furthermore, the sum of a Brownian function and a Cantor staircase with $\delta < H$ yields $D = 2-H$ and $\lambda = \delta$, hence $1 < 2-H < 2-\lambda$, which suggests that a curve's D can lie anywhere between 1 and $2-\lambda$. This result is proven in Love & Young 1937 and Besicovitch & Ursell 1937. See also Kahane & Salem 1963, p. 27, Theorem II.

Similarly, let X(t) and Y(t) be Lipschitzian with the exponents λ_1 and λ_2. The heuristic suggests that covering the graph of the vector function of coordinates X(t) and Y(t) requires at most $r^{\lambda_1+\lambda_2-3}$ boxes of side r, hence $1 \le D \le 3-(\lambda_1+\lambda_2)$. For example, the graph of a vector Brownian function is of dimension $D = 2$.

As to the trail obtained by projection on the (x,y) plane, the heuristic is more complicated. In the case $\lambda_1 = \lambda_2 = \lambda$, it suggests that one needs $1/r$ boxes of side r^λ, hence $D = \min(2, 1/\lambda)$. In the general case, one finds

$$1 \le D \le 2 - \min\{0, (\lambda_1+\lambda_2-1)/\max(\lambda_1,\lambda_2)\},$$

which is confirmed by Love & Young 1937.

Returning to the simplest and most important case $\lambda_1 = \lambda_2 = \lambda$, the graph in space, of dimension $3-2\lambda$, can have a greater dimension than its projection, of maximal dimensions $\{\min 2, 1/\lambda\}$. A first exception to this rule is encountered for $\lambda = 1$ (rectifiable case). Another more interesting exception concerns the Peano-Brown case $\lambda = 1/2$.

In all other cases, projection of the spatial graph upon the (x,y) plane changes its dimensions by a variable and complicated amount. The effect of projection on the (x,t) or (y,t) plane is also complicated and moreover is different. Both complications are due to the fact that the graph is self affine but not self similar.

We encounter no such problem with the trail {X(t), Y(t), Z(t)}, assuming the coordinates to be independent fractional Brownian functions with the same H. This trail is self similar and isotropic. Isotropy is impossible under mere self affinity. For this trail, the effect of projections is the same as for ordinary dimension: the plane projections are of dimension $\min(2,1/H)$ and the linear projections of dimension $\min(1,1/H)=1$. The Lipschitz heuristic agrees with this result.

MEDIAN AND SKIP POLYGONS

Many Koch curves with a uniform r, especially in the Peano-Koch limit $D=2$, avoid self intersection but not self contact. Most fortunately, an illustration in Peano 1908 points to one way out of this difficulty. (There is nothing in the text in question to suggest that this tiny illustration is worth particular attention, but in fact it differs from the illustration by Moore to which it claims to refer.) A second different way of avoiding self contacts may be read into the illustration in Hilbert 1891.

MEDIAN POLYGONS. (A neologism.) The method inspired by Peano consists in replacing each of the approximating polygons of a Koch curve by the *median polygon* that joins in sequence the midpoints of the sides of the original poly-

gon. Since the original approximation sides are all of length r^m, the median's sides are of the form μr^m with μ of the order of magnitude of 1. For example, in a square lattice in which the original polygon bends at every vertex, $\mu=1/\sqrt 2$. In general square lattices, $\mu=1$ or $1/\sqrt 2$. In general triangular lattices, $\mu=1$, $\sqrt 3/2$, or $1/2$.

As to the number of sides, it is N^m for the approximate polygon, and its values for the median depends on certain details to which we shall return. In any event, it can be said to be of the form N^m-1+"2 halves." Applying the definition of fractal dimension mechanically, one is led to consider the ratio

$$\log(N^m-1+"2\text{ halves"})/\log(\mu r^m)$$

We observe that

$$\lim_{m\to\infty}\log(N^m-1)/\log(\mu r^m)=\log N/\log r$$

Hence in cases when the median is self similar, the operation "replace a Koch curve by its median" will leave D unchanged.   ◁There seems little doubt that the result holds in all cases. ▶

In one case of special interest, the midpoint of $[0,1]$ is also the midpoint of one of the sides of the standard polygon. A necessary condition is that N is odd, in particular that $N\geq 3$. In that case, the median polygon drawn in stage m includes all the vertices of the median polygon drawn in stage m−1. Every point on the final Koch curve is either the vertex of all the median polygons for m≤k, or a limit of vertices. Furthermore, in all instances I have examined thus far, the heuristic argument concerning D is justified because the median polygon is itself "very nearly" a Koch curve with the same N and r as the original. The term "very nearly" refers to the following two features. First, the end points of a Koch curve remain fixed while it is being built, but the end points of its median move around. However, it suffices to consider a portion of the median polygon between two midpoints. For example, one can string together the "tail half" of one median polygon and the "front half" of another. In this fashion, the term "2 halves" that had been left undefined becomes equal to 1. Second, the Koch construction implicit in the median polygon may well involve several different standard polygons with the same N, thus leading to an additional generalization of the original snowflake algorithm.

◁ When the initial finite Koch polygon includes overlapping sides, the median may have self contacts.

◁ When the original construction involves several distinct values of r, the median may self intersect.

◁ The Peano-Sierpiński curve should be mentioned here (see, for example, Gardner 1976). It is not clear why it was initially introduced, but its mth finite approximation is most compactly described as the median polygon of a *nonconnected* polygon that covers the plane increasingly tightly as m→∞. To draw this last polygon, one starts with a unit square with lines 2^{-m} apart, and one preserves the grid lines for which either the abscissa or the ordinate is an odd integer divided by 2^m. ▶

SKIP POLYGONS. (A neologism.) The Peano-Hilbert curve is neither a Koch curve nor the median of one, and there is no indication how it was designed. However, it turns out that it too can be obtained as a transform of a Peano-Koch curve. The point of departure is described under PEANO CURVES, 3. N=2, C. The transform consists, first, in skipping the even-numbered stages (the only ones considered by Peano), and second, in skipping all even-numbered vertices. This second part of the transform appears to be of general interest as a method for avoiding self contacts.

◁ Alternatively, the Hilbert curve can be said to involve a variant of the median. It starts with 4^m squares of side 2^{-m}, each of which is connected to two neighbors except that an initial and a final square are connected to only one neighbor. Hilbert's curve links the mid-points of these *squares* in sequence. ►

PARADIMENSION

In E-dimensional Euclidean spaces, sets such that $D_T < D = E$ are fractals but raise special problems. Fortunately, in all the specific cases I tackled an alternative dimension-like exponent is available, and I propose that it be called *paradimension* (a neologism). Since it seems to lack generality, it is best described through examples.

CANTOR SETS OF POSITIVE MEASURE ON THE LINE. In Chapter IV and in COV-ERING OF A SET, 3, we build a zero-measure set S on [0,1] through its complementary intermissions. However, subtraction of intermissions can also lead to a set for which the linear measure is positive, hence $D=1$. For example, such is the case with intermissions λ_n of the order of $(n/\sigma)^{-1/\Delta}$, but such that the series $\Sigma\lambda_n$ converges to a value below 1. A necessary condition is $\Delta < 1$, that is $\Delta < D = E$. In that case, the Δ ruling the structure of S is a Besicovitch Taylor exponent but is not a Hausdorff Besicovitch dimension. It may be called a *paradimension*.

◁When λ_n for all n is smaller than the function $(n/\sigma)^{-D}$ corresponding to some $D \epsilon [0,1]$, one can use the following simple process: drop an open interval of length λ_1 in the middle of [0,1], then two intervals of lengths λ_2 and λ_3 in the middles of the remainders, and so on. ►

Similar comments apply to the studies of lunar craters or meteorites in Chapter VII; when the respective exponents 2γ or 3γ are below 2 or 3, they are examples of paradimensions.

LEBESGUE OSGOOD CURVES WITH D=2 ON THE PLANE. We saw at the end of Chapter II how such curves are drawn by cutting out of the plane a tree (or a river network) such that, as one proceeds to the branches' tips, the widths λ decrease faster than the lengths Λ. One can see that, in order to achieve $D=2$, the lengths Λ_0, Λ_1 and Λ_2 before and after a point of branching must on the average satisfy $\Lambda_0^2 = \Lambda_1^2 + \Lambda_2^2$. But the corresponding relation for the λ must be $\lambda_0^2 > \lambda_1^2 + \lambda_2^2$. For example, one may have on the average $\lambda_0^\Delta = \lambda_1^\Delta + \lambda_2^\Delta$ with $\Delta < 2$. In that case, the structure of S is determined by Δ, which again will be called paradimension.

When $E=2$ and $\Delta=1$, the relationship $\lambda_0 = \lambda_1 + \lambda_2$ expresses the conservation of a classical expression, namely of the branches' cumulative width. Therefore, it is desirable in all cases to view the above relation between λ_0, λ_1 and λ_2 as expressing some form of nonstandard conservation. This is achieved by viewing Δ as some kind of dimension.

APPLICATIONS. Many empirical

measurements can be viewed as involving a paradimension.

For example, much of the geometry of natural (wooden) trees can be shown to follow from these two assumptions: (a) branch tips nearly fill 3-space, so that $D=3$ or at least $D=3-\varepsilon$, and (b) cross-section is conserved in branching, so that $\Delta=2$. The geometry of lines is to be explored fully elsewhere.

The relationship postulated under (b) is about the same as the relationship between the widths of branches in self similar trees that nearly fill the plane. This near identity may explain why the Figures on top of Plate 65 and on Plate 67 resemble wooden trees.

In branching off of mammalian arteries, it was also argued in Chapter II that tips fill up space so that $D=3$ or at least $D=3-\varepsilon$. However, an empirical law due to Groat implies $\Delta=2.7$.

Finally, the conservation of the quantity λ^{Δ} with $\Delta=3/2$ plays a central role in the analysis of neurons by the method of Rall 1959. See also Jack et al. 1975. However, it has not been established whether fractal concepts are of use in this context, so the fact that Δ resembles a paradimension may be coincidental.

WARNING. One main reason for including the present overly sketchy entry is to serve as warning that it is unlikely that the concept of fractal dimension could possibly include every dimension-like exponent encountered in mathematics. And the same is surely true of such exponents in physics.

■ PARETO LAW AND DISTRIBUTION – Pareto distribution is a synonym of hyperbolic distribution of probability. It is favored by economists and other social scientists. Pareto law is used either as a synonym of Pareto distribution or to designate the law of distribution of income (see Chapter X). Both meanings are examples of power laws.

PEANO CURVES

This synonym of *plane-filling curve* involves at least two sources of possible ambiguity.

First of all, one can view such a curve as a "trail" (see BROWNIAN SETS), but in this sense a Peano curve is really a domain of the plane fully defined by its boundary. ("The term *space-filling curve* seems to be one which should be dismissed;" Young & Young 1906.) In all the classical examples, said boundary is a Euclidean shape but in most of the new examples to be studied below it is itself a fractal curve. The latter situation is doubtless typical in some senses.

Alternatively, one may want to know in which way this domain is filled, that is, one may be interested in the curve's two coordinate functions $X(t)$ and $Y(t)$. There are many alternative methods for thus filling a domain of given boundary.

Secondly, there is the issue of multiple points: there is no agreement on the question of whether or not a finite approximation to a Peano curve is allowed to self intersect and, more important, to cover its domain repeatedly.

This entry will first dwell on questions relative to multiple points, then will relate Peano curves to other curves encountered in this Essay. Then we shall list some specific Peano-Koch curves and, finally, discuss their relationships with various representations of numbers.

MEDIAN AND SKIP POLYGONS includes additional material on Peano curves.

1. MULTIPLE LIMIT POINTS

The fact that a Peano curve must have multiple points is discussed in DIMENSION (TOPOLOGICAL) and in Chapter II, where I give an (after the fact) reason for considering the need for double points as being quite obvious. The argument will be repeated, then extended to triple points. Next we shall tackle quadruple points.

The intuitive proof that a Peano curve must have double points is based on the observation that *any* such curve can be restated in the form of two complementary networks: rivers and drainage divides (see the caption of Plate 61). In this representation, any two points facing each other across a river are confounded into a double point, and double points must similarly be encountered on drainage divides. After the Peano curve is parameterized intrinsically (by the method described in the third subentry below) the *area* of any basin in our river network becomes simply equal to the *length* of the segment joining the preimages of the points that face each other across the basin's mouth. Each basin can be parameterized by the smaller (or the larger) of the "times" when the first of these facing points are crossed.

Pressing this interpretation, we see that a river network necessary involves forks. This remark proves heuristically that a Peano curve cannot avoid multiple points of order at least three, in fact that such points must be everywhere dense. The conclusion is indeed correct, and would deserve to be better known.

◁ It is discussed in Sierpiński 1974–, **II**, pp. 116 ff, who says it was asserted by Hilbert 1891 on the basis of an incorrect observation, then by Lebesgue in 1911 without proof, in conjunction with the "bricks" discussed under DIMENSION (TOPOLOGICAL). Sierpiński also refers to Pólya and to Mazurkiewicz. Lebesgue's proof appeared in 1921; see Lebesgue 1972–, **IV**, pp. 168 and 198 ff). ▶

As to contact points of fourth and higher orders, they correspond to the subdivision of a river into three or more branches. To the question of whether or not such points are unavoidable, intuition no longer provides a firm answer. We may be glad, therefore, to know that Lebesgue showed them to be acceptable but entirely avoidable.

We now approach an entirely different aspect of multiple points. One often observes pairings of Peano curves filling the same boundary, for example the Peano Cesàro curve of Plate 61 and the Peano-Hilbert curve. In this case, and doubtless in many others, it is possible to go from one to the other by massive rerouting. The basic stage can be illustrated in terms of a road crossing which a diagonal barrier has made into two roads in contact. By switching the barrier to the other diagonal, one obtains a different arrangement of roads in contact. ◁ This operation is vaguely reminiscent of genetic crossover and recombination. ▶ Naturally, rerouting causes the river and drainage divide networks to be scrambled.

2. PEANO CURVES' COORDINATES ARE NONDIFFERENTIABLE

In a Peano curve, the coordinate functions $X(t)$ and $Y(t)$ are continuous but not differentiable. Early aspects of this result are found in Peano 1890 and Moore 1900, and the aspect described in Lax 1973 is relative to the Peano Pólya function, which is a limit of Plate 65 and a generalization of the Peano Cesàro curve. The set of nondifferentiability of the Pólya function is curiously involved in its detail. It involves many sets of measure O which cry out for someone to evaluate their D's.

Peano's result may be viewed as an early counterpart of a now familiar fact about Brownian motion: the plane Brownian trail is essentially a Peano curve and its Brownian line-to-line coordinate functions are continuous but nondifferentiable. The $X(t)$ and $Y(t)$ functions of the Peano Moore curve are known to be Lipschitzian of exponent $\lambda = 1/2$ and there is no doubt (but no proof is known to me) that their level sets are of dimensions $1/2$ and their whole graphs are of dimension $3/2$. (In the variant of the Peano Moore curve constructed by Kline 1945, the coordinates are Lipschitzian with the exponents $\lambda \in]0,1[$ and $1-\lambda$.)

The converse probabilistic problem was implicitly approached in Chapter IX in our study of fractional Brownian planar trails. We saw that when the exponent H of the coordinate functions satisfies $H>1/2$, the trail's dimension is

1/H<2, so that the trail does *not* cover the plane. When H<1/2, the trail covers the plane repeatedly.

The nonprobabilistic converse of Moore's remark is delicate because it involves number theoretical difficulties (of the kind mentioned in DIMENSION (FOURIER)). However, Salem & Zygmund 1945 generalized by Kahane, Weiss & Weiss 1963 show that the trajectory of a point having coordinates of the type of cosine and sine Weierstrass functions can cover a domain of the plane repeatedly. This result should no longer be viewed as merely a curiosity and may reward a fresh look.

3. EXAMPLES OF REGULAR PEANO KOCH CURVES

As mentioned on page 58, the Peano Moore curve is (after the fact) a Koch curve. The same holds true of the Peano Cesàro curve (Plate 61). Recently, the dragon curves to be examined below under N=2 signaled a renewal of interest in the invention of new Peano constructs. The latter come just in time to provide fresh and beautiful material for the applications discussed in this Essay. But one should also be grateful for the interesting questions of pure mathematics which they raise. In particular, they suggest it is rewarding to seek Peano curves with finite approximations having various desirable properties.

The Peano Gosper curve (Plates 62-63) is an example of a new Peano Koch curve, and its major contribution from the mathematical (as opposed to purely esthetic) viewpoint is that it has no finite self contact. After the fact, and once one has become aware of this possibility, it seems that a nonself contacting Peano Koch curve has already been drawn by Peano himself (see MEDIAN AND SKIP POLYGONS and the case N=9 later in this subentry). We shall find examples of avoidance of self contact for N=3, 5, 7, 9, 13, 17, ... but not (yet?) for N=2, 4, 6, 10,11, 15, ...

An attempt to classify the known Peano curves, with stress on the Peano Koch curves, has rapidly led to the identification of many new variants of interest. The present status of the taxonomy is sketched, keeping to the case when the standard polygon's sides are equal. The simplest are polygons such that their sides fall along a lattice. Such polygons exist if N is the sum of two squares of integers.

N=2. The only standard polygon with N=2 and r=1/√2 is made of two perpendicular segments of slopes 45° and −45°.

A. The original Koch construction which places the standard polygon to one side of the preceding approximation leads very rapidly to self intersection, so the limit is very complicated (it was the object of profound study in Lévy 1938). However, a generalized Koch construction may avoid self intersection in a number of different ways.

B. Suppose the initial shape is [0,1]. One method for avoiding self intersection consists in treating differently the odd- and even-numbered stages. During an odd-numbered stage, let the standard polygon be placed to the *left* of all the sides of the polygon with which the stage began; we say it is λ. During an even-numbered stage, let the standard polygon be ρ, that is, placed to the *right*. This procedure yields an interpolate of the procedure due to Cesàro which is examined under N=4. The limit's boundary is of dimension D=1: a right isosceles triangle covered from one end of the hypothenuse to the other.

C. Another method of avoiding self intersection is obtained by interpolating the original Peano algorithm (see N=4 below). In the interpolation's first stage, the algorithm is ρ;λ;λ;ρ. In the second stage, it is ρ,ρ;λ,λ;λ,λ;ρ,ρ. In the third, it is λ,ρ,ρ,λ;ρ,λ,λ,ρ;ρ,λ,λ,ρ;λ,ρ,ρ,λ, and so on. The sequence of even-numbered stages brings us back to Peano's original. The sequence of odd-numbered stages, if modified by skipping all the even-numbered vertices, avoids self contact,

and is identical to the classic construction of Hilbert 1891 (MEDIAN AND SKIP POLYGONS). In any event, the limit's boundary is a square of dimension D=1.

D. *The dragon curve.* Still another method proceeds as follows. In each construction stage, the standard polygon is alternatively ρ and λ. When the initial step on each stage is to ρ, the limit is J.E. Heightway's *dragon curve*, publicized in M. Gardner 1967 and investigated in Davis & Knuth 1970. (See also Ball & Coxeter 1974, p. 166.)

Finite approximations of the original dragon curve are open ended but can readily be closed by selecting a closed initial polygon. By starting with [0,1] followed by [1,0], one obtains a shape that may be called *twindragon,* two dragons joined belly to belly. ◁ Let me observe that the twindragon is identical to the domain drawn in Knuth 1968–, **II,** Section 4.1 – as amended in the forthcoming second edition, a portion of which Dr. Knuth has kindly let me see. ► I proved (but the proof is too lengthy to be given here) that if D denotes the fractal dimension of the boundary (dragon's "skin"), $2^{D/2}$ must be the root of the equation $x^3 - x^2 - 2 = 0$; hence, D=1.5236.

The dragon is easy to generalize, in fact there is a generalized dragon corresponding to every combined index of the form $0,i_1i_2i_3$, where $i_n = 0$ or 1. When $i_n = 0$, the nth stage standard polygon starts as ρ; when $i_n = 1$, it starts as λ. When $i_{2n} = 1$ and $i_{2n+1} = 0$, the generalized dragon is merely a right isosceles triangle covered in a nonclassical way: from one 45° corner (Davis & Knuth 1970) to the 90° corner. Thus, a dragon's boundary may be a standard curve or a fractal.

In finite approximations to dragons, every interior vertex is a point of self contact. The best way to avoid them is to draw median polygons.

N=3. The literature includes two possibilities obtained by Davis & Knuth 1970 as variants of the dragon and called *terdragons* by them. Both use a standard polygon shaped like a Z with equal sides making angles of 60°. This polygon is alternately placed as ρ and λ. The two variants use different combined indexes $0,i_1i_2...$ (as defined above under $N = 2$). Again, every variant's boundary is a Koch curve. Its standard polygon corresponds to $N = 4$ and $r = 1/4$ with sides having the following slopes with respect to [0,1]:0°,60°,0°,−60°. These polygons are placed alternatively as ρ and λ with initial positions following the above index $0,i_1i_2...$. Hence in every terdragon, the boundary (skin) shares the dimension $\log 4/\log 3$ of the triadic Koch curves. If one starts with an initial equilateral triangle, one obtains *terdragon triplets* that fill without gap a shape resembling a Koch snowflake in which bays or promontories have been scrambled.

All terdragons suffer from finite self contacts. Again, their median polygons avoid this difficulty. They are the simplest (smallest N) nonself contacting approximate polygons of Peano Koch curves. The orientation index varies according to the initial terdragon but in all cases the standard polygon is an L with an angle of 60° and an upright bar twice as long as the horizontal. Alternatively, given the finite approximations to three terdragon triplets that follow each other along an equilateral triangle, one can combine the back half of the median polygon of one of them with the front half of the median polygon of its immediate follower. This construction, as specialized to the terdragon, (combined index 0.0000), was known to R. W. Gosper, who followed a different path to its discovery.

N=4. This is the lowest N of the form $N_0{}^m$, with N_0 and m integers. A Peano curve for N_0 is also a Peano curve for $N = \Sigma_{0 \leq m \leq N_0} N_0{}^m m_m$ with m_m integers. Such reinterpretations need not be mentioned except for the present $N = 4$, in which case they are of historical interest.

A. A first possibility follows Peano's original method. Translated from the author's excessively concise algebra, it seems to involve the Koch construction as generalized in Plate 57, with [0,1] as the initial polygon, and a standard polygon for which the sides' angles are 90°, 0°, 0°, and −90°. In this construction, finite overlap goes beyond *points* of self contact and some sides are counted twice. Thus we deal with yet another extension of the Koch method. Self intersection is however avoided because the repeats of the same side give rise to "growth" in opposite directions, so the next construction stage separates them.

B. In a possibility noted by Cesàro 1905, the initial polygon is [0,1] followed by [1,0] and the slopes of the standard polygon's sides are 0°, 90°, −90°, and 0°. The result is the same as when the procedure described under N=2, B is reinterpreted by skipping the odd-numbered stages. Again, finite overlap extends beyond points of contact. A graph of the Peano Cesàro curve is readily obtained from one half of the black portion of Plate 61 by rotating it either way by 45°.

C. A possibility (which escaped the attention of the classical writers) results from the fact that any dragon's construction (see under N=2) can be reinterpreted by skipping the odd-numbered construction stages. In the corresponding standard polygons, the slopes of the sides are either 0°, 90°, 0°, and −90°, or 90°, 0°, −90°, and 0°.

N=5. A Peano Koch curve devoid of finite self contact (apparently new but inspired by the Peano Gosper curve; see Mandelbrot 1977m) uses a standard polygon in which the slopes of the sides are obtained by adding $\theta = \tan^{-1}(1/2)$ to 0°, 90°, 0°, −90°, and −90°. The boundary, which I call *quintet*, is a Koch curve with the dimension

$$D = \log 3 / \log \sqrt{5} = \log 9 / \log 5 = 1.3652.$$

The forms of the wrapping and the filling are determined by the value of a real binary index $0,i_1 i_2...$ that combines the orientation indexes of the successive construction stages.

N=7. We have the Peano Gosper curve of Plates 62-63 and variants corresponding to other indices.

N=8. One example that does not result from the fact that $8 = 2^3$ involves the angles 0°, 90°, 0°, 0°, −90°, 180°, −90°, and -90°. Self contact is not avoided.

N=9. The classical possibility is the Peano Moore curve, which does *not* result from the fact that $9 = 3^2$ and uses an entirely different grid. Moore 1900 predated Koch and did not describe his method in terms of standard polygons, but in effect he placed such polygons alternatively as ρ and λ. The standard polygons were the same as in Plate 59 (boundary dimension $D = 1$).

The median polygon of the construction on Plate 59 is itself a Peano Koch approximate polygon. Its standard polygon is one fourth of a bent cross ($N=9$ and $r=1/3$) with the following slopes: 90°, 0°, −90°, −90°, 0°, 0°, 0°, 90°, 180°. The rule that describes the alternation of the polygon and of its inverse could hardly be guessed at in advance.

The choice of the overall index does not affect the boundary of a curve of Peano Moore type, but it affects its median. For example, Moore's original is more complicated than the variant of Plate 59: in order to be able to view its median polygon as an approximate Koch curve, it is necessary to allow for two distinct standard polygons.

N=13, 17, 29, 41, AND SO ON. The construction described under N=5 is easy to generalize to all values of N that are the sum of two squares of mutually prime integers and at the same time are of the form a^2+4b where both a and b≤a are integers.

4. CANTOR'S AND OTHER NUMBER REPRESENTATIONS

Irrespective of the values of D, a Koch curve of known N is parameterized most conveniently when t is represented in the initial positional number system of base N. The number of sides in the polygon being denoted by j, t runs over $[0,j]$;

if j=n, t can simply run over $[0,1]$. The Koch map of t = 0. $a_1a_2...$ is then

$$\Sigma_{1 \le m < \infty} V(a_1 a_2 ... a_m)\gamma^{-m}.$$

This representation is in Cesàro 1905. Here each of the vectors which play the roles of digits results by rotation and symmetry from one of N standard vectors that join the origin of the standard polygon to its N other vertices.

Now add the condition D=2. The need for a nontrivial algorithm to rule the V's was the main contribution of Peano 1890, who stressed that the resulting correspondence between the line and the plane is profoundly distinct from any correspondence obtained by way of a positional number system. The latter are indeed one-to-one and cannot be continuous, while a Peano Koch mapping is continuous and must be many-to-one.

The matter deserves amplification. In a positional number system that maps a line on a plane, the base and/or the digit types must be nonstandard: negative, or imaginary, or complex. (Digit types are the possible values of the digits, and digit tokens are the digits as encountered in a number's representation.) For example, consider Cantor's classical one-to-one correspondence between the line and the plane − about which Cantor wrote to Dedekind, "I see it but I do not believe it" (Dauben 1974). This correspondence is best thought of as involving a positional system with base $N = 4$ and digit types 0, 1, i or i+1. The customary description is different; it involves the base 2 and maps the point

$$t = 0.\ \alpha_1\alpha_2\alpha_3\alpha_4...$$

on the point of coordinates

$$x = 0.\ \alpha_1\alpha_3...$$
$$y = 0.\ \alpha_2\alpha_4...$$

However, each couple of binary digits $\alpha_1\alpha_2$ is a single digit in the base 4, say, a. To map $t = 0.a_1a_2...$ on ΣV_m^{-m}, Cantor uses the following code:

$a = 0$ (00 binary) $\rightarrow V = 0$;
$a = 1$ (01 binary) $\rightarrow V = i$;
$a = 2$ (10 binary) $\rightarrow V = 1$; and
$a = 3$ (11 binary) $\rightarrow V = i+1$.

It is characteristic of a positional number system that each V_m only depends on one a_m. It follows that the correspondence is necessarily discontinuous. Peano's contribution was to insure continuity by injecting into the V's various rotations and symmetries dependent upon the past α's.

The base of a positional number system can also be imaginary, as described in Knuth 1968–, II, Section 4.1. Consider first the example demonstrated privately by R. E. Maas in which $\gamma = 1+2i$ so that the base is $N = |\gamma|^2 = 5$ and the digit types are 0, i, −1, −i, 1. To the point $t = 0.\ a_1a_2...$ this system associates the point $\Sigma V_m\gamma^{-m}$. Again the factors V only depend on a_m, using the following code:

$a = 0 \rightarrow V = 0$;
$a = 1 \rightarrow V = i$;
$a = 2 \rightarrow V = -1$;
$a = 3 \rightarrow V = -i$; and
$a = 4 \rightarrow V = 1$.

The resulting points $\Sigma V_m\gamma^{-m}$ can be shown to fill the boundary of the quintet curve described in the preceding subentry for the case $N = 5$. In the same spirit one can say retrospectively that Cantor's correspondence had been a way of filling the boundary of Peano's original curve.

The third main category of number representations concerns the case when γ is imaginary or complex but the digit types are real. Using $\gamma = 1/\sqrt{N}$ and the digits 0 to $N-1$, one has a one-to-one but discontinuous map of [0,1] upon a rectangle of sides N and $\sqrt{N}$.

More interesting by far is the example due to W. Penney where $\gamma = i-1$ and the digits are 0 or 1 (Knuth 1968–, **II**, Section 4.1). The map is again one-to-one but discontinuous and the boundary of the map of [0,1] happens to be identical to that of the twindragon. Hence the relationship between this correspondence and the twindragon Peano Koch map of $[0,1]$ followed by $[1,0]$ is the same as the relationship between Cantor's and Peano's original maps.

Finally, observe that each Peano Koch curve and each positional number representation involves a self similar tesselation – that is a hierarchy of tiles each of which can be tiled by reduced replicas of itself. In many instances, the tesselations are determined by the boundary, but the case of two joined dragons involves two different tesselations.

To obtain a positional number system, the values of γ and the $|\gamma|^2$ digit must be chosen carefully. If, for example, one tries $\gamma = i + 1$ and the digits are 0 or 1, the correspondence $\Sigma V_m 2^{-m} \rightarrow \Sigma V_m \gamma^{-m}$ turns out to map $[0,1]$ on the twindragon (again!). However, it is both discontinuous and many-to-one.

POTENTIALS, CAPACITIES AND THE DIMENSION

1. INTRODUCTION AND CONJECTURE

The Hausdorff Besicovitch dimension is greatly extended in scope by the fact that it plays a central role in the theory of classical potentials. And also of gen-eralized (Marcel Riesz) potentials using kernels of the form $|u|^{-F}$, where $F \neq E - 2$. In particular, the special value $D = 1$ is intimately linked with the Newtonian potential in 3-space.

This link (to be amplified in the next subentry) suggests the following conjecture concerning various cosmological theories that predict $D = 1$, such as the Fournier and Jeans-Hoyle theories sketched in Chapter IV. It should be possible to rephrase them as corollaries of Newtonian gravitation. Furthermore, the observed departure of the value of D from 1 should be traceable to non-Newtonian affects – presumably relativistic ones.

2. HAUSDORFF BESICOVITCH DIMENSION AND POTENTIALS

The only known contact of potential theory with the study of stellar clustering has been very informal and brief and occurred well before potential theory was generalized. The Olbers paradox, indeed, has a counterpart in the case of gravitation. Suppose that $E = 3$, that the mass $M(R)$ within a sphere of radius R around the origin ω is proportional to R^D with $D = 3$, and that the potential refers to the Newtonian kernel R^{-F} with $F = 1$. The total potential at ω due to all the masses being of the form $\sim \int R^{-F} R^{D-1} dR =$

XII □ □ □ MATHEMATICAL LEXICON AND ADDENDA

$\int R\, dR$, it would diverge at infinity. It was suggested by Seeliger that one way of eliminating this paradox is to keep $D = 3$ but make $F > 3$, so the potential ceases to be Newtonian. The Fournier-Charlier model shows one can keep $F = 1$ and make $D < 1$. For the general integral $\int R^{D-1-F} dR$, the condition of convergence at infinity is clearly $D < F$. And the condition of convergence at the origin is $D > F$. This argument, therefore, establishes a one-to-one link between D and F, and in particular it relates $D = 1$ to $F = 1$.

This link was restated more tightly by Pólya & Szegö and put in final form in Frostman 1935. The major advance was that the result was made to apply not to a single origin ω but to all points in a (compact) set S. Consider a unit mass distributed on S so that the little domain around $\mathbf{u}$ contains the mass $d\mu(\mathbf{u})$. At the point $\mathbf{t}$, the kernel $|\mathbf{u}|^{-F}$ yields the potential function

$$\Pi(\mathbf{t}) = \int |\mathbf{u}-\mathbf{t}|^{-F} d\mu(\mathbf{u})$$

The physical concept of capacity, borrowed from electrostatics, was transformed by de la Vallée Poussin into a method for measuring the "contents" of sets. The idea is that if S has a high capacity $C(S)$, the total mass $\int d\mu$ can be so shuffled as to insure that the maximum potential is small. More precisely, one takes the supremum of the potential over all points $\mathbf{t}$, then the infimum of the result with respect to all the distributions of a unit mass over S, and finally on sets

$$C^{-1}(S) = \inf\,[\sup_{\mathbf{t}}\Pi(\mathbf{t})].$$

There is a simple relationship between $C(S)$ and the Hausdorff Besicovitch dimension of S. When $F > D$, the capacity of S vanishes, meaning that even the "most efficient" distribution of mass leads to a potential that is infinite somewhere. When $F < D$, on the other hand, the capacity of S is positive. Thus the Hausdorff Besicovitch dimension is also a capacity dimension in the sense due to Pólya & Szegö. This identity was proved in full generality in Frostman 1935.

◁Equivalent definition: $C^{-1}(S)$ is the infimum among all the distributions of mass supported by S, of the energy defined by the double integral

$$\int\int |\mathbf{t}-\mathbf{u}|^{-F} d\mu(\mathbf{s})d\mu(\mathbf{t}). ▸$$

Among nonelementary recent mathematical treatments of potential theory, I favor duPlessis 1970. Its Chapter 3 on the conductor problem and capacity summarizes (luckily, with no change in notation) the more detailed treatment in Landkof 1966-1972. The capacity is a kind of measure, but the described relation between capacity and Hausdorff measure in the dimension D is quite involved; see S. J. Taylor 1961.

3. "ANOMALOUS DIMENSION"

The kernels $|\mathbf{u}|^{-F}$ with $F \neq E-2$ are associated in the physicists' minds with an imbedding space having the "Euclidean" dimension $E* = 2-F$. (I do not believe this usage is meant to imply any actual generalization of E to nonintegers.) Given, a), the link between D and E (firmly established by Frostman), and b), the role of D in describing stellar clustering (established in Chapter V of this Essay), the terminology of anomalous dimension leads to the following statements. A fractal dimension $D=0$ for stellar matter is the contrary of anomalous, but the observed fractal dimension $D \sim 1.3$ seems to involve an embedding space of anomalous dimension.

POWER LAWS

Self similarity and self affinity inevitably lead to theories rife with power laws. One may say, only half in jest, that the mathematics underlying this Essay is best characterized as making the best possible use of the first lesson in calculus, concerned with powers, and as avoiding exponentials.

For example, the probability distribution of a self similar random variable X must be of the form $Pr(X>x) = x^{-D}$, which is commonly called *hyperbolic* or *Pareto* distribution.

MOMENTS. Its most striking feature is that moments $\langle X^h \rangle$ of an order $h > D$ are divergent. The "loss" of moments of order 5 or above would pass unnoticed. Moments of order 3 or 4 are used in evaluating the skewness and the kurtosis but one readily lives without them. Life without variance is harder, and life without expectation demands psychological adjustment.

EXAMPLES. For a few phenomena the necessity of a law closely akin to a power has been established from basic principles. The Holtsmark law of spectral line widening and of stellar attraction (Feller 1971, pp. 172-174, and Chandrasekhar 1943, p. 70) was for a long time the only known example. Now another example has been identified in molecular biology, Mandelbrot 1974d.

There is a whole field of physics in which exponents also predominate, namely, the theory of phase transitions and critical phenomena. We saw an example in Chapter VII and many others can be found, for example, in Stanley 1971. Formally, it is closely related to Bernoulli percolation theory (Chapter VIII), and there can be no doubt that the geometry of other critical phenomena is also rife with fractals.

■ RANDOM – The term *random element* is awkward and so are many of its kin. Many of them could be improved by using *random* as a noun. I am sorry that I did not dare, in this Essay, to follow my own suggestion (and that, in the French version, I did not attempt to bring *randon* and *à randon* back into the French language from whence they came).

■ SCALING – A generic term that includes the specific concepts discussed under AFFINITY AND SIMILARITY.

STABLE RANDOM ELEMENTS IN THE SENSE OF LEVY

GENERAL REFERENCES. Feller 1966 (Volume II) is handy and complete, but hard to use. Lamperti 1966 is a good introduction. Gnedenko & Kolmogorov 1954 was very complete in its time and continues to be recommended. See also Lukacs 1970. The original and great treatises by Lévy (1925, 1937-1954) exhibit the distinctive characteristics of his style (see Chapter XI).

1. THE GAUSSIAN IS STABLE

The Gaussian distribution is known to have the following property. Let G_1 and G_2 be two independent Gaussian random variables satisfying the conditions

$$\langle G_1 \rangle = \langle G_2 \rangle = 0;$$
$$\langle G_1{}^2 \rangle = \sigma_1{}^2, \langle G_2{}^2 \rangle = \sigma''{}^2.$$

Then their sum $G_1 + G_2$ satisfies

$$\langle G_1 + G_2 \rangle = 0;$$
$$\langle (G_1 + G_2)^2 \rangle = \sigma_1{}^2 + \sigma_2{}^2,$$

and $G_1 + G_2$ is itself Gaussian. Thus the Gaussian property is left invariant by the addition of independent random variables. In other words, the functional equation

$$(L) : s_1 X_1 + s_2 X_2 = sX,$$

combined with the auxiliary relationship

$$(A2) = s_1{}^2 + s_2{}^2 = s^2$$

has the Gaussian as a possible solution. In fact, except for scale, the Gaussian is the only distribution satisfying both (L) and (A2). Furthermore, if (L) is combined with the alternative auxiliary relation $\langle X^2 \rangle < \infty$, the Gaussian is again the unique solution.

The condition (L) has been the object of profound study by Lévy 1925, who also burdened it by calling it *stability*. This use of an otherwise overworked term is regrettable, but it is unfortunately bound to stick.

2. OTHER STABLE VARIABLES

Since the auxiliary condition $\langle X^2 \rangle < \infty$ is ordinarily taken for granted, it is not appreciated that it can have important consequences, and the Gaussian is widely believed to be the only stable distribution. Such is definitely not the case. The fact was first recognized in Cauchy 1853, p. 206, where extensive use is made of a certain random variable first considered by Poisson. It is such that

$$Pr(X>x) = Pr(X<-x) = 1/2 - \pi^{-1} tan^{-1}x,$$

and therefore such that the density is

$$1/[\pi(1+x^2)].$$

Cauchy showed this variable to be what is now called stable. It is the unique solution of the combination of (L) with the alternative auxiliary relation

$$(A1) = s_1 + s_2 = s.$$

It may be observed that, for this Cauchy variable, $\langle X^2 \rangle = \infty$, in fact $\langle X \rangle = \infty$. Hence, in order to be able to express the obvious notion that the product of X and a nonrandom s has a scale equal to s times the scale of X, one must measure scale by some characteristic quantity other than the root mean square. It may, for example, be the distance between the median value M, and the quartile Q. M is the value such that $Pr(X>M) = Pr(X<M)$ and hence in the present case it is equal to 0

and the quartile Q is such that $Pr(X>Q) = 1/4$.

◁ The Cauchy variable most often serves as a counterexample; the first instance of this use is found in Bienaymé 1853, pp. 321-323. ▶

Cauchy also considered the generalized auxiliary relation

$$(AD) = s_1{}^D + s_2{}^D = s^D$$

and asserted on the basis of purely formal calculation that for every D the conditions (L) combined with (AD) has one solution, the random variable of density

$$\pi^{-1} \int_{0<x<\infty} exp(-u^D)cos(ux)du.$$

This last expression is the Laplace transform of $exp(-u^D)$. ◁ Cauchy was very familiar with the use of the Laplace-Fourier transform in probability, but Cauchy's successors forgot all about it, and it was only revived in the 1920s by Lévy, who denoted it by the overworked term, *characteristic function.* ▶

Cauchy's assertion remained unnoticed and purely formal until Pólya and Lévy showed that in the case $0<D<2$, it is indeed justified. However, in the case $D>2$, the assertion is invalid because the above-written formal density takes on negative values, which is an absurdity. Lévy showed moreover that, if X is allowed to be nonsymmetric stable, the combination of (L) and (AD) has other solutions. When $D < 1$, the most asymmetric variable takes positive values only.

3. CENTRAL LIMIT THEOREMS AND ROLE OF HYPERBOLIC VARIABLES

Given an infinite sequence X_n of random variables, the central limit problem inquires whether or not it is possible to select the weights a_n and b_n so that the sum $a_N \Sigma_{1 \leq n \leq N} X_n - b_N$ has a nontrivial limit for $N \to \infty$.

When the X_n are independent and identically distributed with $\langle X_n^2 \rangle < \infty$, the answer is, classically, to the affirmative, and the limit must be Gaussian.

When the last condition is replaced by $\langle X_n^2 \rangle = \infty$, the answer is more complex: (a) the selection of a_N and b_N is not always possible; (b) when it is possible the limit is stable non-Gaussian; (c) in order that the exponent of the limit be D, a sufficient condition on the X_n is that they follow the hyperbolic distribution of exponent D, and the necessary condition is barely less demanding.

4. STABLE LINE-TO-LINE FUNCTION

This is a random function having independent increments and such that $X(t) - X(0)$ is stable. The scaling factor $a(t)$ that makes $[X(t) - X(0)]a(t)$ independent of t must take the form $a(t) = t^{-1/D}$. This process generalizes the ordinary Brownian motion to $D \neq 2$.

The most striking property of $X(t)$ is that it is discontinuous and includes jumps. When $D < 1$, it includes nothing but jumps; the number of those occurring between times t and $t + \Delta t$ and having an absolute value exceeding u is a Poisson random variable and its expectation is equal to $|\Delta t| u^{-D}$.

The partial numbers of positive and negative jumps can be written as $q |\Delta t| u^{-D}$ and $(1-q) |\Delta t| u^{-D}$. Obviously, the isotropic case corresponds to $q = 1/2$. And the case $q = 1$ involves positive jumps only. The corresponding line-to-line function is called *stable subordinator* and serves to define the Lévy staircase (see CANTOR AND LEVY STAIRCASES).

As $u \to 0$, we observe that $u^{-D} \to \infty$. Therefore, the total expected number of jumps is infinite – however small the length of Δt. Nevertheless, the fact that the associated probability is infinite ceases to seem paradoxical when one notes that almost surely the jumps for which $u < 1$ add to a finite cumulative total. This conclusion becomes natural after it is noted that a small jump's expected length is finite. Indeed, it is less than the cumulation of the expected jump lengths, which is

$$\int_{0 < u < 1} D u^{-D-1} u \, du = D \int_{0 < u < 1} u^{-D} du < \infty.$$

When $1 < D < 2$, on the contrary, this last integral diverges, suggesting that the total contribution of the small jumps is infinite. This last inference is indeed correct and, as a result, the structure of $X(t)$ is more complex; it includes a continuous term and a jump term, both infinite but having a finite sum. It is interesting to recall a fact used in Chapter IV (a reference is given in 6. DIMENSIONS below); when $E \geq 2$, the trail of a stable Lévy $X(t)$ in E-space has a fractal dimension equal to D. As a result, even though the intuitive mental association of a dimension in excess of 1 with linelike continuity cannot be maintained because of the jumps in $X(t)$, the presence of a continuous component endows $X(t)$ with a degree of substance in excess of anything compatible with $D < 1$.

5. STABLE LEVY VECTORS

Let the functional equation (L) in the definition of stability be changed by making X into a random vector **X**. Given a unit vector **V**, it is clear that the combined equations (L) and (AD) have an elementary solution that is the product of **V** by a scalar stable variable. Lévy 1937-1954 has shown that the most general solution is merely the sum of elementary solutions that correspond to all directions in space and are weighted by a distribution over the unit sphere. These contributions may be finite in number, or

infinite but discrete, or infinitesimal. In order that the vector **X** be isotropic, the elementary contributions must be distributed uniformly over all directions.

A stable Lévy vector function of time admits the same sort of decomposition as a stable scalar function: it is the sum of jumps following the hyperbolic distribution. The jump directions are ruled by a distribution over the unit sphere.

6. DIMENSIONS

In the non-Gaussian case, the earliest calculation of a dimension was performed by McKean 1955 and by Blumenthal & Getoor 1960c, 1962. A further reference, with full bibliography is Pruitt & Taylor 1969.

7. SPACE-TO-LINE STABLE RANDOM FUNCTIONS

The construction of the space-to-line Brownian function given by Chentsov 1957 was generalized to the stable case in Mandelbrot 1975b.

STATIONARITY AND KIN: CONDITIONAL STATIONARITY AND SPORADIC FUNCTIONS

The notions of *random process* and *statistical stationarity* generalize two properties of the Bernoulli process of coin tossing.

1. RANDOM FUNCTIONS AND STATIONARITY

A random function $X(t)$ is defined mainly through the joint probability distributions of the values $X(d_m)$ it takes at an arbitrary number of instants d_m. The function is said to be stationary if the joint distribution of the $X(t+d_m)$ is independent of the value of t.

Thus, one begins by attaching a measure μ to the point X of coordinates $X(t+d_m)$. The set Ω of all possible X is given the measure $\mu(\Omega)$, with $1 < \mu(\Omega) < \infty$, and the measure of the X within some domain A is given the measure $\mu(A)$ with again $0 < \mu(A) < \infty$. The probability of A is defined as $\mu(A)/\mu(\Omega)$.

EXAMPLES OF STATIONARITY. (1) Independent values. (2) Markov processes. (3) Stationary Gaussian processes as defined by the spectral density $\Phi(f)$; this $\Phi(f)$ may be infinite for some f – for example, $\Phi(0) = \infty$ for the affine noises, as long as $\int_0^\infty \Phi(f)df < \infty$.

EXAMPLES OF NONSTATIONARITY. (1) Brownian function $B(t)$ conditioned by $B(0) = 0$. (2) The function $Z(t)$ defined as equal to 1 wherever $B(t) = 0$ and to 0 otherwise.

2. INTUITIVE STATIONARITY

It was shown in the section on COUNTERINTUITIVE INSTANCES OF STATISTICAL STATIONARITY in Chapter X that, compared to the intuitive notion of stationarity, the above definition is of excessive generality.

3. SPORADIC FUNCTIONS AND CONDITIONAL STATIONARITY

On the other hand, compared to the needs of many applications, the same definition proves of insufficient generality. For example, the theory of fractal noises (Chapter IV) suggests that one consider a modification of the above nonstationary process $Z(t)$, wherein one does not assume a zero at $t=0$, merely a zero anywhere between $t=0$ and $t=T$. The result is another random process depending on T as a parameter. This process is still not stationary, but it satisfies the following restricted condition of stationarity. The joint distribution of the vector $X(t+d_m)$ is independent of t as long as all the instants $t+d_m$ lie between 0 and T. Thus the nonstationary process $Z(t)$ includes latently a whole class of random processes, each satisfying a restricted form of stationarity.

These processes are so intimately interrelated that to say they form a class is somehow insufficient. Preferring a single compact notion to cover them, I have suggested in Mandelbrot 1967b the notion of sporadic process, one that differs from a random process by the fact that the measure $\mu(\Omega)$ is infinite. The use of

$\mu(\Omega) = \infty$ for random *variables* goes back at least to Rényi 1955. To prevent $\mu(\Omega) = \infty$ from leading to catastrophe, one assumes that such a process is never observed directly, only as conditioned by some event such that $0 < \mu(C) < \infty$. Mandelbrot 1967b describes how sporadic functions allow one to exorcize some instances of infrared catastrophe.

4. DIFFERENT TRANSITIONS TO NONSTATIONARITY

There exist many different families of random functions or surfaces parameterized by a single β, which are stationary for β below some critical β_0 and nonstationary for $\beta > \beta_0$. It may well be that different physical transition phenomena correspond to different forms of the above mathematical transition. First example: the Gauss Markov processes of spectral density $1/(B + f^2)$ are stationary but their limit for $\beta \to 0$ is Brownian motion which is nonstationary. Second example: the Gaussian functions or surfaces defined as having the spectral density $f^{-\beta}$, and restricted to a lattice to avoid an ultraviolet catastrophe for $\beta < 1$. In this case, $\beta_0 = 1$, that is, the transition scans for an $1/f$ noise.

SPLITTING OF LEVEL SETS INTO IS-LANDS AND CONTINENTS. The structure of the zerosets has already been examined for two examples of the latter family. The stationary case $\beta = 0$ corresponds to the Bernoulli percolation studied in Chapter VIII. Here a transition from "island" to "continent" is observed when the "sealevel" drops below a critical value, but the stationary "continent" that emerges when $p > p_0$ is uncharacteristically spindly. The nonstationary cases $1 < \beta < 3$ correspond to the models studied in Chapter IX: Brownian for $\beta 2$ and fractional Brownian otherwise. Here a change in sealevel involves no transition because there is no intrinsic distinction between continents and islands. Let β increase through $\beta = 1$, the sealevel being set such that $\beta < 1$ involves islands. The transition one observes differs profoundly from that relative to Bernoulli percolation.

The sole purpose of this subentry is to stress the multiplicity of behaviors one can encounter in this context.

■ **TERAGON** – From the Greek *teras,* "a wonder or a monster," and *gonia,* "angle." The prefix *tera* is also used in the metric system to denote 10^{12}.

Teragon is proposed, as a neologism, to denote polygons having an extremely large number of angles and sides, especially polygons used as approximations of fractal curves.

In the present Essay, this term (like

the *noun* random) has been avoided so as to minimize neologisms, but it seems appropriate to put it on the record.

WEIERSTRASS FUNCTIONS AND KIN

This entry describes a classical construct that was originally introduced as a mathematical monster and continued to be viewed as such, yet that actually requires a small modification to yield an excellent rough model of a wide class of natural phenomena.

Let γ and w be real numbers satisfying $\gamma > 1$ and $1/\gamma < w < 1$. Weierstrass forms the series

$$\Sigma_{0 \leq n < \infty} \, w^n \exp(2\pi i \gamma^n t).$$

It obviously converges absolutely for every value of t (indeed, it is bounded by $\Sigma w^n = (1-w)^{-1}$). The limit, to be denoted by $W_0(t)$, is the first published example of a function that is continuous but nowhere differentiable.

Originally, γ is an odd integer, making $W_0(t)$ a periodic function. However, Hardy 1916 shows that the properties of continuity and nondifferentiability continue to hold when γ is any real > 1.

If one allows $w < 1/\gamma$, the function completely changes in character and becomes differentiable.

For earlier mentions of applications of the function, see the quotations from Hadamard and Perrin in Chapter I, and L. F. RICHARDSON in Chapter XI.

CELLERIER FUNCTION. This is the (excluded) limit case of the Weierstrass function, corresponding to $w = 1/\gamma$. This function's spectrum is approximately f^{-1}. Cellérier's story is discussed in Chapter XI under WEIERSTRASS.

ULTRAVIOLET AND INFRARED CATASTROPHES AND THEIR MUTUAL RELATIONSHIPS IN THE SELF SIMILAR CASE. We know from Chapter II that nondifferentiability had given rise to bitter controversies linked with its being presumed quite unnatural. Much later and quite independently, physicists discovered the so-called $1/f$ noises (see AFFINE NOISES), which also became the object of controversy because of their being hard to fit into the prevailing mathematical formalisms. The first such noises were described circa 1920. *Postscript:* Actually around 1850; see JOSEPH EFFECT in Chapter XI.

Superficially, these two lonely hearts' desires were divergent, since the Weierstrass function addressed itself to a high-frequency effect, an ultraviolet catastrophe, while $1/f$ noises encountered low-frequency difficulties, infrared catastrophe. But we shall see momentarily that the Weierstrass function can readily be made to involve low frequency difficulties, while the $1/f$ noises can be extended in the opposite direction. This having been done, they meet and match. More precisely, the randomized form of an extended Weierstrass function matches some aspects of certain $1/f$ noises in Euclidean time; for other aspects, see CANTOR AND LEVY STAIRCASES.

SPECTRAL PROPERTIES. The match is best achieved through the spectrum. The Weierstrass function has a line spectrum. For each frequency of the form $f=\gamma^n$, it is a spectral line of energy (amplitude squared) equal to w^{2n}. Hence, the total energy in frequencies $f \geq \gamma^n$ is $w^{2n} + w^{2(n+1)} + \ldots = w^{2n}(1-w^2)^{-1} = f^{-2H}(1-w^2)^{-1}$, with the exponent

$$H = -\log w / \log \gamma.$$

From $w<1$, it follows that $H>0$, and from $1/w<\gamma$, it follows that $H<1$. Any value of H between 0 and 1 can be achieved by proper choice of γ and w.

The cumulative spectrum f^{-2H} is also encountered for processes with the continuous spectral density f^{-2H-1}. For example, the Weierstrass cumulative spectrum can take the form f^{-1}, which is encountered for ordinary Brownian motion of spectral density f^{-2}. One difference, which is important but cannot be discussed here, is that in one case the spectrum is absolutely continuous and in the other case it is discrete and lacunary (Zygmund 1959 **I,** p. 202).

LOW FREQUENCY CUTOFF. Another difference, which we shall momentarily eliminate, is that the Brownian cumulative spectrum f^{-1} extends all the way down to $f=0$, while the Weierstrass spectrum is bounded below by $f=1$. This bound is related to the fact that originally γ was an integer and the Weierstrass function was periodic. Periodicity clashes with the self similarity embodied in the spectrum f^{-2H}.

EXTENDED FORM WEIERSTRASS FUNCTIONS. To avoid the low frequency cutoff, form the expression $\Delta W_0(t) = W_0(0) - W_0(t)$ and change it by letting n run from $-\infty$ to ∞. The added terms converge, and their limit is continuous and differentiable. This procedure yields the extended function

$$W(t) = \sum_{-\infty < n < \infty} w^n [1 - \exp(2\pi i \gamma^n t)].$$

Like its Weierstrass function component, the extended function is continuous but nowhere differentiable.

For this extended function, the cumulative spectrum is proportional to f^{-2H} for all values of $f>0$.

◁ The extended function must be known but my search for literature on its account has been unsuccessful. ▶

SELF AFFINITY WITH RESPECT TO THE FOCAL TIME t=0. The spectrum of the extended function is self similar; it involves harmonics without a fundamental. The extended function's value at time $t\gamma^m$ is

$$\Sigma_{-\infty < n+m < \infty} w^{-m} w^{n+m} [1 - \exp(2\pi i\gamma^{n+m}t)] .$$

Hence W satisfies the relationship

$$W(t\gamma^m) = w^{-m} W(t) .$$

In other words, the extended function $w^m W(\gamma^m t)$ is independent of m. Alternatively, as long as h is of the form γ^m, $h^{-H} W(ht)$ is independent of h; that is, $W(t)$ is self affine with respect to r's of the form γ^{-m}, together with the exponent H and the focal time 0. In the Brownian spectrum case, one has $H = 1/2$, as for ordinary Brownian motion.

GAUSS RANDOM FUNCTIONS WITH A WEIERSTRASS SPECTRUM. The next step toward realism consists in making $W(t)$ into a random function. The simplest and most intrinsic method consists in starting with the extended Weierstrass function and then multiplying all its Fourier coefficients by independent complex Gaussian factors of zero mean and unit variance. The real and imaginary parts of the result will be called Gauss-Weierstrass functions. In several ways, they are approximate fractional Brownian motions. When the values of H are matched, their spectra are as close to being the same as allowed by the fact

that one is discrete and the other continuous. Moreover, the result of Orey 1970 and Marcus 1976 remains applicable and shows that their level sets have the same fractal dimension.

SIMULATION. The Gauss Weierstrass algorithm is especially convenient from the viewpoint of actual simulation of a sample of duration L. It suffices to settle on a value of γ and then to truncate the low frequencies at the value nearest to some prescribed multiple of the outer scale L. Similarly, when a function is examined at discrete intervals of η, one will truncate the high frequencies at some multiple of this inner scale η. As L is multiplied by $\mu = \gamma^m$, the Weierstrass algorithm requires m additional terms, three if $\mu = 2$ and $\gamma = 2^{1/3}$. On the other hand, the number of high-frequency spectral lines remains independent of L, the lowest wavelengths being $\eta/2^{-1/3}$, η and $\eta 2^{1/3}$.

By way of contrast, suppose the fractional Brownian motion is being approximated by a fractional Brown-Fourier series. Then a doubling of L requires that the number of harmonics also be doubled, with most new harmonics going into changing mid- and high-frequency properties, in pure waste.

In the Gauss-Weierstrass approach, the approximation is improved by choosing γ closer to 1. The choice of γ is a matter of compromise between accuracy and cost.

◁ **AN INTRIGUING RELATIONSHIP BE-TWEEN THE WEIERSTRASS SPECTRUM AND MUSICAL SCALES**. The notion of a tempered musical scale implies a discrete spectrum with frequencies spread logarithmically, more or less uniformly, on both sides of some arbitrarily prescribed frequency for the note A. For example, the twelve-tone scale corresponds to $\gamma = 2^{1/12}$. As a result, each musical instrument uses a relatively high proportion of the low frequencies within its overall frequency band, but the proportion of actually used frequencies decreases rapidly as one proceeds to high notes. If such a scale is extrapolated to inaudibly high and low frequencies, the support of its spectrum becomes identical to that of the Weierstrass function with the same value of γ. On the other hand, the spectral support of a musical scale differs profoundly from that of a periodic function, in which the frequencies are *linearly* uniform.

◁ Consequently, in order to add low frequencies to a piece of music, it suffices to add new instruments capable of the desired low tones. Had the musical scale been linearly rather than logarithmically uniform, the high and middle tones

would also have had to be changed. The contrast between these two methods for the synthesis of complex sounds is reminiscent of the contrast (see the preceding subentry) between the two basic methods of synthesis of broad band noise. In both cases the Weierstrass method is the better one.

◁ Nevertheless, it raises problems. The Euler-Fourier theorem states that a periodic function can be represented as a series of *linearly* spaced harmonics, but on the other hand the bulk of high harmonics of accepted notes of low instruments are not themselves accepted notes. One cannot repeal the theorem, but undesirable linear harmonics can be minimized by training.

ZEROSETS OF RELATED FUNCTIONS. The Rademacher functions are squared-off variants of the sine functions of the form $\sin(2\pi\gamma^n t)$ in which γ has been set to 2. Wherever the sine is positive (respectively, negative or vanishing), the Rademacher function is set equal to 1 (respectively, to -1, or 0). (Zygmund 1959 **I**, p. 202.) The natural generalization of the Weierstrass function is a series in which the nth term is the product of w^n by the nth Rademacher function. The spectral exponent is again $H = -\log w / \log 2$. Obviously, the sum is a discontinuous function. Intuitively, the precedent of fractional Brownian motion suggests that the sets where the Weierstrass-Rademacher function crosses a prescribed values are of dimension $1-H$. This is indeed the result derived by Beyer 1962 under the restriction that $1/H$ is an integer. It would be desirable to extend it to other values of w, also to study the Rademacher-type functions with $\gamma \neq 2$.

In the same spirit, Singh 1935 refers to numerous papers on the zerosets of variants of the Weierstrass Fourier function. In some cases, the D is easy to evaluate. The topic deserves a fresh look.

BIBLIOGRAPHY

Reference to the bibliography is through the author's name and date. Multiple dates in one item refer to multiple editions, translations, or reprints. Dates followed by – refer to the first volumes in multivolume collections. In cases of ambiguity, the date is followed by a letter, which is sometimes sequential but most often related to the publication's title or to the name of the serial in which it appeared. This new convention is intended as a mnemonic device.

Because the serials referenced in this list belong to different disciplines, their titles are given without abbreviation.

A few general references are included, but the list does not remotely attempt to be balanced or complete on this account.

Abell, G.O. 1965. Clustering of galaxies. *Annual Reviews of Astronomy and Astrophysics* **3**, 1-22.

Arthur, D.W.G. 1954. The distribution of lunar craters. *Journal of the British Astronomical Association* **64**, 127-132.

Azbel, M.Ya. 1964. Energy spectrum of a conduction electron in a magnetic field. *Soviet Physics JETP* **19**, 634-645.

Bachelier, L. 1900. *Théorie de la spéculation.* Thesis for the Doctorate in Mathematical Sciences (defended March 29, 1900). *Annales Scientifiques de l'Ecole Normale Supérieure* **III-17**, 21-86. Translation in Cootner (Ed.) 1964.

Bachelier, L. 1914. *Le jeu, la chance et le hasard.* Paris: Flammarion.

Ball, W.W.Rouse & Coxeter, H.S.M. 1974. *Mathematical Recreations and Essays.* 12th edition. Toronto University Press.

Balmino, G., Lambeck, K. & Kaula, W.M. 1973. A spherical harmonic analysis of the Earth's topography. *Journal of Geophysical Research* **78**, 478-481.

Barber, M.N. & Ninham, B.W. 1970. *Random and Restricted Walks: Theory and Applications.* New York, London, Paris: Gordon & Breach.

Batchelor, G.K. 1953. *The Theory of Homogeneous Turbulence.* Cambridge University Press.

Batchelor, G.K. & Townsend, A.A. 1949. The nature of turbulent motion at high wave numbers. *Proceeding of the Royal Society of London* A **199**, 238-255.

Berger, J.M. & Mandelbrot, B.B. 1963. A new model for the clustering of errors on telephone circuits. *IBM Journal of Research and Development* **7**, 224-236.

Berman, S.M. 1970. Gaussian processes with stationary increments: local times and sample function properties. *Annals of Mathematical Statistics* **41**, 1260-1272.

Besicovitch, A.S. 1935. On the sum of digits of real numbers represented in the dyadic system (On sets of fractional dimensions II). *Mathematische Annalen* **110**, 321-330.

Besicovitch, A.S. & Taylor, S.J. 1954. On the complementary interval of a linear closed set of zero Lebesgue measure. *Journal of the London Mathematical Society* **29**, 449-459.

Besicovitch, A.S. & Ursell, H.D. 1937. Sets of fractional dimensions (V): On dimensional numbers of some continuous curves. *Journal of the London Mathematical Society* **12**, 18-25.

Beyer, W.A. 1962. Hausdorff dimension of level sets of some Rademacher series. *Pacific Journal of Mathematics* **12**, 35-46.

Bidaux, R., Boccara, N., Sarma, G., Sèze, L., de Gennes, P.G. & Parodi, O. 1973. Statistical properties of focal conic textures in smectic liquid crystals. *Le Journal de Physique* **34**, 661-672.

Bienaymé, J. 1853. Considérations à l'appui de la découverte de Laplace sur la loi de probabilité dans la méthode des moindres carrés. *Comptes Rendus* (Paris) **37**, 309-329.

Billingsley, P. 1967. *Ergodic Theory and Information.* New York: Wiley.

Blumenthal, L.M. & Menger, K. 1970. *Studies in Geometry.* San Francisco: W.H. Freeman.

Blumenthal, R.M. & Getoor, R.K. 1960c. A dimension theorem for sample functions of stable processes. *Illinois Journal of Mathematics* **4**, 308-316.

Blumenthal, R.M. & Getoor, R.K. 1960m. Some theorems on stable processes. *Transactions of the American Mathematical Society* **95**, 263-273.

Blumenthal, R.M. & Getoor, R.K. 1962. The dimension of the set of zeros and the graph of a symmetric stable process. *Illinois Journal of Mathematics* **6**, 370-375.

Bondi, H. 1952; 1960. *Cosmology.* Cambridge University Press.

Borel, E. 1912-1915. Les théories moléculaires et les mathématiques. *Revue Générale des Sciences* **23**, 842-853. Translated as Molecular theories and mathematics. *Rice Institute Pamphlet* **1**, 163-193. See also Borel 1972–, **III**, 1773-1784.

Borel, E. 1922. Définition arithmétique d'une distribution de masses s'étendant à l'infini et quasi périodique, avec une densité moyenne nulle. *Comptes Rendus* (Paris) **174**, 977-979.

Borel, E. 1972–. *Oeuvres de Emile Borel.* Paris: Editions du CNRS.

Bouligand, G. 1928. Ensembles impropres et nombre dimensionnel. *Bulletin des Sciences Mathématiques* **II-52**, 320-334 & 361-376.

Bouligand, G. 1929. Sur la notion d'ordre de mesure d'un ensemble plan. *Bulletin des Sciences Mathématiques* **II-53**, 185-192.

Boyd, D.W. 1973. Improved bounds for the disk packing constant. *Aequationes Mathematicae* **9**, 99-106.

Bragg, W.H. 1934. Liquid crystals. *Nature* **133**, 445-456.

Browder, F.E. 1976. Does pure mathematics have a relation to the sciences? *American Scientist* **64**, 542-549.

Brush, S.G. 1968. A history of random processes. I. Brownian movement from Brown to Perrin. *Archive for History of Exact Sciences* **5**, 1-36. Also in Brush 1976, 655-701.

Brush, S.G. 1976. *The Kind of Motion We Call Heat, a History of the Kinetic Theory of Gases in the 19th Century.* Amsterdam: North-Holland.

Cantor, G. 1883. Grundlagen einer allgemeinen Mannichfältigkeitslehre. *Mathematische Annalen* **21**, 545-591. Also in Cantor 1932. Trans. H. Poincaré, as Fondements d'une théorie générale des ensembles. *Acta Mathematica* **2**, 381-408.

Cantor, G. 1932-1966. *Gesammelte Abhandlungen mathematischen und philosophischen Inhalts.* Ed. E. Zermelo. Berlin: Teubner. Hildesheim: Olms (reprint).

Carathéodory, C. 1914. Über das lineare Maß von Punktmengen – eine Verallgemeinerung des Längenbegriffs. *Nachrichten der K. Gesellschaft der Wissenschaften zu Göttingen. Mathematisch-physikalische Klasse* 404-426. Also in Carathéodory 1954–, **4**, 249-275.

Carathéodory, C. 1954–. *Gesammelte mathematische Schriften.* Munich: Beck.

Cartier, P. 1971. Introduction à l'étude des mouvements browniens à plusieurs paramètres. *Séminaire de Probabilités V* (Strasbourg). Lecture Notes in Mathematics **191**, 58-75. New York: Springer.

Cauchy, A. 1853. Sur les résultats les plus probables. *Comptes Rendus* (Paris) **37**, 198-206.

Cellérier, Ch. 1890. Note sur les principes fondamentaux de l'analyse. *Bulletin des Sciences Mathématiques* **14**, 142-160.

Cesàro, E. 1905. Remarques sur la courbe de von Koch. *Atti della Reale Accademia delle Scienze Fisiche e Matematiche di Napoli* **XII**, 1-12. Also in Cesàro 1964–, **II**, 464-479.

Cesàro, E. 1964–. *Opere Scelte.* Rome: Edizioni Cremonese.

Chandrasekhar, S. 1943. Stochastic problems in physics and astronomy. *Reviews of Modern Physics* **15**, 1-89. Reprinted in *Noise and Stochastic Processes.* Ed. N. Wax. New York: Dover.

Charlier, C.V.L. 1908. Wie eine unendliche Welt aufgebaut sein kann. *Arkiv för Matematik, Astronomi och Fysik* **4**, 1-15.

Charlier, C.V.L. 1922. How an infinite world may be built up. *Arkiv för Matematik, Astronomi och Fysik* **16**, 1-34.

Chentsov, N.N. 1957. Lévy's Brownian motion for several parameters and generalized white noise. *Theory of Probability and its Applications.* **2**, 265-266.

Chorin, A.J. 1975. Lectures on turbulence theory. Berkeley: Publish or Perish.

Clayton, D.D. 1975. *Dark Night Sky, a Personal Adventure in Cosmology.* New York: Quadrangle/New York Times Book Co.

Cootner, P.H. (Ed.) 1964. *The Random Character of Stock Market Prices.* Cambridge, Mass.: MIT Press.

Corrsin, S. 1959d. On the spectrum of isotropic temperature fluctuations in an isotropic turbulence. *Journal of Applied Physics* **22**, 469-473.

Corrsin, S. 1959b. Outline of some topics in homogeneous turbulence flow. *Journal of Geophysical Research* **64**, 2134-2150.

Corrsin, S. 1962. Turbulent dissipation fluctuations. *Physics of Fluids* **5**, 1301-1302.

Dauben, J.W. 1974. Denumerability and dimension: the origins of Georg Cantor's theory of sets. *Rete* **2**, 105-133.

Dauben, J.W. 1975. The invariance of dimension: problems in the early development of set theory and topology. *Historia Mathematicae* **2**, 273-288.

Davis, C. & Knuth, D.E. 1970. Number representations and dragon curves. *Journal of Recreational Mathematics* **3**, 66-81 & 133-149.

de Chéseaux, J.P.L. 1744. Sur la force de la lumière et sa propagation dans l'éther, et sur la distance des étoiles fixes. *Traité de la comète qui a paru en décembre 1743 et en janvier, février et mars 1744.* Lausanne et Genève: Chez Marc-Michel Bousquet et Compagnie.

de Gennes, P.G. 1974. *The Physics of Liquid Crystals.* Oxford: Clarendon Press.

de Gennes, P.G. 1976. La percolation: un concept unificateur. *La Recherche* **7**, 919-927.

Denjoy, A. 1964. *Hommes, formes et le nombre.* Paris: Albert Blanchard.

Denjoy, A. 1975. Evocation de l'homme et de l'œuvre. *Astérisque* **28-28.** Ed. G. Choquet. Paris: Societé Mathématique de France.

de Vaucouleurs, G. 1956. The distribution of bright galaxies and the local supergalaxy. *Vistas in Astronomy* **II**, 1584-1606. London, New York: Pergamon.

de Vaucouleurs, G. 1970. The case for a hierarchical cosmology. *Science* **167**, 1203-1213.

de Vaucouleurs, G. 1971. The large scale distribution of galaxies and clusters of galaxies. *Publications of the Astronomical Society of the Pacific* **73**, 113-143.

de Wijs, H.J. 1951 & 1953. Statistics of ore distribution. *Geologie en Mijnbouw* (Amsterdam) **13**, 365-375; **15**, 12-24.

Domb, C. 1964. Some statistical problems connected with crystal lattices. *Journal of the Royal Statistical Society* **26B**, 367-397.

Domb, C. & Green, M.S. (Eds.) 1972–. *Phase Transitions and Critical Phenomena.* New York: Academic Press.

Domb, C., Gillis, J. & Wilmers, G. 1965. On the shape and configuration of polymer molecules. *Proceedings of the Physical Society* **85**, 625-645.

Domb, C. & Hioe, F.T.. 1969. Mean square interchain distances in a self avoiding walk and correlations in a self avoiding walk. *Journal of Chemical Physics* **51**, 1915-1928.

duBois Reymond, P. 1875. Versuch einer Classification der willkürlichen Functionen reeller Argument nach ihren Änderungen in den kleinsten Intervallen. *Journal für die reine und angewandte Mathematik* (Crelle) **79**, 21-37.

Dugac, P. 1973. Elements d'analyse de Karl Weierstrass. *Archive for History of Exact Sciences* **10**, 41-176.

Dugac, P. 1976. Notes et documents sur la vie et l'œuvre de René Baire. *Archive for History of Exact Sciences* **15**, 297-384.

duPlessis, N. 1970. *An Introduction to Potential Theory.* New York: Hafner.

Eggleston, H.G. 1953. On closest packing by equilateral triangles. *Proceedings of the Cambridge Philosophical Society* **49**, 26-30.

Federer, H. 1969. *Geometric Measure Theory.* New York: Springer.

Feller, W. 1949. Fluctuation theory of recurrent events. *Transactions of the American Mathematical Society* **67**, 98-119.

Feller, W. 1951. The asymptotic distribution of the range of sums of independent random variables. *Annals of Mathematical Statistics* **22**, 427.

Feller, W. 1950-1957-1968. *An Introduction to Probability Theory and Its Application.* (Vol. 1.) New York: Wiley.

Feller, W. 1966-1971. *An Introduction to Probability Theory and Its Application.* (Vol. 2.) New York: Wiley.

Fournier d'Albe, E.E. 1907. *Two New Worlds: I The Infra World; II The Supra World.* London: Longmans Green.

Fréchet, M. 1941. Sur la loi de répartition de certaines grandeurs géographiques. *Journal de la Société de Statistique de Paris* **82**, 114-122.

Friedlander, S.K. & Topper, L. 1961. *Turbulence: Classic Papers on Statistical Theory.* New York: Interscience.

Frostman, O. 1935. Potentiel d'équilibre et capacité des ensembles avec quelques applications à la théorie des fonctions. *Meddelanden fran Lunds Universitets Mathematiska Seminarium* **3**, 1-118.

Gamow, G. 1954. Modern cosmology. *Scientific American* **190** (March) 54-63. See also in Munitz (Ed.) 1957, 390-404.

Gangolli, R. 1967. Lévy's Brownian motion of several parameters. *Annales de l'Institut Henri Poincaré* **3B**, 121-226.

Gardner, M. 1967. Mathematical games. *Scientific American* (March, April and June).

Gardner, M. 1976. Mathematical games. *Scientific American* (December).

Gelbaum, B.R. & Olmsted, J.M.H. 1964. *Counterexamples in Analysis.* San Francisco: Holden-Day.

Gerver, J. 1970. The differentiability of the Riemann function at certain rational multiples of π. *American Journal of Mathematics* **92**, 33-55.

Gillispie, C.C. (Ed.) 1970-1976. *Dictionary of Scientific Biography.* Fourteen volumes. New York: Scribner's.

Gnedenko, B.V. & Kolmogorov, A.N. 1954. *Limit Distributions for Sums of Independent Random Variables.* Trans. K.L. Chung. Reading, Mass.: Addison Wesley.

Grant, H.L., Stewart, R.W. & Moilliet, A. 1959. Turbulence spectra from a tidal channel. *J. Fluid Mechanics* **12**, 241-268.

Grenander, U. & Rosenblatt, M. 1957 & 1966. *Statistical Analysis of Stationary Time Series.* New York: Wiley.

Hack, J.T. 1957. Studies of longitudinal stream in Virginia and Maryland. *United States Geological Survey Professional Papers* **294-B.**

Hadamard, J. 1912. L'œuvre mathématique de Poincaré. *Acta Mathematica* **38**, 203-287. Also in Poincaré 1916–, **XI**, 152-242. Or in Hadamard 1968, **4**, 1921-2005.

Hadamard, J. 1968. *Oeuvres de Jacques Hadamard.* Paris: Editions du CNRS.

Haggett, P. 1972. *Geography: A Modern Synthesis.* New York: Harper & Row.

Hardy, G.H. 1916. Weierstrass's non-differentiable function. *Transactions of the American Mathematical Society* **17**, 322-323. Also in Hardy 1966–, **IV**, 477-501.

Hardy, G.H. 1966–. *Collected Papers.* Oxford: Clarendon Press.

Harris, T.E. 1963. *Branching Processes.* New York: Springer.

Hartmann, W.K. 1977. Cratering in the solar system. *Scientific American* (January) 84-99.

Harvey, W. 1628. *De motu cordis.* Trans. Robert Willis, London, 1847, as *On the Motion of the Heart and Blood in Animals.* Excerpt in *Steps in the Scientific Tradition: Readings in the History of Science.* Ed. R.S. Westfall et al. New York: Wiley.

Hausdorff, F. 1914-1949. *Grundzüge der Mengenlehre.* New York: Chelsea (reprint).

Hausdorff, F. 1919. Dimension und äußeres Mass. *Mathematische Annalen* **79**, 157-179.

Hawkes, J. 1974. Hausdorff measure, entropy and the independence of small sets. *Proceedings of the London Mathematical Society* (3) **28**, 700-724.

Hawkins, G.S. 1964 Interplanetary debris near the Earth. *Annual Review of Astronomy and Astrophysics* **2**, 149-164.

Hawkins, T. 1970. *Lebesgue's Theory of Integration: Its Origins and Development.* Madison: The University of Wisconsin Press.

Hermite, C. & Stieltjes, T.J.. 1905. *Correspondance d'Hermite et de Stieltjes.* 2 vols. Ed. B. Baillaud & H. Bourget. Paris: Gauthier-Villars.

Hilbert, D. 1891. Über die stetige Abbildung einer Linie auf ein Flächenstück. *Mathematische Annalen* **38**, 459-460. Also in Hilbert 1932-1965, **3**, 1-2.

Hilbert, D. 1932-1965. *Gesammelte Abhandlungen.* Berlin: Springer. New York: Chelsea (reprint).

Hiley, B.J. & Sykes, M.F. 1961. Probability of initial ring closure in the restricted random walk model of a macromolecule. *Journal of Chemical Physics* **34**, 1531-1537.

Hille, E. & Tamarkin, J.D. 1929. Remarks on a known example of a monotone continuous function. *American Mathematics Monthly* **36**, 255-264.

Hofstadter, D.R. 1976. Energy levels and wave functions of Bloch electrons in rational and irrational magnetic fields. *Physical Review B* **14**, 2239-2249.

Holtsmark, J. 1919. Über die Verbreiterung von Spektrallinien. *Annalen der Physik* **58**, 577-630.

Horton, R.E. 1945. Erosional development of streams and their drainage basins; Hydrophysical approach to quantitative morphology. *Bulletin of the Geophysical Society of America* **56**, 275-370.

Howard, A.D. 1971. Truncation of stream networks by headward growth and branching. *Geophysical Analysis* **3**, 29-51.

Hoyle, F. 1953. On the fragmentation of gas clouds into galaxies and stars. *Astrophysical Journal* **118**, 513-528.

Hoyle, F. 1975. *Astronomy and Cosmology. A Modern Course.* San Francisco: W.H. Freeman.

Hurewicz, W. & Wallman, H. 1941. *Dimension Theory.* Princeton University Press.

Hurst, H.E. 1951. Long term storage capacity of reservoirs. *Transactions of the American Society of Civil Engineers* **116**, 770-808.

Hurst, H.E. 1955. Methods of using long-term storage in reservoirs. *Proceedings of the Institution of Civil Engineers* Part I, 519-577.

Hurst, H.E., Black, R.P., and Simaika, Y.M. 1965. *Long-term Storage, an Experimental Study.* London: Constable.

Jack, J.J.B., Noble, D. & Tsien, R.W. 1975. *Electric Current Flow in Excitable Cells.* Oxford University Press.

Jaki, S.L. 1969. *The Paradox of Olbers' Paradox.* New York: Herder & Herder.

Jeans, J.H. 1929-1961. *Astronomy and Cosmogony.* Cambridge University Press. New York: Dover (reprint).

Jona-Lasinio, G. 1975a. The renormalization group: a probabilistic view. *Il Nuovo Cimento* **26B**, 99-119.

Jona-Lasinio, G. 1975b. Critical behavior in terms of probabilistic concepts. *International (Marseilles) Colloquium on Mathematical Methods in Quantum Field Theory.*

Kahane, J.P. 1964. Lacunary Taylor and Fourier series. *Bulletin of the American Mathematical Society* **70,** 199-213.

Kahane, J.P. 1968. *Some Random Series of Functions.* Lexington, Mass.: D. C. Heath.

Kahane, J.P. 1969. Trois notes sur les ensembles parfaits linéaires. *Enseignement mathématique* **15,** 185-192.

Kahane, J.P. 1970. Courbes étranges, ensembles minces. *Bulletin de l'Association des Professeurs de Mathématiques de l'Enseignement Public* **49,** 325-339.

Kahane, J.P. 1971. The technique of using random measures and random sets in harmonic analysis. *Advances in Probability and Related Topics,* **1,** 65-101. Ed. P. Ney. New York: Marcel Dekker.

Kahane, J.P. 1974. Sur le modèle de turbulence de Benoit Mandelbrot. *Comptes Rendus* (Paris) **278A,** 621-623.

Kahane, J.P. & Mandelbrot, B.B. 1965. Ensembles de multiplicité aléatoires. *Comptes Rendus* (Paris) **261,** 3931-3933.

Kahane, J.P. & Peyrière, J. 1976. Sur certaines martingales de B. Mandelbrot. *Advances in Mathematics* **22,** 131-145.

Kahane, J.P. & Salem, R. 1963. *Ensembles parfaits et séries trigonométriques.* Paris: Hermann.

Kahane, J.P., Weiss, M. & Weiss, G. 1963. On lacunary power series. *Arkiv för Mathematik, Astronomi och Fysik* **5,** 1-26.

Kaufman, R. 1968. On Hausdorff dimension of projections. *Mathematika* **15,** 153-155.

Kirkpatrick, S. 1973. Percolation and conduction. *Reviews of Modern Physics* **45,** 574-588.

Kline, S.A. 1945. On curves of fractional dimensions. *Journal of the London Mathematical Society* **20,** 79-86.

Knuth, D. 1968–. *The Art of Computer Programming.* Reading, Mass.: Addison Wesley.

Kohlrausch, R. 1847. Über das Dellmann'sche Elektrometer. *Annalen der Physik und Chemie* (Poggendorf) **III-12,** 353-405.

Kohlrausch, R. 1854. Theorie des elektrischen Rückstandes in der Leidener Flasche. *Annalen der Physik und Chemie* (Poggendorf) **IV-91,** 56-82 & 179-214.

Kolmogorov, A.N. 1931. Über die analytischen Methoden in der Wahrscheinlichkeitsrechnung. *Mathematische Annalen* **104,** 415-458.

Kolmogorov, A.N. 1940. Wiensersche Spiralen und einige andere interessante Kurven im Hilbertschen Raum. *Comptes Rendus (Doklady) Académie des Sciences de l'URSS (N.S.)* **26,** 115-118.

Kolmogorov, A.N. 1941. Local structure of turbulence in an incompressible liquid for very large Reynolds numbers. *Comptes Rendus (Doklady) Académie des Sciences de l'URSS (N.S.)* **30,** 299-303. Reprinted in Friedlander & Topper 1961, 151-155.

Kolmogorov, A.N. 1962. A refinement of previous hypotheses concerning the local structure of turbulence in a viscous incompressible fluid at high Reynolds number. *Journal of Fluid Mechanics* **13,** 82-85.

Kolmogorov, A.N. & Tihomirov, V.M. 1959-1961. Epsilon-entropy and epsilon-capacity of sets in functional spaces. *Uspekhi Matematicheskikh Nauk* (N.S.) **14,** 3-86. Translated in *American Mathematical Society Translations* (Series 2) **17,** 277-364.

Korčak, J. 1938. Deux types fondamentaux de distribution statistique. *Bulletin de l'Institut International de Statistique* **III,** 295-299.

Kraichnan, R.H. 1974. On Kolmogorov's inertial range theories. *Journal of Fluid Mechanics* **62,** 305-330.

Kuo, A.Y.S. & Corrsin, S. 1971. Experiments on internal intermittency and fine structure distribution functions in fully turbulent fluid. *Journal of Fluid Mechanics* **50,** 285-320.

Kuo, A.Y.S. & Corrsin, S. 1972. Experiments on the geometry of the fine structure regions in fully turbulent fluid. *Journal of Fluid Mechanics* **56,** 477-479.

Lamperti, J. 1962. Semi-stable stochastic processes. *Trans. of the American Mathematical Society* **104,** 62-78.

Lamperti, J. 1966. *Probability: a Survey of the Mathematical Theory.* Reading, Mass.: W. A. Benjamin.

Landau, L.D. 1943. On the theory of the intermediate state of superconductivity. *Journal of Physics USSR* **7,** 99. *Journal of Experimental and Theoretical Physics USSR* **13,** 377. Translation in Landau 1965a, 365-379, and in Landau 1965b, 119-137.

Landau, L.D. 1965a. *Collected Papers.* Ed. D. de Haar. Oxford: Pergamon.

Landau, L.D. 1965b. *Men of Physics: Landau I.* Ed. D. de Haar. Oxford: Pergamon.

Landau, L.D. & Lifshitz, E.M. 1953-1959. *Fluid Mechanics.* London: Pergamon. Reading: Addison Wesley.

Landkof, N.S. 1966-1972. *Foundations of Modern Potential Theory.* New York: Springer.

Lavoie, J.L., Osler, T.J. & Tremblay, R. 1976. Fractional derivatives of special functions. *SIAM Review* **18**, 240-268.

Lawrance, A.J. & Kottegoda, N.T. 1977. Stochastic modelling of riverflow time series. *Journal of the Royal Statistical Society* A, **140** (in the press).

Lax, D. 1973. The differentiability of Pólya's function. *Advances in Mathematics* **10**, 456-464.

Leath, P.L. 1976. Cluster size and boundary distribution near percolation threshold. *Physical Review* B **14**, 5046-5055.

Lebesgue, H. 1903. *Sur le problème des aires.* See Lebesgue 1973, IV, 29-35.

Lebesgue, H. 1972–. *Oeuvres scientifiques.* Vol. I to V. Genève: Enseignement Mathématique.

Leibniz, G.W. 1849-1962. *Mathematische Schriften.* Hildesheim: Olms (reprint).

Leopold, L.B. 1962. Rivers. *American Scientist* **50**, 511-537.

Leopold, L.B. & Langbein, W.B. 1962. The concept of entropy in landscape evolution. *U.S. Geological Survey Professional Papers* **500A.**

Leray, J. 1934. Sur le mouvement d'un liquide visqueux emplissant l'espace. *Acta Mathematica* **63**, 193-248.

Lévy, P. 1925. *Calcul des probabilités.* Paris: Gauthier Villars.

Lévy, P. 1930. Sur la possibilité d'un univers de masse infinie. *Annales de Physique* **14**, 184-189. Also in Lévy 1973 **II**, 534-540.

Lévy, P. 1937-1954. *Théorie de l'addition des variables aléatoires.* Paris: Gauthier Villars.

Lévy, P. 1938. Les courbes planes ou gauches et les surfaces composées de parties semblables au tout. *Journal de l'Ecole Polytechnique* Série III **7-8**, 227-291. Also in Lévy 1973 **II**, 331-394.

Lévy, P. 1948-1965. *Processus stochastiques et mouvement brownien.* Paris: Gauthier-Villars.

Lévy, P. 1957. Brownian motion depending on n parameters. The particular case $n=5$. *Proceedings of the Symposia in Applied Mathematics* **VII**, 1-20. Providence, R.I.: American Mathematical Society.

Lévy, P. 1959. Le mouvement brownien fonction d'un point de la sphère de Riemann. *Circolo matematico di Palermo, Rendiconti.* Ser. 2, **8**, 297-310.

Lévy, P. 1963. Le mouvement brownien fonction d'un ou de plusieurs paramètres. *Rendiconti di Matematica* (Roma) **22**, 24-101.

Lévy, P. 1965. A special problem of Brownian motion and a general theory of Gaussian random functions. *Proceedings of the Third Berkeley Symposium in Mathematical Statistics and Probability Theory* **2**, 133-175. Berkeley: University of California Press.

Lévy, P. 1970. *Quelques aspects de la pensée d'un mathématicien.* Paris: Albert Blanchard.

Lévy, P. 1973–. *Oeuvres de Paul Lévy.* Ed. D. Dugué, P. Deheuvels & M. Ibéro. Paris, Bruxelles & Montréal: Gauthier Villars.

Lieb, E.H. & Lebowitz, J.L. 1972. The constitution of matter: existence of thermodynamics for systems composed of electrons and nuclei. *Advances in Mathematics* **9**, 316-398.

Livingston, J.D. & DeSorbo, W. 1969. The intermediate state in type I superconduction. *Superconductivity.* Ed. R.D. Parks. New York: Marcel Dekker.

Lorenz, E.N. 1963. Deterministic nonperiodic flow. *Journal of the Atmospheric Sciences* **20**, 130-141.

Love, E.R. & Young, L.C. 1937. Sur une classe de fonctionnelles linéaires. *Fundamenta Mathematicae* **28**, 243-257.

Lukacs, E. 1960-1970. *Characteristic Functions.* London: Griffin. New York: Hafner.

Lydall, H.F. 1959. The distribution of employment income. *Econometrica* **27**, 110-115.

Maitre, J. 1964. Les fréquences des prénoms de baptême en France. *L'Année sociologique* **3**, 31-74.

Mandelbrot, B.B. 1951. Adaptation d'un message à la ligne de transmission. I & II. *Comptes Rendus* (Paris) **232**, 1638-1640 & 2003-2005.

Mandelbrot, B.B. 1955b. On recurrent noise limiting coding. *Information Networks, the Brooklyn Polytechnic Institute Symposium,* 205-221. Ed. E. Weber. New York: Interscience.

Mandelbrot, B.B. 1960e. The Pareto-Lévy law and the distribution of income. *International Economic Review* **1**, 79-106.

Mandelbrot, B.B. 1961e. Stable Paretian random functions and the multiplicative variation of income. *Econometrica* **29**, 517-543.

Mandelbrot, B.B. 1962e. Paretian distributions and income maximization. *Quarterly Journal of Economics of Harvard University* **76**, 57-85.

Mandelbrot, B.B. 1963b. The variation of certain speculative prices. *The Journal of Business of the University of Chicago* **36**, 394-419. Reprinted in Cootner (Ed.) 1964.

Mandelbrot, B.B. 1963e. New methods in statistical economics. *Journal of Political Economy* **71**, 421-440.

Mandelbrot, B.B. 1964. Derivation of statistical thermodynamics from purely phenomenological principles. *Journal of Mathematical Physics* **5**, 164-171.

Mandelbrot, B. 1965c. Self similar error clusters in communications systems and the concept of conditional stationarity. *IEEE Transactions on Communications Technology* **COM-13**, 71-90.

Mandelbrot, B.B. 1965h. Une classe de processus stochastiques homothétiques à soi; application à la loi climatologique de H. E. Hurst. *Comptes Rendus* (Paris) **260**, 3274-3277.

Mandelbrot, B.B. 1965z. Information theory and psycholinguistics. *Scientific Psychology: Principles and Approaches,* 550-562. Ed. B.B. Wolman & E. N. Nagel. New York: Basic Books. Also, *Language, Selected Readings.* Ed. R.C. Oldfield & J.C. Marshall. London: Penguin. Also, with substantial appendices, *Readings in Mathematical Social Science.* Ed. P. Lazarfeld and N. Henry. Chicago, Ill.: Science Research Associates (1966: hardcover). Cambridge, Mass.: M.I.T. Press (1968: paperback).

Mandelbrot, B.B. 1967b. Sporadic random functions and conditional spectral analysis; self-similar examples and limits. *Proceedings of the Fifth Berkeley Symposium on Mathematical Statistics and Probability* **3**, 155-179. Ed. Lucien LeCam & J. Neyman. Berkeley: University of California Press.

Mandelbrot, B.B. 1967s. How long is the coast of Britain? Statistical self-similarity and fractional dimension. *Science* **155**, 636-638.

Mandelbrot, B.B. 1967i. Some noises with $1/f$ spectrum, a bridge between direct current and white noise. *IEEE Transactions on Information Theory* **13**, 289-298.

Mandelbrot, B.B. 1967t. Sporadic turbulence. *Proceedings of the International Symposium on Boundary Layers and Turbulence Including Geophysical Applications.* Supplement to *The Physics of Fluids* **10**, S302-S303.

Mandelbrot, B.B. 1968p. Les constantes chiffrées du discours. *Encyclopédie de la Pléiade: Linguistique,* 46-56. Paris: Gallimard.

Mandelbrot, B.B. 1969e. Long-run linearity, locally Gaussian process, H-spectra and infinite variances. *International Economic Review* **10**, 82-111.

Mandelbrot, B.B. 1970p. Discussion of a paper by Prof. N.F. Ramsey. *Critical Review of Thermodynamics,* 230-232. Ed. E.B. Stuart et al. Baltimore, Md.: Mono Book.

Mandelbrot, B.B. 1970y. *Statistical Self Similarity and Very Erratic Chance Fluctuations.* Trumbull Lectures, Yale University (to be published).

Mandelbrot, B.B. 1971f. A fast fractional Gaussian noise generator. *Water Resources Research* **7**, 543-553.

Mandelbrot, B. 1972j. Possible refinement of the lognormal hypothesis concerning the distribution of energy dissipation in intermittent turbulence. In Rosenblatt & Van Atta (Eds.) 1972, 333-351.

Mandelbrot, B.B. 1972c. Statistical methodology for nonperiodic cycles: from the covariance to R/S analysis. *Annals of Economic and Social Measurement* **1**, 257-288.

Mandelbrot, B.B. 1972z. Renewal sets and random cutouts. *Zeitschrift für Wahrscheinlichkeitstheorie und verwandte Gebiete* **22**, 145-157.

Mandelbrot, B.B. 1973f. Formes nouvelles du hasard dans les sciences. *Economie Appliquée* **26**, 307-319.

Mandelbrot, B.B. 1973j. Le problème de la réalité des cycles lents, et le syndrome de Joseph. *Economie Appliquée* **26**, 349-365.

Mandelbrot, B.B. 1973v. Le syndrome de la variance infinie, et ses rapports avec la discontinuité des prix. *Economie Appliquée* **26**, 321-348.

Mandelbrot, B.B. 1974c. Multiplications aléatoires itérées, et distributions invariantes par moyenne pondérée. *Comptes Rendus* (Paris) **278A**, 289-292 & 355-358.

Mandelbrot, B.B. 1974d. A population birth and mutation process, I: Explicit distributions for the number of mutants in an old culture of bacteria. *Journal of Applied Probability* **11**, 437-444.

Mandelbrot, B. 1974f. Intermittent turbulence in self-similar cascades: divergence of high moments and dimension of the carrier. *Journal of Fluid Mechanics* **62**, 331-358.

Mandelbrot, B.B. 1975b. Fonctions aléatoires pluri-temporelles: approximation poissonienne du cas brownien et généralisations. *Comptes Rendus* (Paris) **280A**, 1075-1078.

Mandelbrot, B. 1975f. On the geometry of homogeneous turbulence, with stress on the fractal dimension of the iso-surfaces of scalars. *Journal of Fluid Mechanics* **72**, 401-416.

Mandelbrot, B.B. 1975m. Hasards et tourbillons: quatre contes à clef. *Annales des Mines* (November), 61-66.

Mandelbrot, B. 1975o. *Les objets fractals: forme, hasard et dimension.* Paris & Montreal: Flammarion.

Mandelbrot, B.B. 1975u. Sur un modèle décomposable d'Univers hiérarchisé: déduction des corrélations galactiques sur la sphère céleste. *Comptes Rendus* (Paris) **280A**, 1551-1554.

Mandelbrot, B.B. 1975w. Stochastic models for the Earth's relief, the shape and the fractal dimension of the coastlines, and the number-area rule for islands. *Proceedings of the National Academy of Sciences USA* **72**, 3825-382

Mandelbrot, B.B. 1976c. Géométrie fractale de la turbulence. Dimension de Hausdorff, dispersion et nature des singularités du mouvement des fluides. *Comptes Rendus* (Paris) **282A**, 119-120.

Mandelbrot, B.B. 1976o. Intermittent turbulence & fractal dimension: kurtosis and the spectral exponent 5/3+B. In Temam (Ed.) 1976, 121-145.

Mandelbrot, B.B. 1977h. River networks through certain plane-filling mazes.

Mandelbrot, B.B. 1977m. The fractal geometry of percolation, polymers and almost everything else. In *Statistical Mechanics and Statistical Methods in Theory and Application.* Ed. U. Landman. New York: Plenum.

Mandelbrot, B. & McCamy, K. 1970. On the secular pole motion and the Chandler wobble. *Geophysical Journal* **21**, 217-232.

Mandelbrot, B. & Van Ness, J.W. 1968. Fractional Brownian motions, fractional noises and applications. *SIAM Review* **10**, 422.

Mandelbrot, B. & Wallis, J.R. 1968. Noah, Joseph and operational hydrology. *Water Resources Research* **4**, 909-918.

Mandelbrot, B. & Wallis, J.R. 1969a. Computer experiments with fractional Gaussian noises. *Water Resources Research* **5**, 228.

Mandelbrot, B. & Wallis, J.R. 1969b. Some long-run properties of geophysical records. *Water Resources Research* **5**, 321-340.

Mandelbrot, B. & Wallis, J.R. 1969c. Robustness of the rescaled range R/S in the measurement of noncyclic long run statistical dependence. *Water Resources Research* **5**, 967-988.

Manheim, J.H. 1964. *The Genesis of Point-set Topology.* Oxford: Pergamon. New York: Macmillan.

Marcus, A. 1964. A stochastic model of the formation and survivance of lunar craters, distribution of diameters of clean craters. *Icarus* **3**, 460-472.

Marcus, M.B. 1976. Capacity of level sets of certain stochastic processes. *Zeitschrift für Wahrscheinlichkeitstheorie verw. Gebiete* **34**, 279-284.

Marsden, J.E. & McCracken, M. 1976. *The Hopf Bifurcation and Its Application.* Applied Mathematical Sciences Series, **19**. New York: Springer.

Marstrand, J.M. 1954a. Some fundamental geometrical properties of plane sets of fractional dimension. *Proceedings of the London Mathematical Society* (3) **4**, 257-302.

Marstrand, J.M. 1954b. The dimension of Cartesian product sets. *Proceedings of the Cambridge Philosophical Society* **50**, 198-202.

Mattila, P. 1975. Hausdorff dimension, orthogonal projections and intersections with planes. *Annales Academiae Scientiarum Fennicae, Series A Mathematica* **I**, 227-244.

Max, N.L. 1971. *Space Filling Curves.* 16 mm color film. International Film Bureau, Chicago, Ill. Accompanying book (preliminary edition), Educational Development Center, Newton, Mass.

McKean, H.P., Jr. 1955. Sample functions of stable processes. *Annals of Mathematics* **61**, 564-579.

McKean, H.P., Jr. 1963. Brownian motion with a several dimensional time. *Theory of Probability and its Applications* **8**, 357-378.

McKean, H.P., Jr. 1975. Brownian local times. *Advances in Mathematics* **15**, 91-111.

Melzak, Z.A. 1969. On the solid packing constant for circles. *Mathematics of Computation* **23**, 169-172.

Menger, K. 1943. What is dimension? *American Mathematical Monthly* **50**, 2-7.

Menschkowski, H. 1967. *Probleme des Unendlichen.* Braunschweig: Vieweg.

Minkowski, H. 1901. Über die Begriffe Länge, Oberfläche und Volumen. *Jahresbericht der Deutschen Mathematikervereinigung* **9**, 115-121. Also in Minkowski 1967 **2**, 122-127.

Minkowski, H. 1967. *Gesammelte Abhandlungen.* New York: Chelsea (reprint).

Monin, A.S. & Yaglom, A.M. 1963. On the laws of small scale turbulent flow of liquids and gases. *Russian Mathematical Surveys* (translated from the Russian). **18**, 89-109.

Monin, A.S. & Yaglom, A.M. 1971 & 1975. *Statistical Fluid Mechanics, Volumes 1 and 2* (translated from the Russian). Cambridge, Mass.: MIT Press.

Montroll, E.W. & Scher, H. 1973. Random walks on lattices. IV. Continuous-time walks and influence of absorbing boundaries. *Journal of Statistical Physics* **9**, 101-135.

Moore, E.H. 1900. On certain crinkly curves. *Transactions of the American Mathematical Society* **1**, 72-90.

Munitz, M.K. (Ed.) 1957. *Theories of the Universe.* Glencoe: The Free Press.

Nelson, E. 1967. *Dynamical Theories of Brownian Motion.* Princeton University Press.

Newman, J.R. 1956. *The World of Mathematics, a Small Library.* New York: Simon & Schuster.

North, J.D. 1965. *The Measure of the Universe.* Oxford: Clarendon Press.

Novikov, E.A. & Stewart, R.W. 1964. Intermittency of turbulence and the spectrum of fluctuations of energy dissipation (in Russian). *Isvestia Akademii Nauk SSR; Seria Geofizicheskaia* **3**, 408-413.

Nye, M.J. 1972. *Molecular Reality. A Perspective on the Scientific Work of Jean Perrin.* London: Macdonald. New York: American Elsevier.

Obukhov, A.M. 1941. On the distribution of energy in the spectrum of turbulent flow. *Comptes Rendus (Doklady) Académie des Sciences de l'URSS (N.S.)* **32**, 22-24.

Obukhov, A.M. 1962. Some specific features of atmospheric turbulence. *Journal of Fluid Mechanics* **13**, 77-81. Also in *Journal of Geophysical Research* 67, 3011-3014.

Olbers, W. 1823. Über die Durchsichtigkeit des Weltraums. *Astronomisches Jahrbuch für das Jahr 1826 nebst einer Sammlung der neuesten in die astronomischen Wissenschaften einschlagenden Abhandlungen, Beobachtungen und Nachrichten,* **150**, 110-121. Berlin: C.F.E. Späthen.

Oldham, K.B. & Spanier, J. 1974. *The Fractional Calculus.* New York: Academic Press.

Orey, S. 1970. Gaussian sample functions and the Hausdorff dimension of level crossings. *Zeitschrift für Wahrscheinlichkeitstheorie und verwandte Gebiete* **15**, 249-156.

Osgood, W.F. 1903. A Jordan curve of positive area. *Transactions of the American Mathematical Society* **4**, 107-112.

Painlevé, P. 1895. Leçon d'ouverture faite en présence de Sa Majesté le Roi de Suède et de Norwège. Manuscript first printed in Painlevé 1972– **1**, 200-204.

Painlevé, P. 1972–. *Oeuvres de Paul Painlevé.* Paris: Editions du CNRS.

Paley, R.E.A.C. & Wiener, N. 1934. *Fourier Transforms in the Complex Domain.* New York: American Mathematical Society.

Partridge, E. 1958. *Origins.* New York: Macmillan.

Peano, G. 1890. Sur une courbe, qui remplit une aire plane. *Mathematische Annalen* **36**, 157-160. Translation in Peano 1973.

Peano, G. 1908. *Formulario mathematico. Vol. 5* Turin: Bocca, pp. 239-240. Translation in Peano 1973, 148-149.

Peano, G. 1973. *Selected Works.* Ed. H.C. Kennedy. Toronto University Press.

Peebles, P.J.E. 1974c. Statistical analysis of catalogs of extragalactic objects. IV. Cross-correlation of the Abell and Shane-Wirtanen catalogs. *The Astrophysical Journal Supplement Series No. 253* **28**, 37-50.

Peebles, P.J.E. 1974i. The gravitational-instability picture and the nature of the distribution of galaxies. *The Astrophysical Journal* **189**, L51-L53.

Peebles, P.J.E. 1975. Statistical analysis of catalogs of extragalactic objects. VI. The galaxy distribution in the Jagellonian field. *The Astrophysical Journal* **196**, 647-652.

Peebles, P.J.E. & Groth, E. 1975. Statistical analysis of catalogs of extragalactic objects. V. Three-point correlation function for the galaxy distribution in the Zwicky catalog. *The Astrophysical Journal* **196**, 1-11.

Perrin, J. 1906. La discontinuité de la matière. *Revue du Mois* **1**, 323-344.

Perrin, J. 1909-1910. Mouvement brownien et réalité moléculaire. *Annales de chimie et de physique* **VIII 18**, 5-114. Trans. F. Soddy, as *Brownian Movement and Molecular Reality.* London: Taylor & Francis. Trans. J. Donau, as *Die Brownsche Bewegung und die wahre Existenz der Moleküle.* Dresden: Steinkopff.

Perrin, J. 1913-1916. *Les Atomes.* Paris: Alcan. 1913 is the date of this first edition. It was reprinted in 1970 by Gallimard, thus superseding several revisions that had aged less successfully. 1916 is the date of *Atoms,* the English translation by D.L. Hammick; London: Constable. New York: Van Nostrand.

Peyrière, J. 1974. Turbulence et dimension de Hausdorff. *Comptes Rendus* (Paris) **278A**, 567-569.

Poincaré, H. 1916–. *Oeuvres de Henri Poincaré.* Paris: Gauthier Villars.

Pontrjagin, L. & Schnirelman, L. 1932. Sur une propriété métrique de la dimension. *Annals of Mathematics* **33**, 156-162.

Pruitt, W.E. 1975. Some dimension results for processes with independent increments. *Stochastic Processes and Related Topics,* **I**, 133-165. Ed. M.L. Puri. New York: Academic Press.

Pruitt, W.E. & Taylor, S.J. 1969. Sample path properties of processes with stable components. *Zeitschrift für Wahrscheinlichkeitstheorie und verwandte Gebiete* **12**, 267-289.

Rall, W. 1959. Branching dendritic trees and motoneuron membrane resistivity. *Experimental Neurology* **1**, 491-527.

Rayleigh, Lord 1880. On the resultant of a large number of vibrations of the same pitch and arbitrary phase. *Philosophical Magazine* **10**, 73. Also in Rayleigh 1899-1964 **1**, 491-.

Rayleigh, Lord 1899-1964. *Scientific Papers.* Cambridge University Press. New York: Dover (reprint).

Rényi, A. 1955. On a new axiomatic theory of probability. *Acta Mathematica Hungarica* **6**, 285-335.

Rényi, A. 1970. *Foundations of Probability.* San Francisco: Holden Day.

Richardson, L.F. 1922-1965. *Weather Prediction by Numerical Process.* Cambridge University Press. (The dates refer to the original and to a reprint, which also contains a biography as part of a new introduction by J. Chapman.)

Richardson, L.F. 1926. Atmospheric diffusion shown on a distance-neighbour graph. *Proceedings of the Royal Society of London. A* **110**, 709-737.

Richardson, L.F. 1960a. *Arms and Insecurity: a Mathematical Study of the Causes and Origins of War.* Ed. N. Rashevsky & E. Trucco. Pittsburgh (now, Pacific Grove): Boxwood Press.

Richardson, L.F. 1960s. *Statistics of Deadly Quarrels.* Ed. Q. Wright & C.C. Lienau. Pittsburgh (now, Pacific Grove): Boxwood Press.

Richardson, L.F. 1961. The problem of contiguity: an appendix of statistics of deadly quarrels. *General Systems Yearbook* **6**, 139-187.

Richardson, L.F. & Stommel, H. 1948. Note on eddy diffusion in the sea. *Journal of Meteorology* **5**, 238-240.

Rogers, C.A. 1970. *Hausdorff Measures.* Cambridge University Press.

Rosenblatt, M. & Van Atta, C. (Eds.) 1972. *Statistical Models and Turbulence.* Lecture Notes Physics **12.** New York: Springer.

Ross, B. (Ed.) 1975. *Fractional Calculus and Its Applications.* Lecture Notes in Mathematics **457.** New York: Springer.

Ruelle, D. 1972. Strange attractors as a mathematical explanation of turbulence. In Rosenblatt & Van Atta (Eds.) 1972, 292-299. New York: Springer.

Ruelle, D. 1975. The Lorenz attractor and the problem of turbulence. *Conference on Quantum Dynamics Models and Mathematics.* Bielefeld. Also in Temam (Ed.) 1976, 146-158.

Ruelle, D. & Takens, F. 1971. On the nature of turbulence. *Communications on Mathematical Physics* **20,** 167-192 & **23,** 343-344.

Saaty, T.L. & Weyl, F.J. 1969. *The Spirit and Uses of the Mathematical Sciences.* New York: McGraw-Hill.

Saffman, P.G. 1968. Lectures on homogeneous turbulence. *Topics in Nonlinear Physics* Ed. N.J. Zabusky. New York: Springer.

Salem, R. & Zygmund, A. 1945. Lacunary power series and Peano curves. *Duke Mathematical Journal* **12,** 569-578.

Scheffer, V. 1976. Equations de Navier-Stokes et dimension de Hausdorff. *Comptes Rendus* (Paris) **282A,** 121-122.

Scheffer, V. 1977. Partial regularity of solutions to the Navier-Stokes equation. *Pacific Journal of Mathematics* (forthcoming).

Scher, H. & Montroll, E.W. 1975. Anomalous transit-time dispersion in amorphous solids. *Physical Review B* **12,** 2455-2477.

Schönberg, I.J. 1938a. Metric spaces and positive definite functions. *Transactions of the American Mathematical Society* **44,** 522-536.

Selety, F. 1922. Beiträge zum kosmologischen Problem. *Annalen der Physik,* Series 4, **68,** 281-334.

Selety, F. 1923a. Une distribution des masses avec une densité moyenne nulle, sans centre de gravité. *Comptes Rendus* (Paris) **177,** 104-106.

Selety, F. 1923b. Possibilité d'un potentiel infini, et d'une vitesse moyenne de toutes les étoiles égale à celle de la lumière. *Comptes Rendus* (Paris) **177,** 250-252.

Selety, F. 1924. Unendlichkeit des Raumes und allgemeine Relativitätstheorie. *Annalen der Physik* Series 4, **73,** 291-325.

Shante, V.K.S. & Kirkpatrick, S. 1971. An introduction to percolation theory. *Advances in Physics* **20,** 325-357.

Sierpiński, W. 1915. Sur une courbe dont tout point est un point de ramification. *Comptes Rendus* (Paris) **160,** 302. More detailed version in Sierpiński 1974-, **II,** 99-106.

Sierpiński, W. 1916. Sur une courbe cantorienne qui contient une image biunivoque et continue de toute courbe donnée. *Comptes Rendus* (Paris) **162,** 629. More detailed version in Sierpiński, 1974-, **II,** 107-119.

Sierpiński, W. 1974-. *Oeuvres Choisies.* Ed. S. Hartman et al. Warsaw: Éditions scientifiques de Pologne.

Sinai, Ja.G. 1976. Some rigorous results in the theory of phase transitions. *Statistical Physics* (IUPAP, 1975), 139-149. Ed. L. Pál & P. Szépfalusy. Budapest: Academy of Sciences.

Singh, A.N. 1935-53. *The Theory and Construction of Nondifferentiable Functions.* Lucknow (India): The University Press. Also in *Squaring the Circle and Other Monographs.* Ed. E.W. Hobson, H.P. Hudson, A.N. Singh & A.B. Kempe. New York: Chelsea.

Spitzer, L. 1968. Dynamics of interstellar matter and formation of stars. *Stars and stellar systems* **7,** 1-63.

Stanley, H.E. 1971. *Introduction to Phase Transitions and Critical Phenomena.* New York: Oxford University Press.

Stanley, H.E., Birgeneau, R.J., Reynolds, P.J. & Nicoll, J.F. 1976. Thermally driven phase transitions near the percolation threshold in two dimensions. *Journal of Physics C (Solid State Physics)* **9,** L553-L560.

Steinhaus, H. 1954. Length, shape and area. *Colloquium Mathematicum* **3,** 1-13.

Stent, G. 1972. Prematurity and uniqueness in scientific discovery. *Scientific American* **227** (December) 84-93.

Strahler, A.N. 1952. Hypsometric (area-altitude) analysis of erosional topography. *Geological Society of American Bulletin* **63**, 1117-1142.

Sulem, P.L. & Frisch, U. 1975. Bounds on energy flux for finite energy turbulence. *Journal of Fluid Mechanics* **72**, 417-423.

Swift, J. 1733. On Poetry, a Rhapsody.

Taylor, G.I. 1935. Statistical theory of turbulence; parts I to IV. *Proceedings of the Royal Society of London* **A151**, 421-478. Reprinted in Friedlander & Topper 1961, 18-51.

Taylor, G.I. 1970. Some early ideas about turbulence. *Journal of Fluid Mechanics* **41**, 3-11.

Taylor, S.J. 1955. The α-dimensional measure of the graph and the set of zeros of a Brownian path. *Proceedings of the Cambridge Philosophical Society* **51**, 265-274.

Taylor, S.J. 1961. On the connection between Hausdorff measures and generalized capacities. *Proceedings of the Cambridge Philosophical Society* **57**, 524-531.

Taylor, S.J. 1964. The exact Hausdorff measure of the sample path for planar Brownian motion. *Proceedings of the Cambridge Philosophical Society* **60**, 253-258.

Taylor, S.J. 1966. Multiple points for the sample paths of the symmetric stable process. *Zeitschrift für Wahrscheinlichkeitstheorie und verwandte Gebiete* **5**, 247-264.

Taylor, S.J. 1967. Sample path properties of a transient stable process. *Journal of Mathematics and Mechanics* **16**, 1229-1246.

Taylor, S.J. 1973. Sample path prospectus of processes with stationary independent documents. *Stochastic Analysis*. Ed. D.G. Kendall & E.F. Harding. New York: Wiley.

Taylor, S.J. & Wendel, J.C. 1966. The exact Hausdorff measure of the zero set of a stable process. *Zeitschrift für Wahrscheinlichkeitstheorie und verwandte Gebiete* **6**, 170-180.

Temam, R. (Ed.) 1976. *Turbulence and Navier Stokes Equations*. Lecture Notes in Mathematics **565**. New York: Springer.

Tennekes, H. 1968. Simple model for the small scale structure of turbulence. *Physics of Fluids* **11**, 669-672.

Tesnière, M. 1975. Fréquences des noms de famille. *Journal de la Société de Statistique de Paris* **116**, 24-32.

Thompson, d'A.W. 1917-1942-1961. *On Growth and Form*. Cambridge University Press. The dates refer to the first, second and abridged editions.

Tietze, H. 1965. *Famous Problems of Mathematics*. New York: Graylock.

Ulam, S.M. 1974. *Sets, Numbers and Universes: Selected Works*. Ed. W. A. Beyer, J. Mycielski & G.-C. Rota. Cambridge, Mass.: M.I.T. Press.

Van der Ziel, A. 1954. *Noise*. New York: Prentice Hall.

van Emde Boas, P. 1969. *Nowhere Differentiable Continuous Functions, with an extended list of references*. Amsterdam: Mathematisch Centrum.

Verveen, A.A. & DeFelice, L.J. 1974. Membrane noise. *Progress in Biophysics and Molecular Biology* **28**, 189-265.

Vilenkin, N.Ya. 1965. *Stories About Sets*. Tr. Scripta Technica. New York, London: Academic Press.

von Koch, H. 1904. Sur une courbe continue sans tangente, obtenue par une construction géométrique élémentaire. *Arkiv för Matematik, Astronomi och Fysik* **1**, 681-704.

von Koch, H. 1906. Une méthode géométrique élémentaire pour l'étude de certaines questions de la théorie des courbes planes. *Acta Mathematica* **30**, 145-174.

von Neumann, J. 1949-1963. Recent theories of turbulence. The dates refer to publication as a report to ONR and in von Neumann, 1961– **6**, 437-472.

von Neumann, J. 1961– *Collected Works*. Ed. A. H. Traub. New York: Pergamon.

von Schweidler, E. 1907. Studien über die Anomalien in Verhalten der Dielektrika. *Analen des Physik* **(4)24,** 711-770.

von Weizsäcker, C.F. 1950. Turbulence in interstellar matter. *Problems of Cosmical Aerodynamics* (IUTAM & IAU). Dayton: Central Air Documents Office.

Voss, R. & Clarke, J. 1975. "1/*f* noise" in music and speech. *Nature* **258,** 317-318.

Wallenquist, A. 1957. On the space distribution of galaxies in clusters. *Arkiv för Matematik, Astronomi och Fysik* **2,** 103-110.

Weierstrass, K. 1872. Über continuirliche Functionen eines reellen Arguments, die für keinen Werth des letzteren einen bestimmten Differentialquotienten besitzen. First published in Weierstrass 1895, **II,** 71-74.

Weierstrass, K. 1895-. *Mathematische Werke.* Berlin: Mayer & Muller.

Wellner, J.A. 1975. Monte Carlo of two-dimension Brownian sheets. *Stochastic Processes and Related Topics,* **II,** 59-75. Ed. M.L. Puri. New York: Academic Press.

Wiener, N. 1948-1961. *Cybernetics.* Paris: Hermann. New York: Wiley (1st edition). Cambridge, Mass.: M.I.T. Press (2d edition).

Wiener, N. 1956. *I am a Mathematician.* Garden City, N.Y.: Doubleday. Cambridge, Mass.: M.I.T. Press.

Wiener, N. 1964. *Selected Papers.* Cambridge, Mass.: M.I.T. Press.

Wiener, N. 1976-. *Collected Works.* Ed. P. Masani. Cambridge, Mass.: M.I.T. Press.

Wigner, E.P. 1960. The unreasonable effectiveness of mathematics in the natural sciences. *Communications on Pure and Applied Mathematics* **13,** 1-14. Also in Saaty & Weyl 1969, 123-140, and in Wigner 1967, 222-237.

Wigner, E.P. 1967. *Symmetries and Reflections: Scientific Essays of Eugene P. Wigner.* Indiana University Press.

Wilson, A.G. 1965. Olbers' paradox and cosmology. Los Angeles, Astronomical Society.

Wilson, A.G. 1969. Hierarchical structures in the cosmos. *Hierarchical Structures,* 113-134. Ed. L. L. Whyte, A. G. Wilson & D. Wilson. New York: American Elsevier.

Wilson, J.T. (Ed.) 1972. *Continents Adrift.* Readings from *Scientific American.* San Francisco: W. H. Freeman.

Yaglom, A.M. 1957. Some classes of random fields in n-dimensional space, related to stationary random processes. *Theory of Probability and Its Applications,* **2,** 273-320. Tr. R.A. Silverman.

Yaglom, A.M. 1966. The influence of fluctuations in energy dissipation on the shape of turbulence characteristics in the inertial interval. *(Doklady Akademii Nauk SSSR* **16,** 49-52. (English trans. *Soviet Physics Doklady* **2,** 26-29.)

Yoder, L. 1974. Variation of multiparameter Brownian motion. *Proceedings of the American Mathematical Society* **46,** 302-309.

Yoder, L. 1975. The Hausdorff dimensions of the graph and range of N-parameter Brownian motion in d-space. *Annals of Probability* **3,** 169-171.

Young, W.H. & Young, G.C. 1906. *The Theory of Sets of Points.* Cambridge University Press.

Zipf, G. K. 1949-1965. *Human Behavior and the Principle of Least-Effort.* Cambridge, Mass.: Addison-Wesley. New York: Hefner.

Zygmund, A. 1959. *Trigonometric Series.* Cambridge University Press.

ACKNOWLEDGMENTS

D'Arcy Thompson describes *On Growth and Form* as having "little need of preface, for indeed it is all 'preface' from beginning to end." The same (as stated in the Introduction) is true of this Essay. Its component chapters are also left without conclusion and the same will be true of the whole – except for one comment. I am as surprised as everyone else that, merely by accepting power function laws and exploring their geometric consequences with care, we find so much to occupy us.

To concoct *"ma macédoine de livre"* has been a long, drawn-out process, during which it has been my pleasure to contract many intellectual and personal debts. Directly or between the lines of this book, in digressions and in the biographical sketches, most of the former have already been recognized. Norbert Wiener and John von Neumann, however, are notable among those who have been slighted by the hazards of citation, because they (especially the latter) exerted their influence less by deed than by example.

A helpful rough translation of the French version was contributed by J. S. Lourie. R. W. Gosper of Stanford allowed me to play with his Peano curve before it was published. Dr. P. L. Renz of W. H. Freeman and Company drew my attention to the Gosper curve and to non-integer based number systems. Dr. J. E. Marsden of Berkeley made useful comments on the text.

At IBM: Dr. M. C. Gutzwiller, Director of the General Sciences Department, made it possible for this work to proceed smoothly. Dr. E. S. Kirkpatrick authorized the use of his percolation programs and halftone fonts. J. L. Oneto (now in Nice) and A. Appel designed the art carried over from the French version. The Research Center library, text processing and graphics staffs were uncommonly helpful.

Last but not least, for support for the work reported in this book and the preparation of the text and illustrations, I am deeply indebted to the Thomas J. Watson Research Center of the International Business Machines Corporation. As Group Manager, Department Director, and now Vice President and Director of Research, Dr. Ralph E. Gomory imagined ways of underwriting my work when it was a gamble, and now of giving it all the support I could use.

INDEX

The letters *a* and *b* following a page number refer, respectively, to the left and right columns of the text. Bold numbers refer to Plates and/or their captions. Roman numerals refer to Chapters wherein many references to that subject are included.

INDEX OF SELECTED DIMENSIONS

The letters *a* and *b* following a page number refer, respectively, to the left and right columns of the text. Bold numbers refer to Plates. These lists do not by any means attempt to be exhaustive.

I: GEOMETRIC SHAPES AND THEIR RIGOROUS DIMENSIONS: EUCLIDEAN (E), FRACTAL (D), AND TOPOLOGICAL (D_T)

When Euclidean dimension is denoted by E, its value is an arbitrary positive integer.

	E	D	D_T	pages
EUCLIDEAN SETS				
Points in finite number	E	0	0	
Straight line, circle, other classical curves	E	1	1	
Planar disc	E	2	2	
Generalized ball	E	E	E	
OTHER NONFRACTAL SETS				
Peano "curve"	2	2	2	**59–63**, 309b
Cantor staircase	2	1	1	100b, **101**, 304b
Squeezed ordinary Brownian trail	1	1	1	89b
Squeezed fractional Brownian trail; $H < 1/E$	E	E	E	233b
NONRANDOM FRACTAL SETS				
Cantor set: the triadic set on the line	1	log2/log3	0	98b, **99**
Cantor sets: other	E	$0 < D < E$	0	**114, 115**, 165b
Koch curve: the triadic snowflake	2	log4/log3	1	**36, 37, 39**, 42a
Koch curve: terdragon's skin	2	log4/log3	1	314a
Koch curve: dragon's skin	2	1.5236	1	313b
Koch curves: other	2	$1 < D < 2$	1	**46–57, 64–67**
Kochlike pattern	2	$0 < D < 2$	0	**125**
Sierpiński gasket and arrowhead curve	2	log3/log2	1	**56**, 186a
Simple curves/Lebesgue-Osgood monsters	2	2	1	**63**, 76b, 308a

	E	D	D_T	pages
RANDOM FRACTALS				
Line-to-E-space (Ordinary) Brownian process:				
trail when E≥2	E	2	1	**11**, 89a
function in $\mathbb{R}^2$	2	1.5000	1	**87**, 90b
function in $\mathbb{R}^{E-1}$ with E > 2	E	1+(E−1)/2	1	305b
zeroset of the function	1	0.5000	0	**87**, 90b
Brownian space (or sphere)-to-line function:	E	E+1/2	E	**210–217, 220**, 284b
zeroset	E	E−1/2	E−1	**8–9**
Burgers scalar turbulence isosurfaces	3	5/2	2	236a
Fractional H, line-to-space Brownian process:				
trail	E	max(E,1/H)	1	**209**, 282b
zeroset	1	1−H	0	**210–215, 218–220**
function	2	2−H	1	**210–215, 221**, 282b
Fractional H, space-to-line Brownian process:				
zeroset	E	E−H	E−1	
Kolmogorov scalar turbulence isosurfaces	3	8/3	2	**222–223**, 236a
Levy stable process:				
trail	2	1/α	1	**137–139**, 394b
zeroset	1	max(0,1−1/α)	0	324b

continued

II: GEOMETRIC SHAPES AND ESTIMATED DIMENSIONS:
EUCLIDEAN (E), FRACTAL (D), AND TOPOLOGICAL (D_T)

The bounds on D that are listed in addition to the more precise estimates are rigorous.

	E	D	D_T	pages
NONRANDOM FRACTALS				
Apollonian gasket	2	1.306951	1	**187,** 188b
(1.300197 < D < 1.314534)				
RANDOM FRACTALS				
Rescaled self avoiding random walk/polygon	2	1.33	1	191a
Rescaled self avoiding random walk	3	1.67	1	191a
River in a Leopold & Langbein network	2	1.28	1	194a
Critical Bernoulli percolation cluster	2	1.77	1	197a

III: NATURAL OBJECTS AND ESTIMATED DIMENSIONS:
EUCLIDEAN (E), FRACTAL (D), AND TOPOLOGICAL (D_T)

For those natural fractals for which a good model is available see the preceding list. The listed values of D are typical values.

	E	D	D_T	pages
NATURAL EUCLIDEAN OBJECTS				
Very fine ball	E	0	0	19b
Very fine thread	E	1	1	19b
Polished sphere	3	2	2	19b
Substantial ball	3	3	3	19b
NATURAL FRACTALS				
Seacoast (Richardson exponent)	2	1.25	1	30b
River network's cumulative bank	2	2.00	1	73b
Individual river's outline (Hack exponent)	2	1.25	1	73a
Vascular system	3	3.00	2	77a
Pulmonary membrane	3	2.90	2	
Tree's bark	3	3.00	2	
Fractal errors	1	0.30	0	104a
Stellar matter in the central zone	3	1.30	0	111b
Turbulence: support of dissipation	3	2.50	2	163b
Word frequencies	1	0.9	0	242b

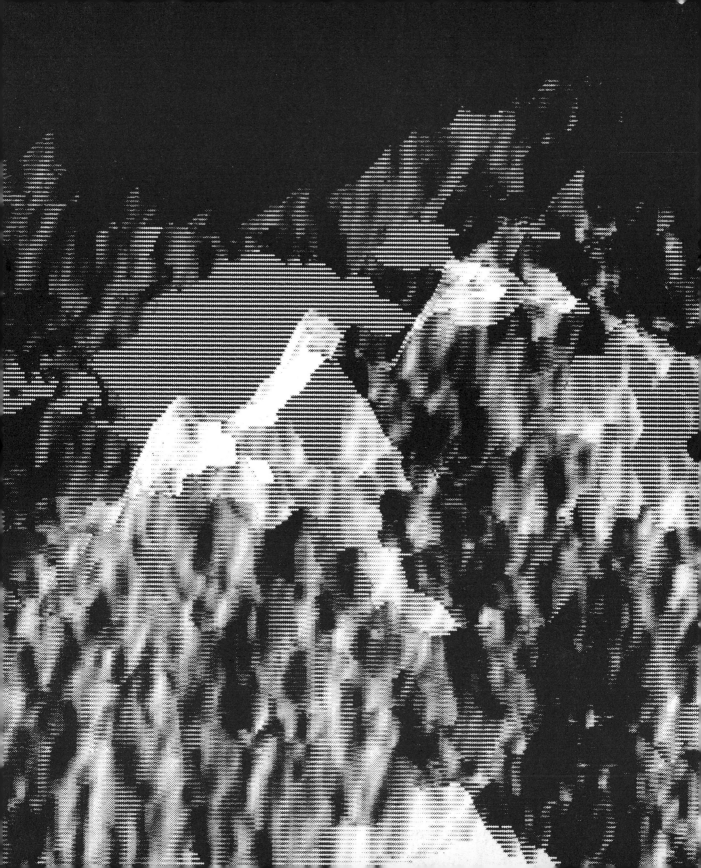